History of the Planetary Systems from Thales to Kepler

J. L. E. Dreyer

Alpha Editions

This edition published in 2020

ISBN : 9789390382446 (Hardback)
ISBN : 9789390382699 (Paperback)

Design and Setting By
Alpha Editions
www.alphaedis.com
email - alphaedis@gmail.com

PREFACE.

IN this book an attempt has been made to trace the history of man's conception of the Universe from the earliest historical ages to the completion of the Copernican system by Kepler in the seventeenth century. Among the various branches of physical science there is no other which in its historical development so closely reflects the general progress of civilisation as the doctrine of the position of the earth in space and its relation to the planetary system. In this we may follow man's gradual emancipation from primitive ideas during the rise of Greek philosophy and science, his relapse into those ideas during the ages following the destruction of the seats of Greek culture, and the rapid advance of knowledge after the revival of learning at the end of the Middle Ages.

What chiefly induced me to write this book was the circumstance that a number of legends on subjects connected with the history of the cosmical systems have been repeated time after time, not only in works on the general history of science and literature like those of Hallam and Draper, but also in books dealing specially with astronomy. Among errors long ago refuted but still frequently produced in print may be mentioned: that Thales knew the earth to be a sphere; that Pythagoras and his school taught the motion of the earth round the sun; that Plato taught the daily rotation of the earth and in his old age inclined to the heliocentric system; that the Egyptians knew that Mercury and Venus move round the sun; that the lunar variation was discovered by Abu'l Wefa; that King Alfonso the X. of Castille found the orbit of Mercury to be an ellipse (which has not been refuted before, so far as I know), and that Cusa and Regiomontanus anticipated Copernicus. On the other hand, some writers are inclined to belittle the knowledge of the Ancients, making out that Plato imagined the earth to be a cube and that the spheres of Eudoxus and the Ptolemaic and Tychonic systems are impossible and absurd.

In order to enable the reader to check every statement made and to form his own opinion on every debatable point, I have given full references to the original authorities. The most recent and best editions have generally been used, though in the case of some of the patristic writers I have only been able to consult old editions.

J. L. E. DREYER.

Armagh Observatory,
December 1905.

CONTENTS.

INTRODUCTION.

THE EARLIEST COSMOLOGICAL IDEAS.

CHAPTER I.

THE EARLY GREEK PHILOSOPHERS.

CHAPTER II.

THE PYTHAGOREAN SCHOOL.

CHAPTER III.

PLATO.

CHAPTER IV.

THE HOMOCENTRIC SPHERES OF EUDOXUS.

CHAPTER V.

ARISTOTLE.

CHAPTER VI.

HERAKLEIDES AND ARISTARCHUS.

CHAPTER VII.

THE THEORY OF EPICYCLES.

CHAPTER VIII.

THE DIMENSIONS OF THE WORLD.

CHAPTER IX.

THE PTOLEMAIC SYSTEM.

CHAPTER X.

MEDIEVAL COSMOLOGY.

CHAPTER XI.

ORIENTAL ASTRONOMERS.

CHAPTER XII.

THE REVIVAL OF ASTRONOMY IN EUROPE.

CHAPTER XIII.

COPERNICUS.

CHAPTER XIV.

TYCHO BRAHE AND HIS CONTEMPORARIES.

CHAPTER XV.

KEPLER.

CHAPTER XVI.

CONCLUSION.

CORRIGENDA.

Page 32, footnote 6. *For* p. 200 *read* p. 206.

Page 64, line 15. *For* comma *read* full stop.

Page 153, footnote. *For* aphelion *read* apogee.

INTRODUCTION.

THE EARLIEST COSMOLOGICAL IDEAS.

For many centuries during the early development of Babylonian civilisation astrology had been eagerly cultivated in the land of the two rivers, before it was found necessary to make a careful study of the motions of the heavenly bodies in order to place the astrological predictions on a more secure base. By degrees the Babylonians acquired a remarkably accurate knowledge of the periods of the sun, moon, and planets, so that they were able to foretell the positions of these bodies among the stars and the recurrence of lunar eclipses, without, so far as we know, formulating any kind of geometrical theory of the celestial motions. But although astronomy as a science thus came into being on the banks of the Euphrates, from whence it eventually exercised a powerful influence on the development of Greek astronomy, the knowledge of the stars was confined to the priests, in whose hands the arrangement of the calendar and the worship of the moon and stars was laid. Outside the ranks of the priesthood there was no attempt at study of any kind, so that speculations on the origin and construction of the world were always interwoven with mythological fancies to the exclusion of independent thought. Astronomy may be said to have sprung from Babylon, but cosmology, distinct from mythological cosmogony, dates only from Greece.

The cosmology of the Babylonians was a compound of the ideas which originally prevailed in the territories around the two ancient sanctuaries, Eridu on the shore of the Persian Gulf and Nippur in northern Babylonia. According to the cosmology of Eridu water was the origin of all things; the inhabited world has sprung from the deep and is still encircled

D.   1

by Khubur, the ocean stream, beyond which the sun-god pastures his cattle. Our knowledge of the cosmology of Nippur is still scanty, but we know that the world had the form of a mountain and that an encircling ocean is not mentioned. In later times, when the dominion of Babylon spread far both east and west, the heaven appeared to be a solid vault, the foundation of which rested on the vast ocean, " the deep " (apsu), which also supported the earth. Above the vault were the upper waters, and above them again "the interior of the heavens," the dwelling of the gods[1], the " sun-illuminated house," from which the sun comes out through a door in the east every morning, and into which it enters every evening through another door. The earth was supposed to be a great mountain, hollow underneath; it had originally been divided into seven zones inside each other, for which afterwards was substituted a division into four quadrants. In the east is the bright mountain or the great mountain of sunrise, in the west the dark mountain or mountain of sunset[2]. The northern part of the earth is unknown and mysterious. Between heaven and earth are the waters of the east and west ocean, which like the south ocean are parts of the apsu. Inside the crust of the earth (above the great hollow interior) is the abode of the dead, the entrance to which is in the west. The vault of heaven is not supposed to be moving, but the sun, moon, and stars are living beings or deities, moving along in paths or orbits. In the earliest times the evening star and the morning star were believed to be the same, but afterwards an endeavour was made to distinguish between Istar of the evening and Istar of the morning, possibly for mythological reasons[3].

In numerous passages of the Old Testament we find ideas as to the construction of the world which are practically the same as those held by the Babylonians. Nothing is said as to

---

[1] In the days of Nippur they were supposed to live on the top of the earth-mountain.

[2] In early (Sumerian) texts there are twin mountains between which the sun passes.

[3] P. Jensen, *Die Kosmologie der Babylonier*, Strassburg, 1890, pp. 253–257; Sayce, *The Religions of Ancient Egypt and Babylonia*, Edinburgh, 1902, pp. 79, 340, 350, 375, 378, 396.

the actual figure of the earth, though the circle of the horizon is alluded to in several places, e.g. "He set a circle upon the face of the deep" (Prov. viii. 27), and "it is He that sitteth upon the circle of the earth" (Is. xl. 22); but the earth is supposed to rest on "pillars" or "foundations," often referred to, e g. in 1 Sam. ii. 8, "for the pillars of the earth are the Lord's, and He hath set the world upon them." On the other hand we read in Ps. cxxxvi. 6 that God "spread forth the earth above the waters," and in Job xxvi. 7, that "He stretcheth out the north over empty space and hangeth the earth upon nothing." Under the surface of the earth is "the great deep," from which fountains and rivers spring, and which plays an important part in the account of the deluge. Under the deep is Sheol, "the land of darkness and the shadow of death" (Job x. 21), under which again Ezekiel apparently supposes "the pit," "the nether part of the earth," to be the destination of the uncircumcised heathen after death (Ezek. xxvi. 20, xxxii. 23). Above the earth is the solid firmament, "strong as a molten mirror" (Job xxxvii. 18), supporting the upper waters "that be above the heavens" (Ps. cxlviii. 4), an idea which is more distinctly set forth in Genesis i. 6, 7: "And God said, Let there be a firmament in the midst of the waters, and let it divide the waters from the waters. And God made the firmament, and divided the waters which were under the firmament from the waters which were above the firmament." The sun and moon are in the same chapter said to be placed "in the firmament," but no particulars are given anywhere as to their motions. The stars are generally referred to as the host of heaven, an expression which was also used in Babylonia, where the moon-god was the "lord of hosts[1]".

Among the Egyptians equally primitive notions prevailed. They imagined the whole universe to be like a large box, nearly rectangular in form, the greatest extent being in the direction from north to south, the direction in which their own country extended. The earth formed the bottom of this box, being a narrow, oblong, and slightly concave floor with Egypt in its centre. The sky stretched over it like an iron

[1] Sayce, l.c. p. 486.

ceiling, flat according to some, vaulted according to others; its earthward face was sprinkled with lamps hung from cords or more generally supposed to be carried by deities, extinguished or unperceived by day but visible to us at night.   This ceiling was at first supposed to be supported by four columns, but afterwards these were superseded by four lofty mountain peaks rising at the four cardinal points and connected by a continuous chain of mountains.   On a ledge a little below the tops of these a great river flowed round the earth, hidden from us towards the north by intervening mountains, behind which the river (Ur-nes) flowed through a valley called Daït which was shrouded in eternal night.   The Nile is a branch of this river, turning off from it at its southern bend.   The river carried a boat in which was a disc of fire, the sun, a living god called Râ, born every morning, growing and gaining strength till noon, when he changes into another boat which carries him to the entrance to Daït, from whence other boats (about which less is known) carry him round to the door of the east during the night.   In later times the book "Am Duat" or the book of the other world gives a detailed account of the journey of the sun-god during the twelve hours of the night, when he in succession passes through and illuminates twelve separate localities of the other world.   The boat is occasionally during the day attacked by a huge serpent, whereby the sun is eclipsed for a short time.   During the summer months the obliquity of Râ's daily course diminishes and he comes nearer to Egypt; during the winter it increases and he goes farther away.   The reason of this change is that the solar bark always keeps close to that bank of the celestial river which is nearest to the abode of men, and when the river overflows at the annual inundation the sun is carried along with it outside the regular bed of the stream and brought yet closer to Egypt.   As the inundation abates the bark descends and recedes, its greatest distance from the earth corresponding with the lowest level of the waters.   The same stream also carries the bark of the moon (Yââhu Aûhû, in some places called the left eye of Horus) which comes out of the door of the east in the evening.   Like the sun, the moon has its enemies; a sow attacks it on the

15th day of each month, and after a fortnight's agony and increasing pallor the moon dies and is born again. Sometimes the sow manages to swallow it altogether for a short time, causing a lunar eclipse.

Of the star-lamps some never leave the sky, others borne by a slow movement pass annually beyond the limits of sight for months at a time. Ûapshetatûi (Jupiter), Kahiri (Saturn), and Sobkû (Mercury) steer their barks straight ahead like Râ and Iââhu, but the red Doshiri (Mars) sails backwards, an assertion which clearly shows that the Egyptians had been specially struck with the length of the retrograde motion of the planet when most conspicuous, i.e. when in opposition to the sun. Bonû (Venus) has a dual personality, in the evening it is Ûati, the star which is first to rise; in the morning it becomes Tiû-nûtiri, the god who hails the sun before his rising. The Milky Way is the heavenly Nile, flowing through the land where the dead live in perpetual happiness under the rule of Osiris[1].

The Egyptians believed that there was a time when neither heaven nor earth existed, and when nothing had being except the boundless primeval water, the Nû, which was shrouded with darkness but contained the germs of the world which at length the spirit of the water called into existence by uttering the word[2]. But this idea can hardly have grown up independently in Egypt, where the desert might rather have been expected to represent the unformed beginning of things. It must have come to Egypt with the Asiatic immigrants, who, probably coming from Babylonia, entered the land from the south and east and transformed the marshes, through which the waters of the Nile made their way, into a rich and highly cultivated country[3].

This idea of the primeval substance being water we find also among the Greeks, but notwithstanding this resemblance we can hardly suppose that the Greek philosophers were to any appreciable extent indebted to the Orientals for any of their

[1] Maspero, *Dawn of Civilization, Egypt and Chaldæa*, London, 1894, pp. 16–19, 85–96. Sayce, l. c. p. 168.
[2] Budge, *Egyptian Ideas of the Future Life*, p. 22.
[3] Sayce, l. c. p. 130.

first notions on the construction of the world. At first Greek
cosmological speculation like that of the eastern nations proceeded on purely mythological lines. The origin of the world
was sought for in a childish manner; supernatural beings were
invented, and earth, sea, and sky were believed to have been
generated by them. Thus Erebus begat with Nyx the æther
and the day, the Earth by itself produced the sea, and with
the Heaven the rivers, &c. But eventually the Greeks shook
themselves free from mythological trammels; they endeavoured
to find the laws which regulated the phenomena of nature
without continual interference from supernatural and capricious
beings, and they reasoned with a freedom of thought to which
their Oriental precursors had never risen.

Greek philosophy does not date further back than the first
half of the sixth century B.C., a date which is but as yesterday
in comparison with the ages which had then elapsed since the
light of civilisation had first commenced to brighten human
existence in Chaldæa, Egypt, and the lands surrounding the
Ægean Sea. Fortunately we possess a splendid source of
information as to the ideas of the universe which prevailed
before the days of the first philosophers, as the Homeric poems
present us with a striking picture of the earth and the heavens
as they appeared to the Greeks in the two or three centuries
immediately preceding the days of Thales. The earth is represented as a flat circular disc surrounded by the mighty river
Okeanos, which starting north of the pillars of Herakles winds
its way north, east, and south of the earth back into itself,
while the heavens like a huge bell cover the whole. The earth
is partly covered by the sea, not only by the Mediterranean but
also by a larger northern sea (traversed by the Argonauts), but
whether this is separated from Okeanos is not stated. In the far
east is the lake of the sun, a large gulf of Okeanos, evidently
the Caspian Sea. South of Egypt and Libya the land reaches
to Okeanos and includes the land of the pygmies, rumours of
which beings must have reached Greece from Egypt, whence
expeditions to the land of the pygmies had been made as early
as B.C. 3300[1]. Beyond Okeanos, to the south-west, is the dark

---

[1] Budge in *Proc. R. Soc.* LXV. p. 347.

and mysterious land of the Kimmerians, who have the un-
enviable distinction of living next door to Erebus, the land of
the dead, which stretches westward from Okeanos in deeper
and deeper gloom, but is otherwise not unlike the earth, having
hills, plains, and rivers, as may be seen from the eleventh book
of the *Odyssey*. In other passages, however, the realm of
Hades is supposed to be beneath the surface of the earth, at
equal distances from the height of heaven above and the depth
of Tartarus below[1]. Over the earth is the region of the ether,
over which again the brazen vault of heaven is thrown, under
which the sun, moon, and stars are moving, rising from Okeanos
in the east and plunging into it again in the west. In the
east, before sunrise, Aurora, preceded by the morning star,
rises from Okeanos, after which Helios makes his appearance
from the lake of the sun. What becomes of the heavenly
bodies between their setting and rising is not stated, but since
Tartarus is never illuminated by the sun[2], they cannot have
been supposed to pass under the earth. Erebus and the land
of the Kimmerians appear to be a kind of appendage to the
earth (like a salad plate), since they are beyond the place
where the sun sets, while thick fog hides even the setting sun
from them[3]. Elysium, the home of the blessed, is also at the
ends of the earth, free from rain and snow, where a gentle west
wind is blowing. In after times these insulæ fortunatæ were
identified with the Canary Islands[4].

The cosmology of Hesiodus is founded on similar ideas, the
broad-breasted earth and the heavens " like unto it " being
developed from Chaos. The dwelling of the dead is subter-

[1] *Il.* VIII. 16, comp. IX. 457, XXII. 482, &c.
[2] *Il.* VIII. 480.
[3] *Od.* XI. 15, 93.
[4] The ignorance of the Homeric poets with regard to the geography of the
West is not to be interpreted as proving that the Greeks had never yet crossed
the seas to the west of Greece. Remains of Mykenean culture in the West
prove that there had been intercourse in early times, but the invasions and
convulsions which brought the Mykenean age to a close about B.C. 1000 severed
the communication between Greece and the West for several centuries; hence
the total silence of the *Iliad* as to western lands. The *Odyssey* probably owes
much to the earliest Milesian and other travellers, who began to go westward
again about the eighth century. See Hall, *The Oldest Civilisation of Greece*,
1901, p. 258.

ranean, above Tartarus where the Titans are imprisoned, and
like Homer Hesiodus makes the depth of Tartarus equal to the
height of the heavens[1], so that the universe is a sphere, divided
by the plane surface of the earth into two hemispheres, while
Chaos is interposed between the lower surface of the earth
and Tartarus[2]. At the western extremity of the earth are the
sources of Okeanos, from whence a branch called the Styx finds
its way to the land of Hades underground.

In one or two places[3] Homer mentions Okeanos as the
origin of everything, and this idea, which as we have seen also
occurs in Babylonian and Egyptian mythology and seemed to
be confirmed by the great extent of the ocean which was
revealed by the advance of geographical knowledge[4], was the
leading one in the cosmology of the earliest philosopher of
Greece and forms a connecting link between the primitive
popular notions and the first attempt at philosophical enquiry.

[1] *Theog.* 720.

[2] Ibid. 814. Opinions have differed as to what Hesiodus meant by chaos;
Plutarch (*Compar. aquæ et ignis*) supposed it to be water.

[3] e.g. *Il.* xiv. 246.

[4] The first Greeks who penetrated through the Pillars of Herakles into the
Atlantic Ocean appear to have been Kolæus the Samian and the Phokeans,
about B.C. 640 (Herod. i. 163, iv. 152).

CHAPTER I.

THE EARLY GREEK PHILOSOPHERS.

OUR knowledge of the doctrines held by the first philosophers of Greece is to a great extent only founded on second or third hand reports. None of the writings of the Ionian philosophers have reached us, and we only possess fragments of those of the other pre-Sokratic thinkers, so that, if we except a few allusions in Plato's works, Aristotle is the earliest author in whose works we find frequent references to the speculations of his predecessors. These references are very valuable, though they must be used somewhat cautiously on account of the polemical tone of Aristotle towards most of the earlier philosophers; but they only make us regret all the more that the book on the history of physics which Theophrastus, Aristotle's principal disciple, is known to have written, is lost with the exception of a few fragments. We have to be thankful for the various compilations of later writers which have been handed down to us and which appear to have been founded on the book of Theophrastus, or rather on later extracts from it, to which were added short accounts of the philosophers who lived after the time of Theophrastus. Of most of these compilers we know next to nothing, hardly even when they lived, except that the earliest flourished in the first century of our era. The *Lives of Philosophers* of Diogenes Laertius (who lived in the second century) is a very carelessly made compilation by a writer who does not always appear to understand what he is writing about, while he is more interested in trivial anecdotes than in the doctrines taught by the philosophers. Still he gives us much valuable information which we should otherwise not have possessed,

and even some (such as his biographical details) which is
independent of Theophrastus. A more important compilation
by another Greek writer is that known by the Latin title
of *Placita Philosophorum* which formerly was attributed to
Plutarch, though it is quite certainly not written by that
great author[1], and is only a reproduction of an older work
which seems to have been lost at an early date. In the form
in which it has been handed down to us the book gives in
separate paragraphs arranged according to subjects the opinions
of philosophers on all conceivable matters, from the rising of
the Nile to the nature of the soul. We possess another
compilation, differently arranged, but founded on the same
source, in the *Eclogæ physicæ* of Stobæus, probably put
together about the fifth century of our era, and generally
agreeing word for word with the *Placita*, though sometimes
one, sometimes the other author goes a little more into detail.
The source of both seems to be a work by a certain Aëtios
mentioned and quoted by Theodoret; and assisted by these and
other quotations Hermann Diels has from the Pseudo-Plutarch
and Stobæus reconstructed the work of Aëtios in a masterly
manner[2]. Several Christian writers found it worth while to
make abstracts from Aëtios in order to obtain weapons against
pagans and heretics. Thus Eusebius copied a considerable part
of the *Placita* into his *Preparatio Evangelica*, while another
writer, now generally acknowledged to be Hippolytus, a well-
known controversial author of the third century, wrote a book
called *Philosophumena*, parts of which are from biographical
sources similar to those used by Diogenes, while others are
founded on Theophrastus but are independent of Aëtios. It
served as an introduction to a *Refutation of all Heresies*, and

---

[1] The philosophical works of Plutarch contain, however, many fragments,
chiefly of Herakleitus, and the *Stromata* attributed to him and preserved by
Eusebius are not without value.

[2] *Doxographi Græci, collegit... H. Diels*, Berlin, 1879. In a lengthy introduction the question of the origin of all the various compilations is most
thoroughly discussed. Translations of the fragments and doxographic paragraphs relating to the Ionian and Eleatic philosophers are found in P. Tannery,
Pour l'histoire de la science Hellène, de Thalès à Empédocle, Paris, 1887, and in
A. Fairbanks, *The First Philosophers of Greece*, London, 1898. The latter also
gives the Greek texts of the fragments.

was till lately ascribed to Origen[1]. Though not derived from the same source, we may also mention in this connection the *Lexicon* of Suidas, a kind of encyclopædia from about the tenth century. In addition to these compilations we have most valuable sources of information about Greek cosmology and astronomy in the work of Theon of Smyrna, written in the second century after Christ and founded on the writings of Adrastus and Derkyllides, now lost[2], as well as in the commentary of Simplicius to Aristotle's book on the Heavens.

The scantiness of our information about the earliest philosophers is nowhere more apparent than in the accounts of the earliest Ionian philosopher, Thales of Miletus, who was born about the year B.C. 640, and died at the age of 78. His cosmical ideas were as primitive as those of Homer. The earth is a circular disc floating " like a piece of wood or something of that kind[3] " on the ocean, the water of which is the principle or origin of everything, and from the evaporation of which the air is formed. Speaking of the " first principles " of early philosophers, Aristotle says[4]: " As to the quantity and form of this first principle there is a difference of opinion; but Thales, the founder of this sort of philosophy, says that it is water (accordingly he declares that the earth rests on water), getting the idea, I suppose, because he saw that the nourishment of all beings is moist, and that heat itself is generated from moisture and persists in it (for that from which all things spring is the first principle of them), and getting this idea also from the fact that the germs of all beings are of a moist nature." Agitation of the water caused earthquakes[5]. The celestial vault limits the world above, while nothing is said about the lower limit or the support of the

[1] The *Philosophumena* are included in the work of Diels, as are also the philosophical history of Pseudo-Galenus and some other extracts from Aëtius.

[2] *Theonis Smyrnæi liber de Astronomia*, ed. Th. H. Martin, Paris, 1849. This is only part of a larger book on the mathematics useful in reading Plato, edited by E. Hiller, Leipzig, 1878.

[3] Arist. *de Cælo*, II. 13, p. 294 a; comp. Seneca, *Quæst. Nat.* III. 13, "terrarum orbem aqua sustineri."

[4] *Metaph.* I. 3, p. 983 b.

[5] Aet. III. 15, Diels, p. 379.

ocean; but apparently the water, being the first principle of everything, does not require anything to support it and was possibly regarded as infinite. Neither are we told what becomes of the stars between their setting and rising, but they were probably believed to pass literally behind and not under the earth.

When we remember that, according to Herodotus (I. 74), Thales was able to predict a solar eclipse, or at least the year in which it happened (probably B.C. 585), it seems at first sight surprising that he had not a more advanced conception of the universe. But it must be borne in mind that this prediction (if it was really made) can only have been the result of some knowledge which Thales had gathered during the lengthy stay which he is supposed to have made in Egypt; and the Egyptians had only borrowed their knowledge of the motion of the sun and moon from the Chaldeans, of whom we know that they were able to some extent to foretell eclipses even at an earlier date. It is, to say the least, very strange that neither Aristotle nor any other astronomical author of antiquity mentions this prediction[1], and of course it is quite out of the question that Thales should have been able to predict how large an eclipse of the sun would be for a particular locality[2]. But whatever announcement he was able to make caused him to be considered a prodigy of wisdom, and centuries after his time, when his undoubted merits as the founder of Greek geometry were still unforgotten notwithstanding the rapid development of this

---

[1] Theon of Smyrna (ed. Martin, p. 322) states very shortly that Eudemus in his history of astronomy had mentioned (among other discoveries) that Thales first "found the eclipse of the sun," but this might simply mean the cause of an eclipse. Diog. Laert. I. 23 tells us that Thales was the first to study astronomy and foretold the eclipses and motion of the sun, as Eudemus relates in his history of astronomy (which is lost). Diogenes is generally very slipshod in his statements; a little below (I. 24) he says that Thales found the sun to be 720 times as great as the moon. This is doubtless a misunderstanding of the fact that the daily path of the sun (360°) is 720 times the moon's apparent diameter. But it is unlikely that this was known at the time of Thales.

[2] He is also said to have been able to foretell the storm which saved the life of Crœsus and even a specially good harvest of olives, by which he realised a large sum of money (Diog. Laert. I. 26). Anaximander is said to have predicted an earthquake (Pliny, *H. N.* II. 191) and Anaxagoras the fall of an aërolite (ibid. II. 149). These stories do not tend to strengthen our belief in the prediction of the eclipse.

science, it became the fashion to consider Thales as the author of many scientific and philosophical truths which in reality were not known till long after his time. Solon was in a similar manner credited with the introduction of many political changes not thought of till much later. We read in the doxography[1] that Thales was aware of the fact that a solar eclipse is caused by the sun being covered by the moon (both of which bodies he held to be of an earthy nature), while he also taught that the moon was illuminated by the sun and became eclipsed when passing through the earth's shadow[2]. These statements must be viewed with some suspicion, particularly as the same compiler in another place[3] tells us that Thales held the earth to be spherical, a statement totally opposed to the same writer's own account (confirmed by other writers) as to the earth being supported by water in the cosmical system of Thales. But when even Homer and Hesiodus were supposed to have known that the earth was a sphere[4], why not Thales also ?

The second philosopher of the Ionian school, Anaximander, a younger contemporary of Thales (about 611–545), did not advance much further in his ideas of the construction of the world. He is credited[5] with having introduced the use of the gnomon among the Greeks, but it had long before his time been in use among the Babylonians[6]. Anaximander gave up the idea that water or any other known substance might be the first principle, and held that this is of the nature of the infinite (τὸ ἄπειρον), that is, matter without any determinate property except that of being infinite. All things are developed out of this and return to it again, so that an infinite series of worlds have been generated and have in turn become again resolved into the abstract mass[7]. The earth is flat or convex

---

[1] Aet. II. 24 and 28, Diels, pp. 353, 358.

[2] The latter only according to Stobæus, Diels, p. 360.

[3] Aet. III. 10, Diels, p. 376.

[4] By Krates of Mallus in the second century B.C., see Susemihl, *Gesch. d. griech. Lit. in der Alexandrinerzeit*, II. p. 5.

[5] By Diog. L. II. 1, and Suidas.

[6] Herod. II. 109.

[7] Aet. I. 3, Diels, p. 277 ; *Strom.* II. Diels, p. 579 ; Cicero, *Nat. Deor.* I. 25, says that Anaximander considered the innumerable worlds to be gods. Zeller, *Phil. d. Griechen*, I. p. 230, thinks that this refers to the stars.

on the surface, but more like a cylinder or a stone column than a disc, the height being one-third of the breadth[1]. We know from Aristotle that Anaximander believed the earth to be in equilibrium in the centre of the world, because it was proper for it not to have a tendency to fall in any particular direction, since it was in the middle and had the same relations to every part of the circumference. The successful sea voyages of the Phokæans had by this time enlarged the Greek horizon, and Anaximander constructed a map of the earth, which is alluded to by Herodotus, who points out the error of assuming the surrounding ocean to be a narrow stream, and Europe and Asia to be of equal size[2]. The heavens were by Anaximander supposed to be of a fiery nature and of a spherical form (a distinct advance), enclosing the atmosphere "like the bark on a tree," and this enclosure forms a number of layers, between which the sun, moon, and stars are situated at different distances, the sun being the most distant and the fixed stars the nearest to us[3]. This shows at once how little the celestial phenomena had been watched, as the frequently occurring occultation of a bright star by the moon must have been unknown to Anaximander; but he has at any rate the credit of having been the first to speculate on the relative distances of the heavenly bodies[4]. With regard to the sun we are to imagine a wheel or ring with a diameter twenty-seven or twenty-eight times as great as the diameter of the earth. The

---

[1] Aet. iii. 10, p. 376; Hippolytus, vi. Diels, p. 559 (comp. p. 218), says that the top was convex, γυρὸν στρογγύλον. Diogenes (ii. 1) is certainly mistaken in stating that Anaximander taught the spherical form of the earth, as this would have been alluded to by Aristotle (*De Cœlo*, ii. 13, p. 295 b) when referring to the ideas of Anaximander about the equilibrium of the earth. Theon (ed. Martin, p. 324) quoting Eudemus, a disciple of Aristotle, says that Anaximander thought the earth was suspended in the air and "moves round the centre of the world." Here again we may be certain that this is neither a rotatory nor a progressive motion, as Aristotle, when denying both these kinds of motion, would not have failed to mention Anaximander in this connection. Neither would Aetius have passed it over. Probably Martin is right when he suggests that the motion alluded to was simply earthquakes (Theon, p. 49).

[2] Herod. iv. 36; Strabo, i. p. 7.

[3] *Strom.* ii. Diels, p. 579; Aet. ii. 15, p. 345; Hippol. *Phil.* vi. p. 560.

[4] According to Eudemus, quoted by Simplicius in his commentary to Aristotle, *De Cœlo*, ii. 10, p. 471 (Heiberg's ed.).

hollow rim of this wheel is filled with fire, which is only seen through a hole in the rim, equal in size to the earth[1]. This explanation was also extended to the moon and stars, the diameter of the lunar ring being nineteen times as large as that of the earth. The moon was self-luminous, and both solar and lunar eclipses were caused by temporary stoppages of the apertures in the rings[2], while the phases of the moon were accounted for by regularly recurring partial stoppages[3]. It is somewhat difficult to make out how Anaximander imagined the solar and lunar rings and the stellar sphere and spheres to be constituted, as the two latter (or at least the stellar sphere) must have been transparent to the sunlight without letting their own "fire" appear except at the apertures. In the case of the moon no doubt its wheel would be inclined to the plane of the sun's wheel (by an angle of 5°, which is the inclination of the lunar orbit), but even so there would be very lengthy solar eclipses twice a year, unless the lunar wheel was perfectly transparent to sunlight. But these matters of detail had probably not been considered by Anaximander, unless perhaps by fire he merely meant a very subtle substance which only became luminous at the apertures[4]. We are also

[1] There is a discrepancy with regard to the size of the solar wheel between Aet. II. 20, p. 348 (Galen, c. 62) and II. 21, p. 351 (Galen, c. 63, p. 626), the former giving 28, the latter 27. The apparent diameter of the sun ought to have been 4° 15′ instead of about 30′, if the figures of Anaximander were right, but no doubt this might have escaped his attention. But on the other hand Hippolytus (*Philos.* VI. Diels, p. 560) says that the circle of the sun is 27 times *that of the moon*, and as he had before him in this case a very reliable extract from Theophrastus, independent of that used for the *Placita* (Diels, p. 153), it is very possible that he is right and that the distance of the sun was $19 \times 27 = 513$ times the diameter of the earth.

[2] Aet. II. 24, 28, 29, Diels, pp. 354–59.

[3] Hippol. *Philos.* VI. Diels, p. 560.

[4] This explanation is suggested by Tannery, *Pour l'hist. de la sc. Hell.* p. 92. These wheels have given commentators a good deal of trouble, e.g. Achilles, *Isagoge in Arati Phæn.* XIX. (Petavii *Doctrina Temporum*, 1703, p. 81): "For as in a wheel the nave is hollow and the spokes part from it towards the rim, so the sun, emitting its light from a hollow, spreads its rays outward and illuminates everywhere." But he has apparently not grasped the idea of the wheel correctly. Martin (*Mém. de l'Acad. des Inscript.* XXIX. pp. 72–86) seems to have been influenced by Achilles. He imagines that the wheels were suggested by the phenomena of solar and lunar halos, that the wheels were therefore not wound round the earth, but that the luminous apertures were at

ignorant whether Anaximander considered the breadth (from north to south) of the rings to be equal to the actual diameters of the sun and moon or much greater, also whether he supposed the wheels to revolve and thereby produce the daily rotation of the heavens and the orbital motions of the sun and moon, but probably the system never advanced beyond a mere sketch and was not worked out in detail.

The cosmical ideas of Anaximenes of Miletus, the third philosopher of the Ionian school, who lived about the middle or second half of the sixth century B.C., were fully as primitive as those of his predecessor. The stars are attached "like nails" to the celestial vault, which is of solid, crystalline material[1], but it is not stated whether he considered it to be a sphere or a hemisphere, though the latter seems most likely, as he supposed that the sun and stars when setting do not go under the earth, but merely pass behind the northern, highest part of it[2]. We are also told that the firmament turns round the earth "like a hat round the head," which looks as if he believed it to be a hemisphere. In the philosophical system of Anaximenes the first cause of all things was air, out of which primary matter all things were formed by compression or rarefaction, so that the flat earth[3] was first produced from air made dense, and from this again by vaporisation fire, out of which the sun, moon, and planets were formed by the revolution of the heavens. The broad earth is supported on air, and the sun, moon, and stars (that is, probably only the planets) are flat bodies[4], prevented

the *centres* of large discs or wheels, the circumferences of which occasionally became visible as halos. But this extraordinary interpretation cannot possibly be reconciled with the clear and unequivocal statements of the doxographers, particularly when we remember the trustworthiness of Hippolytus. Comp. Diels, *Doxographi*, pp. 25 and 156, where it is shown that the comparison of the fire emerging from the holes ὥσπερ διὰ πρηστῆρος αὐλοῦ (*Plac.* II. 20) refers to the pipe of a smith's bellows, so that the holes were certainly not at the centres of the circles.

[1] Aet. II. 14, p. 344.

[2] Hippol. *Philos.* VII. Diels, p. 561, Aet. II. 16, p. 346. Compare Aristotle, *Meteorol.* II. 1, p. 354 a, where this opinion is attributed to "ancient meteorologists."

[3] "Like a table," Aet. III. 10, Diels, p. 377.

[4] Ps.-Plutarch, *Stromata*, III. Diels, p. 580, says that the sun is an earth to which the motion has imparted heat.

from falling by the support of the air, while the density of the air limits their motion in declination[1]. The heat of the sun is caused by the rapidity of its motion, but the stars do not give out any heat owing to their great distance. According to Hippolytus, Anaximenes taught that in addition to the stars there are also in the place occupied by them bodies of a terrestrial nature, carried along with the stars in their motion. Theon of Smyrna says that Anaximenes was aware that the moon borrows its light from the sun, and knew the true cause of lunar eclipses, but this is not mentioned by any other writer[2].

The Ionian school had not advanced very far in the direction of a rational idea of the universe. The earth was flat, the fixed stars were attached to a vault, the planets are barely mentioned, and the nature of the sun and moon very imperfectly understood. But at the other extremity of the Greek world, in the south of Italy, a philosophical school arose in the second half or towards the end of the sixth century, within which by degrees much sounder notions about the heavenly bodies were developed. But it will be more convenient to discuss the opinions of Pythagoras and those of his successors together, and first to finish our review of the cosmical ideas of the remaining pre-Sokratic philosophers. Several of these were to some extent influenced by Pythagoras, and we shall therefore mention here in anticipation that the spherical form of the earth was recognised by the Pythagoreans, most probably already by the founder of the school.

---

[1] Aet. II. 23, p. 352. This seems at least to be the meaning of the statement that the stars are beaten back by the condensed and resisting air. The paragraph is headed Περὶ τροπῶν ἡλίου, i.e. On the solstices, and though only τὰ ἄστρα are mentioned, this word may be supposed to mean the moon and planets which are confined within narrow limits in declination as well as the sun. That the moon in particular has its τροπαί must have been known very early, and it is difficult to see how Zeller (I. 223, note 3) can say, "Von einer den Sonnenwenden zur Seite gehenden Umbiegung der Mondbahn wusste die griechische Astronomie so wenig wie die unsrige." The inclination of the lunar orbit to that of the sun is so small (5°) that the phenomena of "turning back" of sun and moon are very similar.

[2] Theon, ed. Martin, p. 324. Pliny (*H. N.* II. 187) says that the gnomon was invented by Anaximenes, and that he exhibited at Sparta the sundial called Sciotherion. The latter part of the statement may be true.

Almost at the same time as the Pythagorean school there sprang up in the south of Italy another renowned philosophical school, the Eleatic. Its founder was Xenophanes of Kolophon, poet and philosopher, who was born about the year 570 and lived to a great age. He was opposed to the popular polytheistic religion and taught that God is one, though the question whether he ought to be called a pantheist or a monotheist will probably never be settled. The commentator Alexander of Aphrodisias states[1] that according to Xenophanes the first principle is limited and spherical, being homogeneous throughout. Apparently this statement does not rest on any good authority, but in any case it is certainly not to be understood as implying the spherical form of the universe (unless the opinions of Parmenides have been confounded with his predecessor's), and Xenophanes probably merely intended to express poetically that the influence of the deity extends in every direction. There cannot be any doubt as to the opinions of Xenophanes on the form and nature of the world, as they are amply testified to by various writers. According to him the flat earth has no limits, it is "rooted in the infinite," and the air above it is also unlimited[2]. The sun, stars, and comets are fiery clouds formed by moist exhalations which are ignited by their motion[3], and this motion is rectilinear, the circular forms of their daily paths being only an illusion caused by their great distance[4]. The stars are extinguished every morning and new ones formed in the evening, while the sun is likewise formed every day from small fiery particles which

---

[1] Simplicius *in Phys. Aristot.* (Diels, *Doxogr.* p. 481), also Hippolytus, *Philos.* xiv. (Diels, p. 565). The source of both appears to be the pseudo-Aristotelean book *De Melisso, Xenophane et Gorgia*, p. 977 b, 2, and it appears to be totally at variance with the distinct statement of Aristotle (*Metaph.* i. 5, p. 986 b) that Xenophanes had not given any opinion as to the nature of the unity. There must have been considerable difference of opinion among commentators on this subject, as Simplicius also says that Nicolaus of Damascus mentions Xenophanes as saying that the first principle is infinite and immovable.

[2] Arist. *De Cœlo*, ii. 13, p. 294 a. Simplicius, p. 522, 7 (Heib.), thinks that this might merely mean that the earth stretches downward towards the infinite. Compare Achilles (Petavius, iii. p. 76) τὰ κάτω δ' ἐς ἄπειρον ἰκνεῖται, also Aet. iii. 11, p. 377.

[3] Aet. ii. 20, and Stobæus, p. 348; Aet. iii. 2, p. 367.

[4] Aet. ii. 24, p. 355.

are gathered together[1]. The moon is a compressed cloud,
shining by its own light and extinguished every month[2].
There are many suns and moons, according to the various
climates and regions and zones of the earth, and sometimes the
sun arrives at an uninhabited region, and being in an empty
place it becomes eclipsed[3]. All this is extremely primitive,
though we should probably do Xenophanes an injustice if we
suppose that he had formulated a regular theory of the
universe[4], indeed he says in one of the fragments (No. 14) we
possess of his philosophical poem that certainty of knowledge
is unattainable. All the more refreshing is it to meet with the
following observation and rational explanation. Xenophanes
noticed that shells are found on dry land and even on mountains, while in several places, such as the quarries at Syracuse,
imprints of fish and of seals have been found; these imprints,
he says, were made long ago when everything was covered with
mud and the imprints then dried in the mud. He believed
that the earth would eventually sink into the sea and become
mud again, after which the human race will begin anew[5].

Xenophanes was perhaps more of a poet than of a philosopher, and the real founder of the Eleatic school was Parmenides
of Elea, who lived in the early part of the fifth century. He is
said by Diogenes Laertius (IX. 21) to have attached himself to
the Pythagoreans, but he was in his philosophy influenced by
Xenophanes. His doctrines are expounded in a short poem on
Nature, fragments of which are left. In opposition to the
Ionians he does not let everything be derived from a primary

[1] Aet. II. 13, p. 343, and Hippolytus, XIV. p. 565.
[2] Aet. II. 25, and Stob., Diels, pp. 356, 358, 360.
[3] Aet. II. 24, p. 355.
[4] An ingenious suggestion has been made by Berger (*Gesch. d. wissensch.
Erdkunde der Griechen*, II. p. 20) that Xenophanes in talking of the many suns
according to various climates merely expresses himself as we do when we speak
of the Indian sun, the midnight sun, &c., while his mention (Aet. II. 24) of a
solar eclipse which lasted a whole month is turned into an allusion to the long
winter night of the arctic regions! But it does not seem possible to accept these
explanations, as all the other statements attributed to Xenophanes only permit
a literal interpretation and do not in any way point to his having held the
earth to be a sphere and climate to depend on latitude.
[5] Hippolytus, XIV., Diels, p. 566.

substance, nor does he with the Pythagoreans say that everything is number; he only accepts as truth that the ent (or
existent) is, while the nonent (or non-existing) is not, the ent
being "perfect on every side, like the mass of a rounded sphere,
equidistant from the centre at every point," continuous[1]. There
is no such thing as a void, and there is therefore in reality
no change and no motion, as these are not conceivable without
an empty space. With regard to the phenomenal world he
considers the attainment of truth impossible, owing to the
imperfection of our senses, which lends a semblance of plurality
and change to the ent, and he merely claims probability for his
speculations. Parmenides supposes two elements, fire or light,
subtle and rarefied, and earth or night, dense and heavy,
analogous respectively to the ent and the nonent.

Notwithstanding the close connection of his philosophical
doctrines with those of Xenophanes, Parmenides was able to
perceive the spherical form of the earth, and he deserves great
credit for having taken this great step forward, which no
philosopher outside the Pythagorean school was sufficiently
unprejudiced to take till Plato appeared. Theophrastus attributed the discovery to Parmenides and not to Pythagoras[2],
and the former was therefore probably the first to announce it
in writing. He is also said to have been the first to divide the
earth into five zones, of which he made the central, torrid and
uninhabited one, nearly twice as broad as it was afterwards
reckoned to be, extending beyond the circles of the tropics into
the temperate zones[3]. We cannot doubt that the true figure
of the earth was first made clear through the reports of
travellers about certain stars becoming circumpolar when the
observer proceeded to the north of the Euxine, while a very
bright star (Canopus), invisible in Greece, was just visible above
the horizon at Rhodes, and rose higher the further the navi-
gator went south. Travellers had probably also announced the
different length of the day in different latitudes, a fact which

[1] Fragm. 102–109, Fairbanks, p. 96.

[2] Diog. L. VIII. 48; comp. IX. 21, where it is said that Parmenides was the
first to assert the spherical form.

[3] Strabo, II. p. 94 (on the authority of Posidonius); Aet. III. 11, p. 377,
where the extra width of the torrid zone is not mentioned.

has even been supposed to have been known to the writer of the *Odyssey*[1]. Parmenides may, however, also have supposed that the earth ought to be of the same figure as its surroundings[2], as he arranged the universe in a series of concentric layers round the earth. This is the first time we meet with the system of concentric spheres, which afterwards played so important a part in the history of astronomy. The outermost of these layers, "the extreme Olympos," is a solid vault, chained by Necessity to serve as a limit to the courses of the stars[3]. Next to it comes a layer formed of the subtle element, and under this layers (στεφάναι) of a mixed character[4], first the morning and evening star (which he knew to be the same) in the ether, next below which he placed the sun and then the moon, both being of a fiery nature and equal in size. The moon is, however, also said to derive its light from the sun, and to appear "always gazing earnestly towards the rays of the sun" (i.e. having its bright part turned towards the sun); he also says that it shines with borrowed light[5]. The sun and moon have been formed of matter detached from the Milky Way, the sun of the hot and subtle substance, the moon chiefly of the dark and cold[6]. The "other stars in the fiery place which he calls the heavens" (οὐρανὸν) are below the sun and nearer to the earth[7], a strange error into which, as we have seen, Anaximander had also fallen. In the middle is the earth, which Parmenides, like other early philosophers, believed to be in equilibrium there because it had no more tendency to fall in one direction than in another[8].

[1] *Odyss.* x. 82 ff. At Telepylos the shepherd returning at eve meets another driving his flock out at sunrise, and a man who did not require sleep might earn twofold wages. It is of course quite possible that this is pure phantasy and is not the outcome of any knowledge of the greater length of the day in summer in higher latitudes, still the passage was already by Krates of Mallus interpreted in this way. Geminus, vi. 10, ed. Manitius, p. 72. Compare above, p. 13, note 4.

[2] As Aristotle afterwards did, *De Cœlo*, ii. 4, p. 287 a.

[3] Frag. 137–39, Fairbanks, p. 100.

[4] Aet. ii. 7, Diels, p. 335.

[5] Frag. 144–45, Fairbanks, p. 100, comp. Aet. ii. 26, p. 357. On the other hand, Stobæus (Diels, p. 361) says that the dark spots in the face of the moon are caused by dark matter mingled with the fiery, for which reason Parmenides called it the star that shines with a false light.

[6] Stobæus, Diels, p. 349.   [7] Stobæus, Diels, p. 345.

[8] Aet. iii. 15, p. 380.

In the midst of all (and therefore apparently in the centre of the earth) there is a divinity (δαίμων) who rules all, and she generated Eros first of all the gods[1]. This is apparently a reminiscence of an Orphic hymn, in which Hestia, daughter of Kronos, occupies the central position of the eternal fire, and Parmenides perhaps merely introduced it in order to give dignity to the central position of the earth in a poetical manner[2].

A great contrast to the opinions of the Eleatic school, the metaphysical as well as the cosmological, is afforded by the doctrines of Herakleitus of Ephesus, who seems to have flourished about the year 500 B.C. The leading idea in his philosophical system is that nothing is at rest, the sole reality is "Becoming," and to express the never-ending changes in nature he selected fire as the principle, from which all things are derived and into which they must eventually be resolved, after which matter will again be produced as before and a new world will arise[3]. The fire first became water, and from this again the solid earth was formed. There is an unceasing circulation in nature, "upward, downward, the way is one and the same"[4]; in fact, the whole doctrine of Herakleitus centres in this up and down motion of the all pervading essence, and there can therefore be no doubt that he believed the earth to be flat, though this does not seem to be expressly stated anywhere. On rising from the earth moist exhalations are caught in a hollow basin with its cavity turned towards the earth and are ignited as this basin rises from the sea in the east, to be afterwards extinguished when it sets in the west. In this way the sun is produced, and as the sun is constantly renewed, it is

---

[1] Frag. 128–132, Fairbanks, pp. 99–100.

[2] See a memoir by Martin, "Sur la signification cosmographique du mythe d'Hestia dans la croyance antique des Grecques," *Mém. de l'Acad. des Inscr.* XXVIII. 1, pp. 335–353.

[3] Herakleitus appears even to have fixed the period of these renewals of the world, his annus magnus being 10,800 years (Censorinus, *De Die Nat.* c. 18) or 18,000 years (Aet. II. 32, p. 364). The former is probably the right number, as suggested by Tannery (*Pour l'hist. de la sc. Hell.* p. 168), as it is = 30 × 360 or one generation for every day of the year, 30 years being the length of a generation according to Herakleitus.

[4] Frag. 69, Fairbanks, p. 40.

"always young"[1], and it shines more brightly because it moves in purer air, while the moon, which is also a bowl filled with fire, shines more dimly because it moves in thicker air[2]. The stars are fainter owing to their great distance[3]. Eclipses (and probably also the phases of the moon) are caused by the turning of these boat-like basins, whereby their other, non-luminous sides become directed towards us[4]. The sun is only as large as a human foot[5], a curious fallacy which long afterwards was repeated by Epikurus. Day and night, summer and winter depend on the predominance of bright and dark, hot and cold evaporations[6].

The idea of these boats charged with very light material, sailing along the heavenly vault, is quite in keeping with the mythological fables of the Chaldeans and Egyptians mentioned in our introductory chapter. The resemblance may of course be purely accidental, and Herakleitus may never have heard of these myths, but at any rate his hypothesis supplied a plausible popular explanation of the mystery, what became of the sun, moon, and stars between setting and rising, as the boats might gaily pass along the Okeanos and be ready in good time for a fresh ascent.

The views of Empedokles of Agrigentum (about 450 B.C.) were as quaint as those of Herakleitus, and combined Ionian with Pythagorean and Eleatic doctrines. He assumed that there are four primary elements—fire, air, water, and earth— imperishable and unchangeable in quality; and he was the first to fix on these four as such, while his predecessors had only assumed one or two elements[7]. To explain the combinations of the elements he supposes that they are in turn joined to and separated from each other by two motive forces or divine agencies of an attractive and repulsive nature, love and discord, which respectively combine and separate the elements in various proportions and thereby produce all the manifold pheno- mena of nature. Love and discord alternately predominate,

---

[1] Arist. *Meteor.* II. 2, p. 355 a.   [2] Aet. II. 28, Diels, p. 359.
[3] Diog. IX. 10.   [4] Aet. II. 29, p. 359.
[5] Aet. II. 21, p. 351.   [6] Diog. IX. 11.
[7] The word element, στοιχεῖον, is not used by him, it was introduced by Plato.

and the history of the world is thus divided into periods of
different character. Empedokles considered the finite universe
to be spherical, solid, made (at the beginning of the present
period) of condensed air after the manner of crystal[1]. To this
sphere the fixed stars, which were formed of fiery matter
pressed upward by the air, are attached, while the planets
wander freely in space[2]. The moon is air rolled together and
mixed with fire, it is flat like a disc and illuminated by the
sun[3]. Empedokles assumed the existence of two separate
celestial hemispheres[4], one of fire and another of air with only
a little fire, or a day and a night side; and as the sphere was
made to rotate by the push of the fire, the two halves are in
turn above the earth and cause day and night, "the earth makes
night by coming in front of the lights[5]," i.e. by screening the
luminous hemisphere. The ideas of Empedokles about the sun
are peculiar; they are described by Aëtius as if there were two
suns, one in one hemisphere and the other merely its reflection
" from the rounded earth and carried along with the motion of
the fiery atmosphere," and he adds: " Briefly speaking, the sun
is a reflection of the fire surrounding the earth," and in the
next paragraph : " The sun which faces the opposite reflection
is of the same size as the earth[6]." But the doxographic writer
has certainly misunderstood the meaning of Empedokles, as
we read in the *Stromata* of the Pseudo-Plutarch : " The sun is
not of a fiery nature, but a reflection of fire, like that which is
produced in water." This is intelligible enough and agrees
with the statement that the course of the sun is the limit of
the world[7]." The sun is therefore simply an image of the fiery

[1] Aet. II. 11, p. 339. Stobæus (Diels, p. 363) says that Empedokles remarked
that the height of the heavens was less than its breadth, the universe having the
shape of an egg.

[2] Aet. II. 13, pp. 341-2.

[3] Aet. II. 25, 27, 28, pp. 357 b, 358; Ps.-Plutarch, *Strom.* Diels, p. 582,
Achilles, *Isagoge*, XVI. (Petavius, 1703, p. 80).

[4] Ps.-Plutarch, l.c., Aet. III. 8, p. 375.

[5] Plutarch, *Quaest. Plat.* VIII. 3.

[6] Aet. II. 20-21, Diels, p. 350 a, b, 351 b. Compare Plutarch, *Pyth. Orac.*
XII.: " You laugh at Empedokles for saying that the sun arises from the
reflection of the heavenly light from the earth."

[7] Aet. II. 1, p. 328.

hemisphere formed by the crystalline sphere and moving along
in accordance with the displacement of the fiery hemisphere.
We have winter "when the air predominates by condensation
and reaches the upper parts, and summer when the fire predominates and tends to reach the lower regions[1]," so that the
airy and the fiery hemisphere in turn occupy more than half
of the heavenly sphere, and thereby make the sun, the image
of the fiery hemisphere, move south or north according to the
seasons. The turning of the sun at the solstices is also said to
be caused by the enclosing sphere, which prevents the sun
from always moving in a straight line, and by the tropical
circles[2]. It is not easy to see with certainty where Empedokles imagined the axis of rotation of the dark and luminous
hemispheres to be situated, but we can hardly suppose that it
coincided with the axis of the crystalline sphere on which the
fixed stars are situated. Nor does he appear to have mentioned
the annual motion of the sun with regard to the stars, but only
its motion north and south. The daily motion of the sun he
believed to have been so slow at the time when the human
race sprang from the earth, that the day was as long as ten
months are now; afterwards it became reduced to the length
of seven months, and this is the reason why ten-month and
seven-month children can live[3].

Empedokles was aware that solar eclipses are caused by the
moon passing before the sun[4]. The moon he supposed to be
twice as far from the sun as from the earth[5], or, as the sun was
only a reflection from the celestial sphere, the distance of the
moon was one-third of the radius of the sphere. The region
occupied by man is full of evils, and these extend as far as the
moon, but not further, the region beyond being much purer[6].
The planets were fiery masses moving freely in space, obviously
beyond the orbit of the moon[7].

[1] Aet. ɪɪɪ. 8, p. 375.   [2] Aet. ɪɪ. 23, p. 353.
[3] Aet. v. 18, p. 427.   [4] Aet. ɪɪ. 24, p. 354 b.
[5] Aet. ɪɪ. 31, and Stobæus differ, but see Diels, p. 63 and p. 362.
[6] Hippol. *Philos.* ɪv. Diels, p. 559.   Herakleitus was of the same opinion.
[7] It is probably to the planets that Aristotle alludes when saying that
according to Empedokles the spinning round of the heavens is quicker than the
tendency to fall caused by its weight, whereby the motion is preserved. *De
Cœlo,* ɪɪ. 1, p. 284 a; Simplicius, p. 375, 29 (Heib.).

The earth was, according to Empedokles, held in its place by the rapid spinning round of the rotating heavens, "as the water remains in a goblet which is swung quickly round in a circle[1]." In the beginning the pressure of the rotation squeezed the water out of it, from the evaporation of which the air arose[2]. The obliqueness of the axis of the heavens to the horizon he believed to have been caused by the air yielding to the rapid motion of the sun, whereby the north side has been elevated and the south side depressed[3]. The north pole of the heavens must therefore originally have been vertically above the earth, until the latter was tilted so as to bring the northern end up and the southern end down. The earth must therefore necessarily have been supposed to be flat.

Among the earlier systems of philosophy the atomic theory occupies an important place, being by its two leading teachers clearly conceived and worked out into a system of the origin and constitution of the visible universe which compares very favourably with the more primitive notions of previous thinkers and marks a distinct advance towards the enlightened views of the Kosmos afterwards held by Plato and Aristotle. The founder of Atomism was Leukippus, who flourished about the middle of the fifth century, but about whose life nothing is known. He became quite overshadowed by his brilliant disciple, Demokritus of Abdera in Thrace, who further developed the atomic theory, but there is no doubt that its leading features are due to Leukippus. While the Eleatic school had denied motion and the plurality of things, because these were not conceivable without the void, while the void as not being was rejected, the atomic school acknowledged the existence of the void and assumed that matter consists of an infinite number of extremely small, finite, and indivisible bodies which move in the void. By the combination and separation of these atoms the generation and destruction of things are caused. The atoms are indestructible and without beginning, they are similar in quality but not in size and shape, and by the different grouping of the larger and smaller atoms the great variety of things is

[1] Aristotle, *De Cœlo*, ii. 13, p. 295 a.
[2] Aet. ii. 6, p. 334.
[3] Aet. ii. 8, p. 338.

caused. The atoms are under the influence of gravity, that is
to say (according to the views of the ancients on this subject)
they have a tendency to a downward motion, and being of
different sizes they are supposed to fall with different velocities
through the empty space. Demokritus taught furthermore
that the lighter (smaller) atoms during this motion are pushed
aside by the heavier (larger) ones, so that they appear to have
a tendency to move upward. The collisions between the atoms
produced a circular or vortex motion, in which gradually all
atoms in the whole mass take part. In the infinite space the
infinite number of atoms thus produces an infinite number of
worlds, which increase as long as more atoms from outside join
them, and decrease when they lose atoms. The worlds are
therefore subject to continual change, and when two of them
collide they perish. When our own world commenced to be
built up in this manner a kind of skin was formed round it,
which by degrees became thinner, as parts of it owing to the
vortex motion settled in the middle; of these the earth was
formed, while others produced the fire and air which fill the
space between earth and heaven. Of the atoms which were
caught by the spherical membrane, some crowded together into
compounds which were first moist but gradually dried up and
took fire; these are the stars[1]. In the central compound the
lighter parts were forced out and collected as water in the
hollows. The earth, while it was small and light, had at first
moved to and fro, but when it became heavy it settled in the
middle of the world[2]. According to Leukippus the earth
resembles a tympanon, that is to say, it is flat on the surface
but probably with a slightly elevated rim[3]. He accounted for
the inclination of the axis of the heavens to the horizon by
suggesting that the earth has sunk towards the south, because
the north side is frozen and cold, while the south side (of the

---

[1] Diog. L. ix. 31–33.   [2] Aet. iii. 13, p. 378.

[3] Aet. iii. 10, p. 377, Diog. ix. 30. A τύμπανον had a hemispherical bottom,
but Leukippus probably merely meant a thing like a modern tambourine, as
Aristotle (*De Coelo*, ii. 13, p. 294 a), who refers to this idea without mentioning
Leukippus, plainly shows in the sequel that he is talking of a disc and not of a
hemisphere. Otherwise there would be a great difference between the opinions
of Demokritus and Leukippus on this point.

heavens) is heated by the sun and is much thinner[1]. Demokritus compared the earth to a discus, but higher at the
circumference and lower in the middle[2], and supposed that the
lower part of the heavenly sphere is filled with compressed air,
over which the earth fits like a lid, as Aristotle expresses it[3].
At the south side of the heavens the air is weaker and yields
more to the earth, which is heavier and increases more at that
side owing to the great masses of fruit growing there[4], for
which reason the earth is inclined towards the south and the
north pole of the heavens has sunk from the zenith half way
towards the north horizon. This is the same as the explanation
given by Leukippus.

As to the nature and positions of the heavenly bodies there
was some difference of opinion between the two philosophers.
According to Leukippus the orbit of the sun is the most distant
one, and that of the moon the nearest, and between them are
the orbits of the other planets. The moon is eclipsed oftener
than the sun owing to the difference of size of their orbits[5].
Demokritus placed the moon and the morning star nearest to
the earth, then the sun (which he took to be an ignited mass
of stone or iron), then the planets, and finally the fixed stars[6].
The planets appear not to have had any orbital motion, but
merely to have revolved from east to west somewhat more
slowly than the fixed stars[7]. The sun and moon are large, solid
masses, but smaller than the earth; they had originally been
two earths formed like ours and each at the centre of a world;
these had encountered ours, which had absorbed them and

---

[1] Aet. III. 12, p. 377.   [2] Aet. III. 10, p. 377.
[3] *De Cœlo*, II. 13, p. 294 b, where the same opinion is attributed to Anaximenes and Anaxagoras. Aetius (III. 15, p. 380) says however that Demokritus
like Parmenides supposed that the earth remained in equilibrium, because it
was equidistant from every part of the sphere.
[4] Aet. III. 12, p. 377.
[5] Diog. L. IX. 33. In the statement that eclipses are caused by the earth's
inclination to the south there must be a gap in the text.
[6] Aet. II. 15 and 20, pp. 344 and 349; Hippol. 13, Diels, p. 565. According
to Seneca, *Quest. Nat.* VII. 3, Demokritus did not venture to say how many
planets there are.
[7] Aet. II. 16, p. 345. Compare the statement of Lucretius, *De Rer. Nat.* v.
619–26, that the bodies furthest from the heavens (and nearest to the earth)
move slowest.

taken possession of the earth of each[1]. Comets are due to the near approach to each other of two planets[2]. The moon is a solid body and the markings on its face are caused by the shadows of mountains and valleys, a statement which (like that about comets) is also credited to Anaxagoras[3]. Demokritus appears to have had a remarkably correct conception of one celestial phenomenon, the Milky Way, the light of which he is said to have explained as caused by a great multitude of faint stars[4].

Metrodorus of Chios, a disciple of Demokritus, appears like his teacher to have set up some theories of his own on astronomical matters, but they certainly did not mark any advance. With Anaximander he believed the stars and the planets to be nearer to us than the moon and the sun[5]. The fixed stars and planets are illumined by the sun[6]. The earth is a deposit of water, the sun a deposit formed in the air[7]. The air in becoming condensed produces clouds and afterwards water, which puts out the sun; by degrees the sun dries up and transforms the limpid water into stars, and day and night and eclipses are produced when the sun lights up or is put out[8]. The Milky Way is a former path of the sun[9]. These notions are all very primitive and resemble the opinions of Herakleitus and Xenophanes more than those of the Atomic school.

Empedokles, Leukippus, and Demokritus are among the many ornaments of the fifth century B.C., that wonderful

[1] Ps.-Plutarch, *Strom.* VII., Diels, p. 581.
[2] Aet. III. 2, p. 366; Arist. *Meteorol.* I. 6, p. 342 b.
[3] Aet. II. 25, p. 356, and Stobæus, Diels, p. 361 b. Diogenes (IX. 33) says that according to Leukippus the moon has only a slight amount of fire.
[4] Aet. III. 1, p. 365; Macrob. *Somn. Scip.* I. 15. Achilles, XXIV. (Petav. III. p. 86) and Manilius, I. 753, quote the opinion but do not mention Demokritus. On the other hand Aristotle, *Meteor.* I. 8, p. 345 a, says that Demokritus shared the peculiar opinion of Anaxagoras, so perhaps D. did not think that there are more stars crowded together in the Milky Way, but only that we can see them better there.
[5] Aet. II. 15, p. 345.   [6] Aet. II. 17, p. 346.
[7] Ibid. III. 9, p. 376. The earth being heavy sank down, and the sun being light rose up to the upper region like a bladder full of air, and the other stars were placed as if they had been weighed on a balance. Plutarch, *De facie in orbe lunæ*, XV.
[8] Ps.-Plutarch, *Strom.* XI. Diels, p. 582.   [9] Aet. III. 1, p. 365.

century which was ushered in by the life and death struggle
between Asia and Europe, which saw Athens become the
centre of Greek civilisation, adorned by the masterpieces of
Phidias, and witnessing the triumphs of Æschylus, Sophokles,
and Euripides; the century which closed with the political
downfall of Athens and the judicial murder of Sokrates.
During that brilliant period Athens became for the first time
the home of philosophers, while hitherto all the great thinkers
had moved on the outskirts of Greek life, chiefly in Asia Minor
and in southern Italy. The first philosopher of distinction at
Athens was Anaxagoras of Klazomenæ, who was born about
the year 500 and settled at Athens about 456. He differed
from Empedokles and the Atomists by assuming that all
qualitative differences of things already exist in the primary
elements, and he therefore supposes that these exist in an
unlimited number and are not like the atoms of a finite size,
but divisible ad infinitum, while there is no void anywhere.
But while Empedokles could only offer a sort of mythological
explanation of motion and change, and the Atomists believed
a purely mechanical force, gravity, sufficient, Anaxagoras holds
in the history of philosophy a high place as the first to perceive
that matter is not moved by a force inherent in itself; and to
some extent he anticipated Plato and Aristotle by setting up
Mind (νοῦς) as the agent which has produced the world from
the original chaos, by altering matter not qualitatively but
merely mechanically, and by starting that rotation in it which
has resulted in the orderly arrangement of the world which we
see. By this rotation matter was first separated into two great
masses, ether and air, the former warm, light, thin, the latter
cold, dark, heavy. The latter was by the rotation collected in
the middle, and in course of time water was precipitated, and
from this again earth, some of which under the influence of
cold became stones[1]. The earth was at first like mud, but was
dried by the sun, and the remaining water therefore became
salt and bitter[2]. The infinite matter spreads out outside the
world and continues to become included in it[3].

---

[1] Fragm. 1, 2, 8, 9, Fairbanks, pp. 236–42.
[2] Aet. III. 16, p. 381.   [3] Fragm. 6, Fairbanks, p. 240.

A great meteorite which fell at Ægos Potamoi in the year
467 is said to have attracted the special attention of Anaxagoras (he was afterwards even said to have predicted it[1]), and
as it fell in the daytime he supposed it to have fallen from the
sun and was thereby led to believe the sun to be a mass of red-
hot iron greater than Peloponnesus, and therefore not at a
very great distance from the earth[2]. At the time of the
solstices the sun is driven back by the air which its own heat
has thickened, and the same was the case with the moon[3],
while he had no idea of any orbital motion, but only knew the
diurnal motion from east to west[4]. He was the first to think
that the seven planets are arranged in this order: the moon,
the sun, and outside these the remaining five planets[5], an
arrangement which was adopted by Plato and Aristotle. The
stars he supposed to be stony particles torn away from the
circumference of the flat earth and prevented from falling by
the rapidly revolving fiery ether which rendered them luminous
by the friction, but owing to the distance we do not perceive
the heat[6]. When the stars set they pursue their courses under
the earth, which is supported in the centre of the celestial
sphere by the air underneath it[7]. The inclination of the axis
of the heavens to the vertical was caused by a spontaneous
tilting of the earth towards the south after the appearance of
living creatures, in order that there might be hot and cold
climates, inhabitable and uninhabitable regions[8]. The moon

---

[1] Diog. L. II. 10.

[2] Plutarch, *De facie in orbe lunæ*, XIX. Aet. II. 21, p. 351, has "much
greater than Peloponnesus," Hippolytus, VIII. (Diels, p. 562), ὑπερέχειν δὲ τὸν
ἥλιον μεγέθει τὴν Πελοπόννησον.

[3] Aet. II. 23, p. 352; Hippolytus (l.c.) says that the sun and moon are
thrust aside by the air, and the moon alters its course oftener because it cannot
stand the cold, while the sun by its heat thins the air and weakens the
resistance. Herodotus (II. 24) without giving any authority, says that in
winter the sun is driven out of its course.

[4] Aet. II. 16, p. 345.

[5] Proklus *in Timæum*, p. 258 c, on the authority of Eudemus, a disciple of
Aristotle who wrote a history of astronomy.

[6] Plutarch, *Lysander*, XII.; Aet. II. 13, p. 341.

[7] Arist. *De Cælo*, II. 13, p. 294 b ("like a lid"). Simplicius, p. 527 (Heib.),
joins Anaxagoras with Empedokles and Demokritos as having used the illustration of the bowl full of water.

[8] Aet. II. 8, p. 337; Diog. II. 8.

is as large as Peloponnesus, partly fiery, partly of the same
nature as the earth, the inequalities in its "face" being due to
this mixture of material[1], though he is also said to have
believed that there were "plains and valleys" in it[2]. He
knew that it was illuminated by the sun and gave the correct
explanation of its phases[3] (so that he must have recognised the
spherical form of the moon) and of solar and lunar eclipses,
although he thought that the latter were occasionally caused
by other bodies nearer to us than the moon[4]. This latter idea
was perhaps partly borrowed from Anaximenes, who, as we
have seen, assumed the existence of bodies of a terrestrial
nature among the stars, and we shall see in the next chapter
that the Pythagoreans also felt compelled to assume that some
lunar eclipses are due to unknown bodies.

Anaxagoras offered a curious explanation of the Milky Way.
He thought that owing to the small size of the sun the earth's
shadow stretches infinitely through space, and as the light of
the stars seen through it is not overpowered by the sun, we see
far more stars in the part of the sky covered by the shadow
than outside it[5]. An ingenious idea, but Anaxagoras ought
to have seen that if it was correct the Milky Way ought to
change its position among the stars in the course of the year,
an objection which was already made by Aristotle[6]. These
and other speculations on the cause of natural phenomena
appeared too daring to the people of Athens, hitherto un-
accustomed to hearing philosophers reason freely about the
heavenly bodies. The fact that Anaxagoras was a personal
friend of Perikles may also have induced the political enemies
of the latter to hope that it would injure his reputation if a
friend of his could be convicted of a crime against religion.

[1] Aet. II. 30, p. 361. The moon has some light of its own which we see
during lunar eclipses; Olympiodorus *in Aristot. Meteor.* I. p. 200 (Ideler).

[2] Aet. II. 25, p. 356; Hippolytus, l.c.

[3] That he was the first to do so is affirmed by Plato, *Kratyllus*, 409 A,
Plutarch, *Nikias*, XXIII. and Hippol. l.c. Parmenides had however anticipated
him.

[4] Aet. II. 29, p. 360 (Stobæus, on the authority of Theophrastus), Hippol. l.c.

[5] Aet. III. 1, p. 365, and Hippol.

[6] *Meteorol.* I. 8, p. 345 a; comp. Alex. of Aphrod. in Ideler's ed. of the
Met. I. p. 200.

Anyhow Anaxagoras was convicted of "impiety" and died in exile at Lampsakos in 428.

His unpopularity seems to have been shared by his contemporary, Diogenes of Apollonia, the last philosopher of the Ionian school. Like Anaximenes he adopted air as the origin of all substances, which are formed from it by condensation or rarefaction, but he also supposed it to represent the intelligent force which directs these changes. The earth is a flat disc planted in the midst of a vortex produced by heat which settled heavy air there; this became consolidated by cold, while light air took an upward direction and formed the sun[1], but there is still a good deal of air shut up in the earth which causes earthquakes[2]. He adopted the same explanation as Anaxagoras of the inclination of the axis of the sphere. Diogenes was apparently also much impressed by the great meteorite, as he taught that in addition to the visible stars which are like pumice-stone, breathing-holes of the world, giving light because they are penetrated by hot air, there are also dark bodies like stones which occasionally fall down on the earth[3]. He believed the sun to change its course at the solstices owing to the cold air which penetrates it[4].

Very similar opinions were held by Archelaus, who though a pupil of Anaxagoras also adopted air as the primitive substance, out of which the cold and the hot, water and fire were formed. The water collected in the centre, part of it arose as air, part of it became condensed into earth, loose pieces of which formed the stars. The sun is the largest, and the moon the second largest of the heavenly bodies. The earth, which is a   small portion of the universe, is supported by the air is kept in its place by the vortex; it is higher round the l hollow in the middle, which is proved by the fact that does not rise and set simultaneously for all parts of the s it would do if the earth were quite flat. It is ıg to meet with this objection to the flatness of the ›ugh Archelaus evidently did not know that sunrise

Plut. *Strom.*, Diels, p. 583; Diog. ix. 37.
ıeca, *Quæst. nat.* vi. 15.
. ii. 13, p. 341.   [4] Aet. ii. 23, p. 353.

and sunset occur later in western countries than in eastern, since his theory would require the opposite to be the case. Originally the stars turned horizontally round the earth (i.e. the pole was in the zenith), and the sun was therefore not visible from the earth owing to the raised rim, until the heavens became tilted and the heat of the sun was able to dry up the earth[1].

It was characteristic of all the philosophers, whose opinions on the construction of the world we have reviewed, that the explanation of physical phenomena played an important part in their philosophical systems. But the very scanty stock of observed facts on which they were able to build was utterly insufficient to serve as a foundation to their speculations, and the result was that philosophers differed hopelessly from each other as to the explanation of even the principal phenomena. Consequently philosophy for a time came to be looked at with indifference, while a class of men pushed themselves to the front whose sole endeavour was to prepare youths to take their places in the civic life of Athens, to the exclusion of all efforts to seek after truth for its own sake. As regards natural philosophy in particular, the Sophists were absolute sceptics and therefore they do not concern us here; only one of their number, Antiphon (a contemporary of Sokrates), is occasionally alluded to in the doxographic accounts. But he merely echoed now one, now another opinion of some earlier philosopher, as when he with Anaxagoras believed the moon to have a faint light of its own, which he suggested is the more overpowered by that of the sun the nearer the two bodies are to each other, " which also happens to the other stars[2]."

But it is time that we should turn our attention to a most important school in which mathematics and astronomy were assiduously cultivated, and which exercised a powerful influence on the advance of knowledge.

---

[1] Hippolytus, ix., Diels, p. 563 ; Diog. L. ii. 17.
[2] Aet. ii. 28, p. 358.

CHAPTER II.

AMONG the earlier Greek philosophers Pythagoras occupies a peculiar position. Like several others he left no writings of his own, but he founded a philosophical school which lasted upwards of two hundred years, and which attracted an unusual amount of attention, as its members not only formed a sort of religious brotherhood and developed many peculiar opinions, but also threw themselves into politics and therefore became subject to persecution in the south of Italy. As a philosophical school the Pythagoreans, like all the pre-Sokratic schools, mainly aimed at interpreting nature ; and all of them, from the founder of the school and down to the last unknown member of it two hundred years after his time, appear to have been devoted to the cause of science.

Pythagoras was born at Samos about the year 580, settled at Kroton in the south of Italy about 540 or 530, and died there or at Metapontum somewhere about the year 500 or very soon after. He is said by later writers to have travelled a good deal in the East and to have been indebted for much of his knowledge of science to what he had learned during his travels. The earliest notice of his stay in Egypt occurs in a panegyric on an imaginary Egyptian king by the orator Isokrates more than a hundred years after the death of Pythagoras, but it is a question whether an incidental allusion to his having brought Egyptian wisdom to Greece may be accepted as historical evidence, occurring as it does in a work of fiction. Later on his travels to Egypt and Babylon and his studies there were commonly accepted as historical facts and

3—2

are referred to by many writers[1], a circumstance which is doubtless connected with the tendency of the Greeks to associate
the rise of their own civilisation in every possible way with
the older civilisations of the East[2]. But whether Pythagoras
may have laid the first foundation of his great proficiency in
mathematics in Babylonia or Egypt is a problem that does not
concern us here, as we have absolutely no reason to believe
that the strange system of the world which was developed
within the Pythagorean school has in any way been founded
on Eastern ideas. In the course of time important changes
appear to have taken place within the school with regard to
philosophical and scientific doctrines, but we are very imperfectly informed as to the chronology of these changes. It
is therefore very doubtful to whom most of the Pythagorean
doctrines are due, whether to the founder of the school or to his
successors; but whenever a particular Pythagorean is credited
with a doctrine it is reasonable to suppose this to be true,
as there was a general inclination later on to attribute as much
as possible to Pythagoras himself.

The leading idea of the Pythagorean philosophy is that
number is everything, that number not merely represents the
relations of the phenomena to each other but is the substance of
things, the cause of every phenomenon of nature. Pythagoras
and his followers were led to this assumption by perceiving
how everything in nature is governed by numerical relations,
how the celestial motions are performed with regularity,
and how the harmony of musical sounds depends on regular
intervals, the numerical valuation of which they were the first
to determine. It is not here the place to set forth how combinations of odd and even numbers (the perfect and the
imperfect) were supposed to produce everything in the world,
but we must not omit to mention that the world was supposed
to be ruled by harmony, all the different heavenly revolutions
producing different tones, so that each of the planets and the
sphere of the fixed stars emitted its own particular musical

---

[1] For instance, Strabo, XIV. p. 638.
[2] For a full discussion of the myths concerning Pythagoras and his travels
see Zeller, *Die Phil. d. Griechen*, I. pp. 300–308 (5th ed.).

sound, which our ears are unable to hear because we have
heard them from our birth[1], though it was afterwards asserted
that Pythagoras alone of all mortals could hear them. It
seemed, as Plato says[2], " that the ears of man were intended to
follow harmonious movements, just as his eyes were intended
to detect the motions of the heavenly bodies, these two being
sister sciences, as the Pythagoreans declare." This theory of
the harmony of the spheres was elaborated in detail long after
the time of Pythagoras, and we shall revert to it when discussing the opinions of the ancients on the distances of the planets.
We shall here only remind the reader that Pythagoras was
a great mathematician, a fact which is at the root of his
philosophical system[3], and which would doubtless have carried
his followers very far in astronomical research, as their school
continued while it lasted to be the main seat of mathematical
studies, if they had not at an early stage got into a wrong groove.

The Pythagorean school seems to have come to an end in
the course of the fourth century, though religious mysteries,
which had gradually taken the place of philosophical speculation among its members, continued to exist throughout the
Alexandrian period. At the beginning of the first century B.C.
Pythagorean doctrines began again to take their place in the
realm of thought, and now we find for the first time special
opinions as to the construction of the world attributed to
Pythagoras himself, in an account by Alexander Polyhistor,
quoted by Diogenes Laërtius. According to this, Pythagoras
taught that the world is formed of the four elements (earth,
water, air, fire), that it is endowed with life and intellect, and
is of a spherical figure, having in its centre the earth, which is
also spherical and inhabited all over, that there are antipodes
and that what is below as respects us is above in respect to
them[4]. In another place (VIII. 48) Diogenes says that, according to Favorinus, Pythagoras was the first to call the heavens
κόσμος and the earth round (καὶ τὴν γῆν στρογγύλην), though

---

[1] Arist. *De Cœlo*, II. 9, p. 290 b.   [2] *Republ.* VII. p. 530.
[3] Arist. *Metaph.* I. 5, p. 985 b.
[4] Diog. L. VIII. 25. Suidas (under Πυθαγόρας Σάμιος) repeats the account,
using the very same words.

Theophrastus attributed this to Parmenides and Zeno to Hesiodus (!). Aëtius also attributes to Pythagoras a knowledge of the spherical form of the earth, since he states that according to Pythagoras the earth after the analogy of the sphere of the all is divided into five zones[1]. With regard to the heavenly bodies, we are told[2] that Pythagoras was the first to recognise that Phosphorus and Hesperus, the morning and evening stars, were one and the same body, that he held the moon to be a mirror-like body[3], and by Theon of Smyrna that he was the first to notice that the planets move in separate orbits inclined to the celestial equator[4]; while Aëtius says that Pythagoras was the first to perceive the inclination of the zodiacal circle, but that Œnopides claimed the discovery as his own[5].

This is positively all the information to be found in ancient authors as to the astronomical knowledge of Pythagoras. It is significant that Aristotle never mentions him in his book on the heavens, and that Aëtius has so very little to say about him. There is no reason to doubt that he knew the earth to be a sphere, though the fact that Aristotle's disciple Theophrastus attributed the discovery of the true figure of the earth to Parmenides seems to show that while the latter (who flourished about 500 B.C. or a little later) taught it openly and as an important part of his views of the world, Pythagoras accepted the spherical figure from the supposed necessity of the earth and the heavens being of the same shape, without devoting special attention to the construction of the world. And we must doubt the statement of Theon that Pythagoras himself taught that the planets move in separate orbits. Aëtius announces that Alkmæon of Kroton and the mathematicians discovered " that the planets hold an opposite course to the

---

[1] Aet. iii. 14, p. 378.

[2] By Diog. L. viii. 14, Pliny, *H. N.* ii. 37, and Stobæus, Diels, p. 467.

[3] Aet. ii. 25, p. 357.

[4] Theon, ed. Martin, p. 212.

[5] Aet. ii. 12, p. 341. Theon (p. 322), on the authority of Eudemus, calls Œnopides of Chios (about 500–430 B.C.) the discoverer of the zodiacal girdle (διάξωσις), which probably means that he showed the sun's annual path to be a great circle inclined to the equator, as Diodorus Siculus, i. 98, says that Œnopides learned from the Egyptian priests that the sun moves in an inclined orbit and in a direction opposed to the daily motion of the stars.

fixed stars, namely from west to east[1]. In other words,
Alkmæon and the Pythagoreans pointed out that the planets
do not, as the Ionians and other early philosophers thought,
travel from east to west, only somewhat more slowly than the
fixed stars. Now, if Pythagoras himself had formulated the
geocentric system, the earth in the middle, surrounded by the
orbits of the planets at different distances, the discovery of the
independent motion of the planets from west to east would not
have been attributed to one of his younger contemporaries[2],
who, if not exactly a disciple of Pythagoras, was at any rate
very strongly influenced by him; and the founder of the school
can therefore hardly have advanced beyond his predecessors.
It is not at all remarkable that a writer living six hundred
years later should have thought it necessary to suppose that
Pythagoras must have been acquainted with the system of the
world which long before the writer's own time had become
universally accepted; for Pythagoras had by that time to the
adherents of the Neo-Pythagorean school become a mythical
person, an omniscient demigod, from whom all knowledge
possessed by mankind must have descended, and about whose
life the most wonderful stories were circulated, of which earlier
writers had not known anything.

Whether due in the first instance to Pythagoras or to
Parmenides, the doctrine of the spherical figure of the earth
must have made some headway during the first half of the fifth
century. Herodotus mentions that there were people in the
far north who slept six months of the year, and that the
Phœnicians, who were supposed to have circumnavigated
Africa, had the sun on the right while sailing westward[3].
These stories, which Herodotus finds incredible, show that some
people must have been able to perceive the consequences of
the earth being a sphere; and these people must have been
Pythagoreans, for outside their school it is certain that nobody
as yet believed that the earth was spherical. And if this
doctrine took a long time to make itself acceptable, it is no

---

[1] Aet. II. 16, p. 345.

[2] "Alkmæon of Kroton reached the prime of life when Pythagoras was an old man." Arist. *Metaph.* I. 5, p. 986 a.

[3] Herod. IV. 25 and 42.

wonder that the sun and moon were not for a while, even among the Pythagoreans, recognised as spheres; at least we know that Alkmæon believed the sun to be flat and acquiesced in an explanation of lunar eclipses similar to that of Herakleitus[1]; but we have no way of knowing how soon a more rational idea of the figure of the sun and moon began to prevail among the Pythagoreans.

As already mentioned, Pythagoras did not leave any writings, and it was even popularly believed that the members of his school were sworn to secrecy concerning the more important doctrines imparted to them. The Pythagorean philosophy did not become generally known in Greece till after the violent disruption of the school at Kroton, and the first publication emanating from the school is attributed to Philolaus, a native of the south of Italy and a contemporary of Sokrates, who lived at Thebes for a number of years towards the end of the fifth century. Though only fragments are left of his book, there are sufficiently detailed references in the works of other writers to enable us to form a clear idea of the remarkable cosmical system of this philosopher, though we are unable to say with certainty whether it is altogether due to him or had been gradually developed among the followers of Pythagoras. In referring to it, Aristotle does not mention him but only "the Pythagoreans"; but on the other hand, Aëtius always and distinctly credits the system to Philolaus[2], so that it was probably chiefly if not altogether developed by him. By a strange fatality this system of the world has been totally misunderstood by medieval and later writers, and even at the present day the mistake is frequently made that the Pythagoreans taught the motion of the earth round the sun, although it is now nearly a hundred years since Boeckh gave a correct exposition of the system of Philolaus[3].

---

[1] Stobæus, Diels, pp. 352 and 359.

[2] Aet. II. 4, 5, 7, 20, 30, III. 11, Diels, pp. 332 b, 333 a, b, 336–7 b, 349 a, b, 361 b, 377 a. Diog. L. VIII. 85, merely says that Philolaus "was the first to affirm that the earth moves in a circle, though some attribute this to Hiketas of Syracuse."

[3] "De Platonico systemate cœlestium globorum et de vera indole astronomiæ Philolaicæ." Heidelberg, 1810. 4°. Reprinted with new appendices in Böckh's *Kleine Schriften*, Vol. III. Berlin, 1866.

It is not easy to answer the question, how the Pythagoreans came to invent this system of the world. The leading idea of it is that the apparent daily rotation of the starry heavens and the daily motion of the sun are caused by the earth being carried in twenty-four hours round the circumference of a circle. When it had once become clear that sun, moon, and planets describe orbits round the heavens from west to east, it was probably felt objectionable, a flaw in the arrangement, that the whole heavens should rotate in twenty-four hours in the opposite direction. Would it be possible to account for this by another motion from west to east? Somehow the idea of the earth rotating round an axis did not occur to Philolaus, or, if it did, it did not commend itself to him, possibly because there appeared to be no other case of rotation in the world, the moon always turning the same face to us and therefore *not* rotating, as philosophers of that age (and of many succeeding ages) argued[1]. If the moon had an orbital motion, always keeping one side directed to the centre of the orbit, might not the same be the case with the earth? Evidently, if this were the case and the period of the earth's orbital motion were twenty-four hours, an observer on the earth would see the whole heavens apparently turn round, and the sun, moon, and stars rise and set once in the course of a day and a night. And this arrangement would offer the advantage that all motions took place in the same direction, from west to east.

Philolaus and his adherents were perhaps influenced by these considerations, and they considered the nature of the earth too gross to make it fit for the exalted position of occupying the centre of the universe. In this commanding position they placed the "central fire," also described as the hearth of the universe (Ἑστία τοῦ παντός) or the watch-tower of Zeus (Διὸς φυλακή), round which the earth and all the other heavenly bodies moved in circular orbits[2]. The orbit of the earth had of course to be supposed to lie in the plane of its equator, and the fact that nobody had ever seen the central fire

---

[1] Arist. *De Cælo*, II. 8, p. 290 a, 26.

[2] Arist. *De Cælo*, II. 8, p. 293 b; Simplicius, p. 512 (Heib.); Plutarch, *Numa*, c. XI.; Stobæus, Diels, p. 336; Act. III. 11, p. 377.

could be easily explained by assuming that the known parts of the earth, Greece and the surrounding countries, are situated on the side of the earth which is always turned away from the centre of the orbit. It would therefore be necessary to travel beyond India to catch a glimpse of the central fire, and even after travelling so far this mysterious body might still be invisible, as another planet intervenes between it and the earth. About this unseen planet Aristotle says: " They also assume another earth, opposite to ours, which they call the counter-earth (ἀντίχθων), as they do not with regard to the phenomena seek for their reasons and causes, but forcibly make the phenomena fit their opinions and preconceived notions and attempt to construct the universe." In another place Aristotle says[1]: " When they (the Pythagoreans) anywhere find a gap in the numerical ratios of things, they fill it up in order to complete the system. As ten is a perfect number and is supposed to comprise the whole nature of numbers [being the sum of the first four numbers] they maintain that there must be ten bodies moving in the universe, and as only nine are visible they make the antichthon the tenth."

These nine bodies are the earth, the moon, the sun, the five planets, and the sphere of the fixed stars. To complete the number ten, Philolaus created the antichthon, or counter-earth. This tenth planet is always invisible to us, because it is between us and the central fire and always keeps pace with the earth[2]; in other words, its period of revolution is also twenty-four hours, and it also moves in the plane of the equator. It must therefore hide the central fire from the inhabitants of the regions of the earth in longitude 180° from Greece, and there is a curious resemblance between this part of the theory and the ancient idea of the eternal twilight in which the regions far to the west of the Pillars of Herakles were shrouded. The Pythagoreans do not appear to have indulged in speculations as to the physical nature of the antichthon, but it does not seem that they assumed it to be inhabited, as only the region

---

[1] *Metaph.* I. 5, p. 986 a; Alex. Aphrod. *in metaph.*, Arist. *Opera*, v. p. 1513 (§ 198).

[2] Simplicius, *De Cœlo*, II. 13, p. 511 (Heib.); Alexander, l.c.

below the moon and beginning with the earth was considered
to be given up to generation and change and was called the
heavens (οὐρανός), while the kosmos, the place of regulated
motion, embraced the moon, the sun, and the planets; and the
Olympos, the place of the elements in their purity, was the
sphere of the fixed stars[1]. Outside this was the outer fire, and
outside that again the infinite space (τὸ ἄπειρον) or the infinite
air from which the world draws its breath[2].

That the antichthon was a sphere is not expressly stated,
and Boeckh is of the opinion that the Pythagoreans believed
it and the earth to be the two halves of one sphere, cut into
two equal parts along a meridian and separated by a comparatively narrow space, turning the flat sides towards each other
but the convex side of the antichthon towards the central fire,
that of the earth away from it[3]. He acknowledges that Aristotle
certainly understood the two bodies to be two distinct spheres,
but all the same he clings to his own strange idea, apparently
because the conception of a counter-earth probably originally
arose out of the doctrine of Pythagoras that the earth was
round and that there were antipodes. But though this latter
part of the doctrine very probably first suggested to Philolaus
to postulate a tenth planet, why should we suppose that in
doing so he abandoned the grand discovery of Parmenides and
Pythagoras that the earth is a sphere, and most unnecessarily
made it out to be only half a sphere? Is it not more likely
that he left the fundamental part of the cosmical view of the
Pythagoreans intact, and merely added a duplicate earth, that
is, another sphere? The Pythagoreans had plenty of enemies;
some of them would surely in their writings have heaped
ridicule on the earth being sliced in two.

Outside the orbit of the earth came that of the moon, which
body took twenty-nine and a half days to travel round the
central fire, while outside the moon (or above it, according to
the phraseology of the Pythagoreans) the sun and planets

[1] Stobæus, Diels, p. 337; Epiphanius, *Adrers. Hæres.*, Diels, p. 590.
[2] Arist. *Phys.* III. 4, p. 203 a, 6, and IV. 6, p. 213 b, 22. Aet. II. 9, p. 338.
[3] See his earliest paper and his paper "Vom Philolaischen Weltsystem" in
his *Kleine Schriften*, III. p. 335.

described their orbits round the central fire, the sun taking a
year to do so[1]. All the writers of antiquity who mention the
system of Philolaus agree in describing the orbital motion of
the sun[2], and not a single one supposes that the sun and the
central fire are identical, so that there has really been no excuse
for the confusion on this point which has prevailed so long.
With regard to the orbits of the planets authorities differ.
Plutarch states that Philolaus placed the orbits of Mercury and
Venus between those of the moon and the sun, and this is
repeated by several late writers, who, however, do not mention
Philolaus or his system, but assume that the Pythagoreans
placed the earth in the centre[3]. Aëtius says that "some of the
mathematicians" agreed with Plato, others placed the sun in
the midst of all, i.e. with three planets on either side[4]. But
Alexander of Aphrodisias asserts that the Pythagoreans gave
the seventh place among the ten bodies to the sun, and thus
agrees with the detailed account of Stobæus, who says that the
sun came after the five planets, and then came the moon[5].
There was therefore possibly a difference of opinion among the
Pythagoreans on this subject, but as Anaxagoras, Plato, and
everybody else for several centuries adopted the order, moon,
sun, Venus, Mercury, Mars, Jupiter, Saturn, they had probably
borrowed it from the Pythagorean school, and Plutarch and
other late writers perhaps merely credited the Pythagoreans

---

[1] Censorinus (*De die natal.* c. 18–19) states that Philolaus put the solar
year = 364½ days, and adopted a lunisolar period of 59 years in which there
were 21 intercalary lunar months. But it is hardly likely that Philolaus had
not a better knowledge of the length of the year, and he probably merely said
that 729 or 9^3 represents nearly (with an error of about a unit) the number of
months in a great year ($59 \times 12 + 21 = 729$) and also the number of days and
nights in the solar year. See P. Tannery, "La grande année d'Aristarque de
Samos," *Mém. de la Soc. des sc. phys. et nat. de Bordeaux*, 3e Série, IV.
pp. 79–86.

[2] See particularly Stobæus, Diels, p. 337.

[3] Plutarch, *De an. procreat. in Timæo*, XXXI. Pliny, *Hist. Nat.* II. 84. Theon,
ed. Martin, p. 182 (quoting a poem by Alexander of Ætolia, a poet of the third
century B.C., though it is probable that the poem is by Alexander of Ephesus, a
contemporary of Cicero). Censorinus, *De die natali*, c. XIII.

[4] Aet. II. 15, p. 345.

[5] Alex. *in Metaph.* I. 5, Brandis, *Scholia in Arist.* p. 540 b, 2; Stobæus, l.c.
Also Photius, *Bibl.* cod. 259.

with a knowledge of what had since become the prevailing opinion, in accordance with the usual fashion of those days.

It is easy to see that this peculiar system of the world could account for the apparent rotation of the heavens[1], as a point on the earth's equator during the daily revolution of the earth round the central fire would in succession face every point on the celestial equator, thus producing the same effect as if the earth merely rotated on its own axis. Day and night are also easily accounted for : when the inhabited part of the earth swings round in sight of the sun we have day, and when it in turn is carried round to the other side of the central fire, away from the sun, we have night. Similarly the system readily explains the revolutions of the moon and planets and the annual motion of the sun in the zodiac, as well as the changes of the seasons caused by the inclination of the sun's orbit to the equator, or, as the Pythagoreans expressed it, by the fact that the sun moved round the central fire in an oblique circle. Of course the system could not account for the apparent irregularities in the planetary motions any more than a simple geocentric system could : but this was hardly felt to be a serious defect, as these irregularities had not yet been closely followed. A more conspicuous fault was that the daily motion of the earth would make the distances of the sun and moon from the earth vary very considerably in the course of a day, which ought to produce a sensible change in their apparent diameters, especially in the case of the moon, the orbit of which is next to that of the earth. This did not escape the attention of the Pythagoreans, but according to Aristotle they met the objection by pointing out that, even if the centre of the earth were at the centre of the universe the observer would be at a distance from this centre equal to the semi-diameter of the earth, so that even in that case we might have expected parallax and change in the size of the moon. Accordingly the distances of the antichthon and the earth from the central fire were assumed to be very small in comparison with the distances of the moon and the planets, and this was not held to be

---

[1] Arist. *De Cælo*, II. 13, p. 293 a: "The earth being one of the stars is carried in a circle round the centre and produces thereby day and night."

unreasonable, as the idea of the analogy between these distances and the musical intervals which the Pythagoreans entertained must either have been abandoned when the number of planets was raised from seven to ten, or was not fully developed until the Philolaic system had been given up[1].

The central fire was by no means supposed to be the only or even the chief source of light and heat in the universe. The sun borrows light not only from it but also from the fire distributed around the sphere of the visible universe, the ethereal fire, the fire "from above[2]." This light and heat the sun transmits to us after having sifted it through its own body, apparently in order to scatter it to all sides[3]. The sun is therefore also the sole source of the light reflected to us by the moon[4], but the feeble light over the whole surface near new moon must have appeared to the Pythagoreans to be due to the central fire, and it is not impossible that this "ashlight" may have helped to suggest the idea of a luminous body always, as it were, behind the inhabited part of the earth. The moon was held to be a body like the earth, with plants and animals, the latter being fifteen times as powerful as the terrestrial ones and differing from them in the manner of their digestion, and their day is fifteen times as long as ours[5]. Of course the day on the moon is 29½ times as long as our day, but Philolaus

[1] Plutarch (*De animæ procreat.* XXXI.) states that many Pythagoreans, putting the central fire (its radius?) = 1, assumed the distance of the antichthon = 3, of the earth = 9, of the moon 27, of Mercury 81, of Venus 243, and so on, or 3^0, 3^1, 3^2, 3^3....But this is doubtless a later invention, as it is not mentioned by any earlier writer.

[2] This light is said to be conical (μόνον δὲ τὸ ἀνώτατον πῦρ κωνοειδές, Aet. I. 14, p. 312). Schiaparelli (*I Precursori di Copernico*, p. 5) suggests that this looks like an allusion to the zodiacal light, but Zeller (I. p. 435) is more likely right in thinking that this is the Milky Way seen from outside, and we shall see in the next chapter that Plato in the *Republic* refers to it as a column.

[3] Achilles, *Isagoge in Arat.* XIX. (Petavius, 1703, Vol. III. p. 81), Aet. II. 20, p. 349. According to these writers Philolaus described the sun as glassy (ὑαλοειδής). In his transcript, Eusebius (XV. 23) adds the word δίσκος, which is not found either in the Placita or in Stobæus, while the latter elsewhere (Diels, p. 352) says that the Pythagoreans held the sun to be a sphere.

[4] According to Aet. II. 29, p. 360, some later Pythagoreans believed a light to be gradually kindled and to spread over the whole surface of the moon and then die away again, but when these people can have lived is hard to say.

[5] Aet. II. 30, p. 361.

doubtless meant that any point on the moon was illuminated for fifteen times twenty-four hours[1]. Other Pythagoreans were however of the opinion that the markings in the moon were caused by a reflection of our own seas from the lunar surface[2]. Eclipses of the moon were supposed not only to be caused by the passage of the moon through the earth's shadow, but also occasionally by the shadow of the antichthon, and this is the reason why there are more lunar than solar eclipses seen. Some of the Pythagoreans even thought it necessary to assume the existence of several unseen bodies to assist in the production of lunar eclipses, which shows that they had given up the axiom of the ten planets[3]. The same must have been the case with those Pythagoreans who believed comets to be appearances of an additional planet, which like Mercury generally only rose a little above the horizon and therefore escaped notice[4].

We have already mentioned that the sphere of the fixed stars was considered the tenth and last of the bodies which circled round the central fire, and that the fixed stars did actually in the system of Philolaus perform some kind of motion is distinctly affirmed by Simplicius[5]. At first sight the system might seem to be self-contradicting, as the daily motion of the earth round the central fire had been specially designed to explain the apparent rotation of the heavens. But it is quite evident that the motion of the fixed stars was supposed to be a very slow one. The tenth body had to be in motion as well as the other nine, because all divine things move like sun, moon, stars, and the whole heavens, as Alkmæon had said[6]. Now it was obvious that the velocities of the nine bodies

---

[1] Compare the *Epinomis*, p. 978 (a book which, whether written by Plato or not, was certainly influenced by Pythagorean ideas): "Among these the deity has formed one thing, the moon, which at one time appearing greater and at another less, proceeds through her orbit showing continually another day up to fifteen days and nights."

[2] Stobæus, Diels, p. 361. Compare above, p. 38.

[3] Arist. *De Cælo*, II. 13, p. 293 b; Aet. II. 29, p. 360.

[4] Arist. *Meteor.* I. 6, p. 342 b; Aet. III. 2, p. 366, adds that others took a comet to be "a reflection of our vision into the sun, like images reflected in a mirror."

[5] *De Cælo*, p. 512 (Heib.).

[6] Aristotle, *De anima*, I. 2, p. 405 a, 32.

gradually decreased as their distances from the centre increased, consequently some very slow motion of the fixed stars was supposed to be a necessity. To the modern reader this suggests a knowledge of the precession of the equinoxes, as this shows itself as a very slow motion of the stars from west to east, and this explanation was in fact adopted by Boeckh, though he subsequently abandoned it[1]. There seems indeed no reason to believe that any Greek astronomer before Hipparchus had the slightest knowledge of precession, though on the other hand it is not unlikely that the Egyptians may have dimly perceived that those stars, by which they had fixed the directions of certain passages in their monumental buildings, changed their places in the course of centuries, so that those passages no longer pointed to them at their rising or culmination. It is also more than likely that the Babylonians, who at the latest at the end of the eighth century B.C. observed eclipses, must have noticed that the sun at equinox or solstice did not from one century to another stand at the same star[2]. But it is by no means necessary to assume that the Pythagoreans knew anything of this. For in the first place they were not afraid to take for granted anything that suited their philosophical ideas, even though these had not received any support from observations (the central fire and the antichthon), and secondly no observation could disprove the postulated slow motion of the starry sphere. For supposing that this slow rotatory motion amounted to one minute of arc per day, it would only be necessary to assume that the real (not the apparent) period of revolution of the earth round the central fire was twenty-four hours minus four seconds, and an observer would have to take up his station at the centre of motion, at the central fire, before he would be able to say whether the stars stood still and the earth revolved in 24h. 0m. 0s. *or* whether the stars daily moved through an angle of 60″ and the earth in reality revolved in 23h. 59m. 56s. It is therefore not necessary with Boeckh to suppose the motion of the starry sphere to have been an

---

[1] See his "Untersuchungen über das kosmische System des Platon," p. 93 (Berlin, 1852).

[2] Compare Kugler's *Babylonische Mondrechnung*, Freiburg, 1900, p. 103.

exceedingly slow one, as it might in any case escape direct observation.

It is impossible to review this strange system of the world without a certain feeling of admiration. The boldness of conception which inspired the Pythagoreans with the idea that the earth need not necessarily be the principal body, at rest in the centre of the universe, contrasts in a remarkable manner with the prevailing ideas, not only of their own time, but of the next two thousand years. And the manner in which they accounted for the apparent rotation of the heavens is so ingenious that one is inclined to regret that it is so totally erroneous. The system does not appear to have won any adherents outside the philosophical school in which it originated, which is not to be wondered at when we remember how intimately it was connected with the philosophical conception of the nature of numbers which was characteristic of the school of Pythagoras, while it required a very strong faith to accept the doctrine of an unseen central body, of the existence of which there was absolutely no indication anywhere. But the system of Philolaus has all the same been of use to the development of science, not only because it long afterwards through a curious misunderstanding made the Copernican system appear respectable in the eyes of those who only could admire philosophers of classical antiquity, but also because it paved the way for the conception of the earth's rotation on its own axis. It seems that this further step was actually taken by Hiketas, a native of Syracuse and one of the earlier Pythagoreans, but unfortunately we know next to nothing about him. It is not even certain whether he lived before Philolaus or after him; Diogenes[1], after stating that Philolaus first affirmed that the earth moved in a circle, adds: "But others say that Hiketas of Syracuse did this," while the doxography says about him[2]: "Thales and his followers think that there is one earth; Hiketas the Pythagorean that there are two, this one and the antichthon." This passage has certainly become corrupted in the course of repeated compilation and condensation, and in the corresponding passage in the philosophical history of Pseudo-

---

[1] Diog. VIII. 85.   [2] Aet. III. 9, Diels, p. 376.

Galenus the words "some of the Pythagoreans" occur instead
of "Hiketas the Pythagorean[1]." But a very clear statement
about the astronomical opinions of Hiketas is made by Cicero,
who says[2]: "Hicetas of Syracuse, according to Theophrastus,
believes that the heavens, the sun, moon, stars, and all heavenly
bodies are standing still, and that nothing in the universe is
moving except the earth, which, while it turns and twists itself
with the greatest velocity round its axis, produces all the same
phenomena as if the heavens were moved and the earth were
standing still." This testimony of Theophrastus, the principal
disciple of Aristotle, is not contradicted by the statement of
Diogenes[3], and it is certainly of great weight; it seems to show
conclusively that Hiketas taught the rotation of the earth on
its axis in twenty-four hours, which of course involves abandoning the central fire theory. The words "that nothing in the
universe is moving except the earth" are of course absurd, but
they refer only to the explanation of the most striking of all
celestial phenomena, the daily rotation of the heavens, and they
do not weaken the testimony as to the doctrine of Hiketas.
Whether the latter was developed from that of Philolaus or
was possibly anterior to it, is not absolutely certain, but the
former seems more likely. This was certainly the case with
another Pythagorean, Ekphantus, also of Syracuse, who must
have lived later than Philolaus, as his name in the *Placita* is
associated with that of Herakleides of Pontus. We know
almost as little of him as of Hiketas, but as they were both
from the same city Ekphantus may have been a disciple of

---

[1] Galen's *Hist. Philos.* 81, Diels, p. 632. Boeckh suggests that the passage
in Aetius may originally have run thus: "Hiketas the Pythagorean one, but
Philolaus the Pythagorean two." See his *Untersuchungen*, p. 125.

[2] *Acad. prior.*, lib. II. 39, § 123: "Hicetas Syracusius, ut ait Theophrastus,
caelum, solem, lunam, stellas, supera denique omnia stare censet, neque praeter
terram rem ullam in mundo moveri : quæ quum circa axem se summa celeritate convertat et torqueat, eadem effici omnia, quæ si, stante terra, caelum
moveretur. Atque hoc etiam Platonem in Timaeo dicere quidam arbitrantur,
sed paulo obscurius." The last sentence we shall refer to again in the next
chapter.

[3] Because κινεῖσθαι κατὰ κύκλον might also be used to express rotation round
an axis. About the words used by Plato, Aristotle, and Simplicius, to signify
rotation, see Martin, *Mém. de l'Acad. des Inscriptions*, t. XXX. part 2, p. 5.

Hiketas. If what the doxographer[1] says of his opinions is correct he must have been of an eclectic turn of mind, for the doctrine of the indivisible bodies (atoms) is stated to have been taught by him as if he were an adherent of Leukippus and Demokritus. With regard to his cosmical ideas Aetius says: "Herakleides of Pontus and Ekphantus the Pythagorean let the earth move, not progressively ($\mu\epsilon\tau\alpha\beta\alpha\tau\iota\kappa\hat{\omega}s$) but in a turning manner like a wheel fitted with an axis, from west to east round its own centre[2]." This is sufficiently clear, and it may be taken as an indication that some members of the Pythagorean school already in the first half of the fourth century before our era had abandoned the doctrine of the central fire, or rather, that Ekphantus and perhaps others may have modified the original doctrine by substituting for the central fire the fire in the interior of the earth, which revealed itself in volcanic eruptions. There are indeed several circumstances which indicate that the later Pythagoreans did not remain faithful to the original doctrine of Philolaus. We have already mentioned that some of them thought it necessary to introduce additional unseen bodies in space to account for some of the lunar eclipses, which involved giving up the doctrine of the planets being ten in number; and when that doctrine was abandoned the antichthon might also be swept away. Still later, when the Pythagorean philosophy was revived and had to be revised in some points to suit the times, the long abandoned ideas about the central fire and the antichthon had to be explained away or interpreted so as to agree with the geocentric system, as it was supposed to be inconceivable that the divine master or his more immediate followers could have designed a system which had turned out to be untenable. Accordingly Simplicius, after describing the ten bodies of

---

[1] Aet. (Stobaeus), Diels, p. 330.

[2] Aet. III. 13, p. 378. A similar account is given by Hippolytus (*Philos.* xv. Diels, p. 566) in a short and not very clear notice of the opinions of Ekphantus. After stating that we cannot attain a true knowledge of things and that bodies are neither moved by weight nor impact but only by a divine power which he calls mind and soul, Ekphantus is supposed to say that of this the world is a representation, wherefore by the agency of one power [the divine one apparently] it was fashioned of a spherical shape, but the earth at the middle of the world moves round its own centre towards the east.

Philolaus, proceeds thus[1]: "And thus he [Aristotle] understood
the opinions of the Pythagoreans, but those who partook of a
greater knowledge called the fire in the middle the creating
power, which from the middle gives life to the whole earth and
again warms that which has been cooled." And the position
of the earth is explained thus: " But they called the earth a
star because it also is an instrument of time, for it is the cause
of days and nights, for it makes day to the part illuminated by
the sun, but night to the part which is in the cone of the
shadow." Here the motion of the earth is dropped without
substituting rotation for it, and the expression "instrument of
time " (ὡς ὄργανον καὶ αὐτὴν χρόνου) is borrowed from Plato
and shows the anxiety of Simplicius to banish the notion of
the earth being a moving star. Finally he says that the moon
was called the antichthon, because it is an " ethereal earth."

But quite apart from this apology on the part of one of the
very last writers of antiquity we cannot doubt that already
about the middle of the fourth century B.C. a strong opposition
to the system of Philolaus had made itself felt within the
Pythagorean school. But unfortunately we know nothing about
the opinions on the system of the world which prevailed among
the Pythagoreans during the last days of their existence, which
is the more to be regretted as it is quite possible that a more
complete knowledge of the motions of the planets began to be
attained before the extinction of the school towards the end of
the fourth century. We shall return to this subject in our
sixth chapter.

---

[1] Simplicius, *De Cœlo*, p. 512 (Heib.); compare an anonymous scholiast in
Brandis, *Scholia*, p. 504 b, 42–505 a, 5, who says practically the same.

CHAPTER III.

PLATO.

THE speculations of the Pythagoreans, though they had drifted into a wrong channel, had at least accustomed the minds of philosophers to the idea of the spherical earth, and after the time of Philolaus we hear nothing more of the crude notions of earlier ages. The influence of the Pythagoreans was widely felt, and it can be distinctly traced in the views on the system of the world held by the great Attic philosopher, which we shall now consider.

In the writings of Plato (born about 427, died 347 B.C.) we do not find many signs of his having taken any great interest in physical science. The idea, as the pure existence, is to him the sole object of knowledge, while the visible, physical world only to a limited extent partakes of Being, since the formless matter never completely yields to the forming power of the ideas. The opposition between form and matter makes it possible to attain absolute truth only in the eternal and unchangeable ideas, but not in the physical world, the region of the contradictory and incomplete, where at most a high degree of probability may be reached by means of mathematics. But while the opposition of idea and matter is strongly felt if we attempt to investigate the details of the external world, it disappears when we view the latter as a whole; the world is then seen as that wherein the idea rules supreme, as Kosmos, the perfect living Being, formed in the image of the Deity, a divine work of art. In his statements on the construction of the world Plato therefore does not descend to details, and they are not always easy to follow, as they have rather the character of an intellectual play and are frequently interwoven with mythological illustrations from which his philosophical meaning has to be disentangled, though there is in reality a perfect accordance between the plainly worded passages and those clothed in mythological imagery. Even in the only dialogue which is specially devoted to physical questions, the *Timæus*, Plato has mingled myth and science, and partly for this reason there has been considerable difference of opinion on several points of his cosmical system. It is for many reasons desirable to discuss these questions in detail, especially as an uncritical admirer of Plato, O. F. Gruppe, has invented and made himself the champion of a most preposterous theory as to Plato's astronomical knowledge, which to readers unacquainted with Plato's own writings may appear exceedingly plausible[1].

Though there are great difficulties in the way of fixing the order in which Plato wrote his various dialogues, it seems probable that the *Phædrus* is one of the earliest ones[2]. There is nothing astronomical in this dialogue except that the universe is described as a sphere; a distinction being made between the super-celestial space occupied by the eternal ideas, to which the soul wends its way, and the infra-celestial space, the region of sense and appearances. Gruppe, who is anxious to prove that Plato successively passed through all cosmical systems from the Homeric to the Copernican, infers from these poetical descriptions that Plato in the *Phædrus* assumes the heavens to be a material sphere, or rather a bell of crystal, thrown over a flat earth. But there is absolutely no reason to accept this view, as the shape of the earth is nowhere mentioned, and there is nothing in Plato's words to show that he did not believe it to be spherical. And even if we take his words about the heavens literally, the expressions super- and

[1] *Die kosmischen Systeme der Griechen.* Von O. F. Gruppe. Berlin, 1851. Plato's astronomical system is discussed in detail in Th. H. Martin's *Études sur le Timée de Platon*, 2 vols. Paris, 1841, and in a memoir by the same author in the *Mémoires de l'Institut de France, Académie des inscriptions*, t. xxx. (1881), as well as in Boeckh's *Untersuchungen über das kosmische System des Platon*, Berlin, 1852 (in reply to Gruppe's book).

[2] Zeller is of opinion that it cannot have been written later than 392–390. *Phil. d. Gr.* II. a, p. 539 (4th ed.).

infra-celestial might be merely the outcome of Philolaic reminiscences, as we have seen that the Pythagoreans used the words above and below where we should say outside and inside[1]. Another Pythagorean reminiscence might at first sight appear to have dictated the following sentences (*Phædrus*, p. 246): "Zeus, the great ruler of the heavens, steering his winged chariot, sets out first, arranging and taking care of everything, and him follows the host of gods and spirits, in eleven divisions. For Hestia alone remains in the house of the gods. But all the others, who to the number of twelve are ranged in order as ruling gods, set out in the order appointed to each." Of course these twelve gods do not in number correspond exactly to the number of planets (either in the Pythagorean or in any other system), but there can be no doubt that Hestia here means not the central fire but simply the earth, which in Greek literature, from Orphic and Homeric hymns down to Plutarch, is often referred to as Hestia[2]. We are therefore unable to conclude from the *Phædrus* that Plato in his younger days was an adherent of Philolaus, or indeed that this dialogue represents any very decided opinions about the Kosmos.

In the *Phædo*, the dialogue in which Plato has given us a touching picture of the last moments of Sokrates, we find no general theory of the construction of the world, but only an account of the earth, which is said to stand in the midst of heaven, requiring neither air nor any other force to keep it from falling, as it has no cause to incline more in one direction than another. It is round, like a twelve-striped leather ball,

---

[1] *Phædrus*, pp. 247–248. A revolution of the heavens is mentioned, but it means a period of ten thousand years, during which pure souls are free from the body.

[2] Compare Macrobius, *Saturnal.* I. 23 ; Plutarch, *De facie in orbe lunæ*, c. vi. The earth is also referred to as Hestia in the spurious treatise "On the soul of world and nature," alleged to have been written by Timæus the Lokrian (p. 97): " The earth, fixed in the middle, becomes the hearth of the gods " (ἑστία θεῶν). Chalcidius also takes Hestia to mean the earth in the above passage of the *Phædrus*, while he thinks that the dominions of the twelve gods are the sphere of the fixed stars, the seven planets and the four elements. But as Plato would hardly have compared the elements to gods, this explanation seems rather far-fetched. (Chalcidius, CLXXVIII. ed. Wrobel, Leipzig, 1876, p. 227.)

very large, and the people who dwell from Phasis to the Pillars of Herakles only inhabit a small part around the sea, in one of the many hollows in which water, mist, and air have flowed together. " But we do not notice that we live in these hollows of the earth and imagine ourselves living above on the earth, as if one living at the bottom of the sea were to think that he lived at the surface, and because he could see through the water the sun and stars were to take the sea for the heavens[1]." The meaning of this seems to be that the true spherical surface of the earth is at a much higher level than the Mediterranean " hollow," unless the real surface is supposed to be the limit of the atmosphere. Anyhow, we find in this dialogue the earth described as a sphere, unsupported and placed in the centre of the universe.

The daily rotation of the heavens is not alluded to in the two dialogues mentioned, but it is prominently dealt with in the tenth book of the *Republic*, where we find a detailed description of an elaborate machinery for moving the heavenly spheres. Having expounded his conception of the perfect man and the perfect state, and having pointed out the rewards which a just man receives during his life, Plato draws a picture of the rewards and punishments which await man after death. Erus, a warrior, is supposed to have been found lifeless on the field of battle, but when laid on the funeral pyre he revives and relates his experiences during the twelve days his soul was absent from the body. After describing the place of judgment and how the souls of just men ascend to heaven, while those of wicked men have to expiate their misdeeds for a shorter or longer time, he proceeds[2]:

" Everyone had to depart on the eighth day and to arrive at a place on the fourth day after, whence they from above perceived extended through the whole heaven and earth a light as a pillar, mostly resembling the rainbow, only more splendid and clearer, at which they arrived in one day's journey; and there they perceived in the neighbourhood of (κατὰ) the middle of the light of heaven, the extremities of

---

[1] *Phædo*, pp. 108–109.
[2] *Republic*, x. pp. 616–17.

the ligatures of heaven extended; for this light was the band (ξυνδεσμός) of heaven, like the hawsers of triremes, keeping the whole circumference of the universe together."

It is an old idea, going back to Theon of Smyrna[1], that this light was the axis of the celestial sphere. But it is not conceivable that Plato should have described, as existing and visible, a luminous axis which nobody has ever seen. Besides, as remarked by Boeckh, how could this light pass through the earth unless there was a cylindrical hollow space from pole to pole? The hypozomes or hawsers keeping a ship together were outside it, along its whole length; similarly the light keeps the celestial sphere together (as hoops do a barrel) lest the rotation of the sphere should cause it to fall asunder. Already Demokritus had explained the Milky Way to be composed of a vast number of small stars, and Plato therefore here considers it as the outermost layer of the ἀπλανής, farthest from the earth. The souls apparently see the circle of the Milky Way from outside the sphere, so that the circle seen edgeways appears as a cylinder or pillar[2]. By saying that the light extends through heaven and earth we are probably to understand that it passes both above and under the earth. The ligatures (δεσμοί) of the heavens are the solstitial and equinoctial colures intersecting in the poles, which points therefore may be called their extremities (ἄκρα).

The narrative continues: "From the extremities the spinning implement[3] of Necessity is extended by which all the revolutions were kept going, whose spindle and point were both of adamant, but its whirling weight (σφόνδυλος) com-

---

[1] Theon, ed. Martin, p. 194. Martin, ibid. p. 362, accepted this explanation, but in his later memoir, p. 96, he acknowledged it to be wrong.

[2] Boeckh, *Kleine Schriften*, III. p. 301.

[3] ἄτρακτος is the whole of the spinning machine, i.e. the whole heavens, while the spindle (ἠλακάτη) is the axis of the world. Grote (in the paper quoted further on) understood Plato to say that the axis of the world rotated itself and thereby caused the rotation of the heavens, because he supposed ἄτρακτος to be the axis. But Plato does not say one word about the rotation of the axis, and when he gives to each of the eight turning bodies a siren producing one note, he does not give one to the axis. Boeckh, l. c. pp. 312–13. Of course when Plato compares the world to a spinning machine we cannot expect this comparison to suit every detail.

pounded of this and other materials; and the nature of this whorl was of such a kind as to its figure as those we see here. But you must conceive it to be of such a nature as this, as if in some great hollow whorl, hollowed out altogether, there was another similar one, but smaller, within it, adapted to it, like casks fitted one within another; and in the same manner a third, and a fourth, and four others. For the whorls were eight in all, as circles one within another, having their lips (or rims) appearing upwards, and forming round the spindle one united convexity of one whorl; the spindle was passed through the middle of the eight, and the first and outermost whorl [the sphere of the fixed stars] had the widest circumference in the lip, the sixth [Venus] had the second widest, and the fourth [Mars] the third widest, and next the eighth [the moon], and then the seventh [the sun], then the fifth [Mercury], then the third [Jupiter], and the narrowest lip was that of the second [Saturn]."

It is needless to say that the "whorls" represent the spheres of the fixed stars and of the seven planets. But what their "lips" or rims are is not very clear. It is most natural to think of the breadth of the zone within which the occupant of each sphere can appear[1]; and as the fixed stars occur all over the heavens, in all declinations, and Venus can reach a higher declination than any other planet, Plato would have been quite right in classing their "lips" as first and second. But the explanation fails with regard to Mars, and especially Mercury, as the inclination of the orbit of the latter planet at that time was 6° 37′, so that he should not have been put third last. Besides, Plato never alludes to the fact that the planetary orbits have different inclinations to the ecliptic, and in the *Republic* he does not even mention that their paths are not parallel to those of the fixed stars, that is, to the celestial equator. We might also think of the breadth of the intervals between the orbits, but if so we must assume that Plato here adopts a scale of the distance of the planets differing from that mentioned in the *Timæus* and from any other used in antiquity.

---

[1] This explanation was adopted by Martin, *Theonis Smyrnæi Astronomia*, p. 363, but he does not mention it in his later memoir.

It must therefore be left an open question what he meant by these lips[1]. He next continues thus:

" The circle of the largest is of various colours, that of the seventh [the sun] is the brightest, and that of the eighth [the moon] has its colour from the shining of the seventh; those of the second and fifth [Saturn and Mercury] are similar to each other but are more yellow than the rest. But the third [Jupiter] is the whitest, the fourth [Mars] is reddish, the second in whiteness is the sixth [Venus]. The spinning implement must turn round in a circle with the whole that it carries, and while the whole is turning round, the seven inner circles are slowly turned round in the opposite direction to the whole[2]; and of these the eighth moves the swiftest [the moon], and next to it, and equal to one another, the seventh, the sixth, and the fifth [the sun, Venus, and Mercury], and third [in swiftness] is the fourth [Mars], which, as it seemed to them, moved in a rapid motion, completing its circle[3]. The fourth

[1] Another explanation has recently been suggested by Professor D'Arcy Thompson in a paper read before the Brit. Assoc. at Cambridge, 1904. In the sphere-theory of Eudoxus (see next chapter) the third and fourth spheres of each of the five planets have their axes inclined at a certain angle, which represents for the outer planets the length of the retrograde arc, for the inner ones the greatest elongation from the sun. These angles are not given by Eudoxus, but the values computed by Schiaparelli are: Venus 46°, Mars 34°, Mercury 23°, Jupiter 13°, Saturn 6°. The order according to the amount of this angle is therefore the same as that of Plato according to breadth of rim, if we insert the moon and the sun in the series according to their greatest *declination*, 28° and 24°. This explanation is very ingenious, but the analogy between the oscillations of the planets in longitude and those of sun and moon in declination is not a very close one, and it is extremely doubtful whether Schiaparelli's value for Mars (34°) is at all near the value assumed by Eudoxus—if indeed he assumed any particular values.

[2] That is, the planets move slowly along their orbits from west to east while participating in the daily motion of the heavens from east to west.

[3] The last sentence has been misunderstood by several translators. Τὸν τρίτον δέ, φορᾷ ἰέναι ὡς σφίσι φαίνεσθαι ἐπανακυκλούμενον τὸν τέταρτον. I understand this as stating that the third one in speed was the fourth planet, Mars. But Theon of Smyrna, who reproduces this whole paragraph, omits the words τὸν τέταρτον (probably a clerical error caused by the words τέταρτον δέ, with which the next sentence begins) and adds μάλιστα τῶν ἄλλων, which words I believe do not occur in any ms. of the *Republic*. Martin translates this sentence of Theon's thus: Celeritate vero tertium ferri, ut ipsis quidem visum est, quartum, qui retro sese circumfert magis quam ceteri omnes (*Theon*, p. 201). No doubt there is good sense in this from an astronomical point of view, as Mars

[in swiftness] was the third [Jupiter], and the fifth was the second [Saturn], and it [the spinning implement] was turned round on the knees of Necessity, and on each of its circles there was seated a Siren on the outer edge carried round, and each emitting one sound, one tone, but the whole of the eight composed one harmony. There were other three sitting round at equal distances one from another, each on a throne, the daughters of Necessity, the Fates, clothed in white garments, and having crowns on their heads, Lachesis, Klotho, and Atropos, singing to the harmony of the Sirens; Lachesis singing the past, Klotho the present, and Atropos the future. And Klotho at certain intervals with her right hand laid hold of the spinning implement and revolved the outer circle, and Atropos in like manner turned the inner ones with her left hand, while Lachesis touched both of these in turn with either hand."

Although this description of the mechanism of the world forms part of a mythical story, there is no reason to think that Plato wrote it merely in play; on the contrary, the detailed manner in which it is worked out seems to justify the belief that the system really represents Plato's conceptions as to how the heavenly motions might be represented, or, at any rate, the ideas on this matter which were uppermost in his mind when he wrote the *Republic*. And in this place Plato enters rather more into details as regards the planets than he does even in the *Timæus*; in fact this is the only occasion on which he makes more than the merest passing allusion to the outer planets. But though he knows the colour and brightness of each one, his account is very meagre as regards their motions; he tells us that they travel from west to east, but he says nothing about the inclinations of their orbits or whether they travel with uniform velocity. The concluding poetical description of the Sirens chanting to the motion of the planets is

does retrograde more than Jupiter and Saturn, but query, does ἐπανακυκλέω imply a backward motion? In the compound verb, ἐπί would seem to have the force of continuance (goes on completing), comp. *Tim.* p. 40 c, ἐπανακυκλήσεις. Theon and some mss. of Plato leave out the article τὸν before τρίτον, where it does seem out of place and has been the cause of the erroneous translations. Hiller's ed. of Theon (p. 145) gives ὃν φασι instead of ὡς σφίσι.

dictated by the conviction that the universe is ruled by
harmony, that the numerical ratios, which in our earthly world
are represented by musical harmony, also pervade the loftier
ranges of Kosmos. The continuity of the celestial motions is
beautifully expressed by giving the control of them to the
three goddesses of fate: Klotho, who rules the present, directing
the diurnal motion of the whole heavens; Atropos, who rules
the future, controlling the varied planetary motions in the
opposite direction; and Lachesis, goddess of the past, assisting
both operations and thereby securing the participation of the
planets in the daily motion of every part of the universe, the
earth alone excepted.

In the three dialogues mentioned Plato referred only in
passing to the system of the world, but after finishing the
Republic he wrote another dialogue, the *Timæus*, in which he
put together his views on the physical world, and which we
must therefore consider in more detail. This dialogue between
Timæus of Locri in Italy, a Pythagorean philosopher of some
renown, Sokrates, and two other friends, is supposed to take
place on the day following that on which the same party had
discussed the nature of an ideal State. Sokrates first shortly
reviews the political results of the conversation on the previous
day, and Kritias tells a charming old myth about the vanished
island of Atlantis, in order to show that this ideal state of
society had really once upon a time existed among the
Athenians. Before proceeding further in the attempt to trans-
plant the ideal citizens of the *Republic* into real life, the parties
present agree to let Timæus, who has an intimate knowledge
of science, give them a lecture on the origin and construction
of the universe and the formation of man.

Timæus first describes how the Deity or the "framing
artificer" made the world and how he gave it the most perfect
figure, that of a sphere. He gave it only one motion, the one
peculiar to its form, letting it turn uniformly on its axis
without any progressive motion[1]. The soul of the world was
by the Artificer placed in the middle, extending throughout
the whole and spreading over its surface, being both in age and

[1] *Timæus*, p. 34 A.

in excellence prior to its body. The manner in which the soul of the world is formed shows the connection between the cosmical speculations of Plato and those of the Pythagoreans. From one essence indivisible and always *the same,* and from another, *the diverse* or different, divisible and corporal, a third form of essence intermediate between the two was formed. After mingling these and producing one from the three, the Artificer distributed this whole into suitable parts, each composed of a mixture of same, diverse, and essence. We must here pass over Plato's account of the formation of the "soul of the world," which though exceedingly beautiful has no direct connection with his astronomical views, except as regards the distances of the planets, which he supposes proportional to the following figures, derived from the two geometrical progressions 1, 2, 4, 8 and 1, 3, 9, 27.

Moon	1
Sun	2
Venus	3
Mercury	4
Mars	8
Jupiter	9
Saturn	27

By interpolating other numbers between these according to certain rules he forms an arithmetical musical scale of four octaves and a major sixth[1]. But while Plato thus views the "soul of the world" as a harmony of the essences, he does not seem to share in the Pythagorean belief in musical sounds produced by the motion of the planets. He now lets Timæus proceed as follows[2]:

"He [the artificer] split the whole of this composition lengthwise into two halves, laying them across like the letter X; next he bent them into a circle and connected them with themselves and each other, so that their extremities met at the point opposite the point of their intersection, comprehending

---

[1] For further details see Boeckh's memoir "Ueber die Bildung der Weltseele im Timaeos des Platon" (1807), *Kleine Schriften,* III. pp. 109–180, and Martin's *Études,* I. p. 383 sq.

[2] Page 36 B.

them in an uniform motion around the same centre, and he made one of the circles external, the other internal[1]. The motion of the exterior circles he named that of sameness, and that of the interior the motion of the diverse. He caused the circle of the same to revolve along the side of a parallelogram towards the right, and that of the diverse along the diagonal towards the left. And the superiority he gave to the circulation of same and similar, for this one he left undivided, but the inner one he divided into six parts, and forming thus seven unequal circles, arranged by double and triple intervals[2], three of each, he bade these circles move in contrary directions to each other, three at equal velocities[3], the other four with velocities unequal to each other and to the former three, yet in a fixed ratio as to velocity. After the whole composition of the soul [of the world] had been completed according to the mind of the maker he next formed within it all that is bodily, and he fitted them together, joining centre to centre. The soul being interwoven throughout from the middle to the farthest part of heaven and covering this all round externally, and moving itself in itself, made the divine commencement of an unceasing and wise life for evermore. And the body of the world became visible; but the soul, invisible, was made to partake of reason and of the harmony of intelligent beings, made by the most perfect being, and itself the most perfect of created things."

Although Plato here ostensibly deals with the soul of the world, the above really represents his conception of the construction of the universe. The soul as the principle of motion penetrates the whole body of the universe, therefore the motions in the latter are the motions of the soul. Aristotle expresses Plato's conception in the following words[4]: "In a similar manner the *Timæus* shows how the soul moves the

[1] The external circle is the celestial equator, the internal one the zodiac. Proklus, the neoplatonic philosopher, in his voluminous commentary to the *Timæus* (p. 213) gave a perfectly correct exposition of Plato's account of these circles.

[2] That is, by the two progressions 1, 2, 4, 8 and 1, 3, 9, 27.

[3] The sun, Venus, and Mercury.

[4] *De Anima*, I. 3, p. 406 b.

body because it is interwoven with it. For consisting of the
elements and divided according to the harmonic numbers, in
order that it might have an innate perception of harmony and
that the universe might move in corresponding movements,
He bent its straight line into a circle, and having by division
made two doubly joined circles out of the one circle, He again
divided one of them into seven circles in such a manner that
the motions of the heavens are the motions of the soul." And
the principal motion of the soul, the rotation of the " circle of
the Same," is of course the same which the hero of the dialogue
at the beginning of his lecture attributed to the world, when
he described how the Artificer made it " to turn uniformly on
itself" and deprived it of any other motion. The undivided
circle of the Same is therefore the celestial equator, parallel to
which all the stars describe their daily orbits. The circle of
the Diverse intersects it in two diametrically opposite points,
their planes forming an acute angle, like the letter X. While
the circle of the Same revolves laterally towards the right
(that is, from east to west), the motion in the circle of the
diverse is " diagonally " towards the left, in the opposite direction, along the inclined zodiacal circle[1]. The " superiority " is
given to the motion towards the right; that is to say, the daily
rotation of the heavens includes also the seven planets which
are carried along in it, although they at the same time pursue

---

[1] With regard to this distinction between "left" and "right," it looks at
first sight as if Plato were not consistent in his use of these expressions. In
the sixth book of the "Laws" (p. 760) he proposes to let the country be ruled
by the various districts in their turn, "by going to the place next in order
towards the right in a circle, and let the right be that which is in the east."
And the author of the *Epinomis* (p. 987) says that the planetary motions are
ἐπὶ δεξιά. Aristotle (*De Cœlo*, II. p. 284 b, l. 28) says that the right of the
world is the place whence comes the diurnal motion, i.e. the east, and that the
left is the west, towards which this motion is directed. This would be natural
to a Greek, who when taking auspices by watching the flight of birds would
turn to the north, and as a star *sub polo* is moving from left to right, Plato
might well say that the diurnal motion was towards the right. This seems to
me a very simple explanation of the apparent inconsistency of Plato, which has
troubled so many commentators. See Martin's *Études*, II. p. 42; Boeckh's
Untersuchungen über das kosmische System des Platon, pp. 29–32; and Prantl's
edition of Aristotle, II. p. 292. The matter is really of no practical importance
for understanding the Pythagorean and Platonic conception of the universe.

their separate motions along the inclined sevenfold circle of the Diverse, the earth being in the common centre of all these circles. Plato evidently assumes that the seven planets all move in the same plane, and that the line of intersection of this plane with the equator is immovable.

Timaeus next tells his hearers that God resolved to form a movable image of eternity on the principle of numbers, which we call time (p. 38 C). "With this design the Deity created the sun, moon, and the five other stars which are called planets, to fix and maintain the numbers of time. And when he had made these bodies God placed them, seven in number, in the seven orbits described by the revolutions according to the Diverse, the moon in the first orbit nearest to the earth, the sun in the second above the earth, then the morning star and the star sacred to Hermes, revolving in their orbits with the same speed as the sun, but having received a force opposed to it (τὴν δ᾽ ἐναντίαν εἰληχότας αὐτῷ δύναμιν), owing to which the sun and the star of Hermes and the morning star in like manner overtake and are overtaken by each other" (p. 38 D).

The expression "a force opposed" to the sun, in the last sentence, is rather obscure and has given rise to a variety of interpretations from Theon of Smyrna and Proklus down to Martin. Plato evidently refers to the intimate connection between the sun and the two inferior planets which never travel far from it, and this sentence must be considered in connection with the statement already quoted (p. 36 D), that the artificer "bade these circles (of the planets) move in contrary directions to each other." If the latter statement had not been supported by that in 38 D, we might have been inclined to think with Proklus[1] that Plato had perhaps merely made a slip and had meant to say that the planets moved in the direction opposite to that of the daily rotation of the heavens. But the other sentence (about the force opposed to the sun) shows clearly enough that Plato really believed Mercury and Venus to differ in some important manner from the other planets. Theon understood the sentence in question to mean

---

[1] Proklus, *In Timæum*, p. 221 E.

that Plato alluded to the theory of epicycles, which he therefore assumed to be due to Plato, and he was herein followed by
Chalcidius, most of the astronomical part of whose commentary
to the *Timæus* was in fact simply copied from Theon[1]. We
shall farther on discuss the theory of epicycles, and shall here
only mention that it explained the variable velocities of the
planets by assuming that each planet moved on a small circle,
the epicycle, the centre of which moved round the earth on the
circumference of another circle. We shall also see that as the
moon and the sun differ from the five planets in never stopping
and retracing their steps for a while, it is necessary in the
epicyclic theory to assume that the sun and the moon move on
their epicycles from east to west, while the five planets move
on theirs from west to east. An adherent of the epicyclic
theory would therefore have been justified in saying that
Mercury and Venus moved in a direction opposite to that of
the sun's motion, and Theon was not the only one to suppose
this to have been Plato's meaning, since Proklus (who does not
mention Theon) says that "others" interpreted the passage
in this manner[2]. Proklus rejects this explanation altogether,
and points out repeatedly (supported by the opinions of other
leading Neo-Platonists) that Plato never alluded to and did
not know anything about epicycles or excentrics, but only gave
one circle to each planet, and apparently supposed the motion
on this circle to be uniform[3]. Proklus thought, however, that
the "opposite force" referred to the want of uniformity in the
motions of Mercury and Venus (which indeed is very glaring),
and that it was caused by small additional movements produced
voluntarily by the soul of each planet[4]. This feeble explanation may have satisfied the Neo-Platonists, to whom the idea of
a direct interference of supernatural agencies by fits and starts
came natural, but it is hardly necessary to point out how
utterly opposed it is to the Platonic doctrine of Kosmos,

---

[1] Theon, ed. Martin, p. 302; Chalcidius cix, ed. Wrobel, p. 176, reprinted
by Martin, Theon, p. 424.

[2] Proklus, p. 221 E–F.

[3] Ibid. 221 F, 258 E, 272 A, 284 C. He quotes Porphyrius, Iamblichus, and
Theodorus of Asina on p. 258 D–E.

[4] Ibid. 221 E–F, 284 D, compare 259 A.

permeated by the soul of the world and subject only to the laws of harmony. A simpler plan was adopted by Alkinous in his *Introduction* to the doctrines of Plato; he boldly suppresses the inconvenient passage, and merely says that (according to Plato) God placed Mercury and Venus on a circle moving with a velocity equal to that of the sun, but at some distance from it[1].

One thing seems certain enough, which is that Plato knew very little about the motions of Mercury and Venus, except that they are never seen at any great distance from the sun. Martin[2] considers that we are bound to take the expression " a force opposed to the sun " in its literal sense, and to assume that Plato really believed the two planets to move in the direction opposite to that of the sun's annual motion, and that he supposed that this explained why they " overtake each other and are overtaken." Plato would then appear to have overlooked the fact that in that case Mercury and Venus ought to be found at all distances from the sun from 0° to 180°. But as he can hardly be supposed to have been capable of overlooking this, might we not simply interpret the passage as meaning that these two planets were acted on by a force *different* from that which propelled the sun, moon, and three outer planets and enabled them to place themselves at all possible angular distances from each other? We certainly must reject every explanation which (like that of Theon) assumes that Plato had any considerable knowledge of the stationary points and retrograde motions of the planets, as he never makes the slightest allusion to these phenomena[3].

There is, however, a circumstance which may serve as an excuse to Plato, if we cannot get over the word " opposed " in

[1] Chapter xiv.; Burges' translation of Plato, Vol. vi. p. 275. In the *Epinomis*, a book of doubtful authorship but apparently published by Plato's disciple, Philip of Opus, it is said in one place (p. 986 c) that the two planets are " nearly equal in velocity to the sun," in two other places (987 b and 990 b) that they move in the same course as the sun.

[2] *Études*, ii. p. 70, and in his later memoir, p. 36.

[3] Except that he says (p. 39 c), after defining the month and the year, that the courses of the five other planets have not been taken into account except by very few people, as their wanderings are infinite in number and of wonderfu variety.

any other way than by accepting it as a motion from east to west. As seen from the earth Venus takes 584 days to go round the sun, but of this period only 143 days are occupied in passing from the greatest eastern elongation as evening star to the greatest western elongation as morning star. When the planet emerges from the rays of the sun in the evening, it creeps slowly eastward until it reaches the eastern elongation, after which it, still for some time continuing to increase in brightness, starts westward towards the sun at a considerable rate. After appearing on the west side of the sun in the early morning after inferior conjunction, it continues to move westward at great speed, attains its maximum brightness about 36 days after the conjunction, reaches the western elongation, and begins then slowly to wind its way back to the sun. At the times when the planet is most conspicuous it is therefore moving westward, in the direction opposite to that in which the sun is always moving; and only when it is a far fainter object, does it gradually overtake and pass the sun. It is therefore perhaps no wonder that Plato, who had only a very elementary acquaintance with the planetary motions, considered the westward motion of Venus and Mercury the usual and proper one, but did not stop to consider why the planets, when much fainter, overtook the sun.

Plato now continues: " As regards the other planets, however, the task of investigating their revolutions, and the reasons why they were given them, would surpass that of the explanation of the matter which led us to it. These subjects may hereafter perhaps, when we have leisure, meet with the investigation they deserve. When, therefore, each of the stars which should produce time had obtained a motion suitable to it, and their bodies, bound by living chains, had become animated beings and learned their prescribed duties, they pursued their course according to the movements of the Diverse, passing obliquely through the motion of the Same and subordinate to it, one having a larger and the other a smaller circle, and that in the smaller circle moved quicker, that in the larger one more slowly. Owing to the revolution of the Same those revolving the quickest appeared to be overtaken by those moving more slowly, although they themselves overtook these. For while it [the revolution of the Same] turned all their circles [the orbits of the planets] into helices by their moving at one and the same time in opposite directions, it made that one which most slowly moves away from itself, being the fastest, appear to be the nearest[1]."

The last few sentences are somewhat obscurely worded and may require a few words of explanation. When Plato speaks of the velocities of the planets, he does not mean the actual velocities in linear measure, as these cannot be determined without knowing the dimensions of the planetary orbits, quantities which the Greeks had no way of determining, even though Plato, as we have seen, assumed the radii of the orbits proportional to the numbers of the soul-harmony 1, 2, 3, 4, 8, 9, 27. The velocities referred to are the angular velocities or the apokatastatic, as Simplicius calls them[2] with reference to Plato's *Republic* and *Timæus*. In both of these dialogues we find Saturn, describing the largest circle, designated as moving slowest, while the moon, describing the smallest circle, moves quickest, performing in fact a revolution in twenty-seven days, while Saturn takes 29½ years. But let us now fix our attention on the daily rotation of the various heavenly bodies (revolution of the Same) in twenty-four hours. All the fixed stars perform this in the same period, which never changes (hence the expression: circle of the Same or of Sameness), while the planets perform it in different periods. During a day the moon moves about 13 degrees towards the east, therefore she takes fully three-quarters of an hour longer than a fixed star to reach the meridian again after her previous culmination. The sun moves nearly a degree eastward during a day, therefore nearly four minutes more than a sidereal day elapse between two successive culminations of the sun. And Saturn moves only about $\frac{1}{30}$° eastward in a day, consequently it only comes to the meridian eight seconds later every day. If we therefore estimate the speed of the planets, not with regard to their orbital motion

---

[1] 38 E–39 A. I have followed Boeckh's translation of the last two sentences (l. c. p. 35).
[2] Commentary to *De Cælo*, II. 10, p. 470 sq. (Heib.).

but with reference to their daily meridian transit, we see that
if the moon and Saturn start together on a certain day, Saturn,
though the slowest planet, will day by day appear to be more
and more in advance of the moon, so that " owing to the revolution of the Same " Saturn may be considered the fastest and
the moon the slowest planet. This is clearly expressed by
Proklus in his commentary to the *Timæus*: " The ruling
movement of the Same makes the one which is nearest to
it appear to go faster. But that one is nearest to it which
goes least away from it. For supposing the moon and Saturn
to be near the heart of the Lion, then the moon by its proper
motion will leave this star, but Saturn will remain many nights
about the same place[1]."

With regard to the statement that the daily rotation of the
heavens turns the planetary orbits into helices, this is an
obvious consequence of the inclination of these orbits (the
circle of the Diverse) to the celestial equator. While a non-
wandering ($\dot{a}\pi\lambda a\nu\acute{\eta}s$) or fixed star every day describes the same
circle, parallel to the equator, the moon (or any other planet)
does not trace a closed curve on the heavenly vault, as its
distance from the equator changes from day to day; and if we
consider its daily motion, not only with reference to the earth
but also with reference to the fixed stars, we may consider the
apparent orbit a sort of helix[2].

I have entered into all these details and quoted Plato's own
words so largely in order to impress on the reader that the
doctrine of the daily rotation of the heavens in twenty-four
hours round the immovable earth is the fundamental feature
of the cosmical system depicted in the *Timæus*. Not only is it
stated at the outset that the Deity gave the world only one
motion, a revolving one, but the motion of the circle of the
Same runs through the whole account, so that it is absolutely
impossible to separate this idea of the rotation of the heavens
from the description of the other celestial movements, as these
are solely viewed in the light of this rotation. The function of

[1] Proklus, p. 262 F.
[2] Compare Theon of Smyrna, ed. Martin, p. 328, last two lines, where
Derkyllides explains this apparent motion in a helix.

III] *Plato*

the sun in marking day and night is thus described (p. 39 D) in
the passage next after the last paragraph quoted above. " In
order that there might be an evident measure of slowness and
swiftness in the eight revolutions, God kindled a light, which
we now call the sun, in the second of these orbits, in order that
it might shine as much as possible throughout the universe,
and that such living beings as required it might become
acquainted with the number, receiving the knowledge of this
from the revolution of the Same and uniform. Thus then, and
for these causes, arose night and day, as the period of the one
and most intelligent circular motion ; but the month when the
moon, having passed through her orbit, overtakes the sun ; and
the year when the sun has completed her course." Here again
we see the day and night measured by the "revolution of the
Same."

If the astronomical part of the *Timæus* had terminated with
the sentence just quoted nobody would have dreamt of attributing to Plato any cosmical system different to what we have
here stated it to be. But unfortunately there follow a few
more paragraphs of which one has given rise to a great deal of
controversy both among ancients and moderns. Plato first says
(p. 40, A–B) that to each of the divine bodies formed from fire
(the stars) and spherical in shape like the universe were given
two motions, " one on the same spot and uniform (ἐν ταὐτῷ κατὰ
ταὐτὰ), as to one which always thinks the Same about the Same,
the other forward and subordinate to the motion of the Same
and Similar." The first of these two motions seems by all
commentators to have been understood to mean a rotary
motion of each star round its axis, and it is a curious fact
that Plato by purely philosophical reasoning was brought to
the conclusion that the heavenly bodies (apparently both fixed
stars and planets) rotated. The second motion is of course the
diurnal revolution common to all the stars. Immediately after
the accounts of these two motions comes the following much
disputed passage :

" But the earth our nourisher, packed (εἰλλομένην) round
the axis that extends through the universe, He formed as the
guardian and artificer (φύλακα καὶ δημιουργὸν) of night and

day, the first and most ancient of the gods that have been generated within the universe."

In this passage there are two difficulties, the proper translation of the word εἰλομένην (or ἰλλομένην) and the interpretation of the "guardian and artificer."

The earliest allusion to this passage by another author is found in Aristotle's book on the heavens (II. 13, p. 293 b). Having stated his own opinion as to the place of the immovable earth in the centre of the universe, he alludes to the Pythagorean idea of the motion of the earth and continues: " But some say that it lies in the centre and is wound (ἴλλεσθαι) round the axis which is stretched through the universe, as is written in the *Timæus*." Three manuscripts add after the word ἴλλεσθαι the words καὶ κινεῖσθαι, "and is moved." Again at the beginning of the following chapter (14, p. 296 a), Aristotle says that some people consider the earth to be one of the stars, and others place it in the middle and maintain that it is wound and moves (ἴλλεσθαι καὶ κινεῖσθαι) round the central axis; but here he does not mention the *Timæus*.

Simplicius says that ἰλλομένην (with ι) means "bound," and he quotes Homer and Apollonius the poet as having used it thus, while it signifies "impeded" if written with a diphthong, for which he quotes Aeschylus. But he seems to think that it might perhaps be misunderstood so as to signify turning round, and that Aristotle, who discusses all sorts of opinions, has assumed it to mean this, merely to refute it. He mentions that Alexander of Aphrodisias believed Aristotle to have meant rotation though he had not said that Plato had used the word in that sense[1]. Proklus explains the word as " compressed and kept together," and rejects the idea of rotation. He refers to the *Phædo* and to Hestia remaining in the house of the gods as showing that Plato held the earth to be immovable[2]. Chalcidius acknowledges that the word (which he translates by *constrictam*) might mean rotation but thinks it more likely

---

[1] Simplicius, *De Cœlo*, pp. 517–18 (Heib.). Diogenes Laert. (III. 75) is the only other ancient writer who states that according to Plato the earth moves round the centre.

[2] Proklus, p. 281 D. Alkinous (c. xv.) also assumes the earth to be at rest.

(*aliquanto verisimilius*) that the earth adheres to the centre of the world and is at rest, as it was both by Plato and others called Vesta[1]. Cicero, as we have already seen, states that some people believed Plato to have taught the rotation of the earth in the *Timæus*, " *sed paullo obscurius*[2].".

In addition to these writers we may also mention the treatise " On the soul of the world and nature," once supposed to have been written by Timæus of Lokri, whom Plato introduces as the leading person in the dialogue called after him. It is now held to be the production of a later Platonist and to be a mere abstract of Plato's dialogue. This spurious work represents the earth as being at rest, and the doubtful word does not occur in it.

The two principal modern commentators, Martin and especially Boeckh, have maintained that the passage in question cannot possibly refer to the rotation of the earth in twenty-four hours, because Plato so repeatedly and in so much detail has set forth in this very same dialogue that the heavenly sphere (the aplanes) rotates in twenty-four hours. This theory of course excludes the adoption of the earth's rotation, since only one motion can be accepted as an explanation of the diurnal motion of sun, moon, and stars, while the latter would seem to stand still if both the heavens and the earth rotated in the same direction in a day and a night. Boeckh's book is particularly directed against Gruppe's " *Die kosmischen Systeme der Griechen*," a book in which the author absolutely ignores the whole of Plato's account of the motion in the circles of the Same and the Diverse. These words do not occur once in Gruppe's 218 pages, and no reader of his book who was unacquainted with Plato's own work would dream that the *Timæus* has already expounded a perfect astronomical system before the unfortunate passage occurs[3]. An attempt to mediate

---

[1] Chalcidius, cxxii. ed. Wrobel, p. 187.

[2] *Acad. prior.* ii. 39, 123, see above, p. 50, footnote 2. In his translation of the *Timæus* (x) Cicero gives the passage thus : " Iam vero terram altricem nostram, quæ trajecto axe sustinetur...."

[3] It is a pity that a generally well-informed writer like R. Wolf (*Geschichte der Astr.* p. 33) should have blindly copied Gruppe and spoken slightingly of Ideler and Boeckh, whose writings on this subject he has evidently never read.

between the two opposite opinions was made by Grote in a
pamphlet, " Plato's doctrine respecting the rotation of the earth
and Aristotle's comment upon that doctrine[1]." I shall endeavour
to state shortly what appear to me to be the results of all these
discussions.

It is first to be noted that Plato when speaking of the
rotation of heavenly bodies (just before the disputed passage)
does not make use of the verb ἴλλεσθαι but calls that motion
κίνησις ἡ ἐν ταὐτῷ κατὰ ταὐτὰ[2]. With regard to the verb
ἴλλεσθαι or εἰλεῖσθαι, Grote fully accepts the opinion of
Buttmann that the word has only the meaning *to pack, to
fasten*, while that of *turning* or *winding* is altogether foreign to
it and can only be superadded in some cases by the nature of
the case[3]. There is therefore not in the words of Plato any
assertion of a rotation of the earth, the expression used merely
implies that the earth is wrapped or packed or twisted round
the axis of the universe. Nor is there in the words of Aristotle
anything which proves that he understood Plato to refer to
a diurnal rotation, as Grote maintains that he does. Grote
does not insist on this rotation as being a spontaneous act on
the part of the earth, but as the earth is said to be packed
closely round the axis of the world, he holds that it must
necessarily follow any rotatory motion with which this axis
may be endowed. He assumes that this axis was not by Plato
considered as a mere mathematical line but as an axle of some
solid material, round which the matter forming the earth could
be packed or wound. As there is nothing said in the *Timæus*
about the function of this axis, Grote refers to the *Republic*,
where, he maintains, the axis of the world is said to cause the
rotation of the heavens, so that, in other words, it rotates itself.
But when this axis rotates, the closely packed earth must rotate
with it, according to Grote. The difficulty of assuming that
Plato could be so totally blind as to let both the heavenly
sphere and the earth turn in the same direction and in the

[1] London, 1860, 35 pp. 8°, *Minor Works*, p. 237 sq.

[2] In fact, he never uses ἴλλεσθαι in this sense, but in addition to the expression quoted above uses ἐν ταὐτῷ στρέφεσθαι or κύκλῳ κινεῖσθαι or ἐν ἑαυτῷ περιάγεσθαι. *Timæus*, p. 34 A; *Republ.* p. 436 D, E, &c.

[3] Buttmann, *Lexilogus*, Engl. Trans. (6th ed.), p. 262.

same period, does not affect Grote. It is true, he says, that a
person with proper knowledge of astronomy will perceive that
in that case the stars would not rise and set daily, but would
appear immovable; but we must not expect the ancients to
have been so clear sighted (!), and, as a matter of fact, he adds,
Boeckh was the first commentator to perceive that the motion
of "the Same" excluded the possibility of the rotation of the
earth having been taught by Plato.

But in the first place we have no right to borrow the
conception of the nature of the axis from the mythological
account in the *Republic* and dovetail it into the system
expounded in the *Timæus*. Although there is nothing to
prove that Plato in the course of years changed his opinions
about the position of the earth in space, there is no reason why
he should not in a matter of detail have expressed himself
differently when he wrote the two dialogues. Certainly, if it
could be proved that the rotation is taught in the *Timæus*,
there would be a vast difference between the doctrines
contained in the two dialogues, since there is nothing in
the *Republic* which can possibly be contorted into a theory
of the earth's rotation. But we have already mentioned that
Boeckh has clearly proved that the axis of the world in the
Republic is *not* supposed to rotate itself, and though it is no
doubt in that work described as made of a solid material
(adamant), this is only in a mythological tale, in which the
parces and a number of sirens also appear on the stage. In
the *Timæus* there is nothing of all this; the axis is simply
a mathematical line joining the two poles and common to the
earth and the celestial sphere, and this is expressed by Plato by
saying that the earth is packed round the axis of the world.
And with regard to Grote's idea, that Plato might have sup
posed the heavenly sphere, its solid axle and the earth to form
one rigid body and to rotate together, it is simply impossible to
assume that an intellectual giant should have been blind to the
fact that in this case the sun, moon, and stars would never rise
or set, but would all appear immovable; and it is equally
impossible to believe that Aristotle should have commented on
this theory without perceiving that his predecessor had talked

nonsense. The only thing to do, for anyone who believes Plato in this disputed passage to have referred to the diurnal rotation of the earth, is altogether to ignore his detailed account of the rotation of the heavens, and it is this policy which Gruppe has found so convenient.

But even supposing that the word ἴλλεσθαι or εἰλεῖσθαι indirectly implies a turning of some sort, what kind of motion might Plato have had in his mind when he used that expression? It is significant that the passage comes immediately after his statement that all the heavenly bodies had a turning motion. We might suppose that he felt driven to the conclusion that the earth did not form an absolute exception and was not perfectly at rest while everything else was moving. Just as the adherents of Philolaus considered it necessary to endow the sphere of the fixed stars with some slow motion to enable it to be enrolled among the wandering bodies. But this, as it were, involuntarily admitted turning motion of the earth cannot have been intended to explain anything or to account for any observed phenomena, and there is certainly nothing in the words of Aristotle to show that he understood Plato to have had any intention of this kind. It was quite a different question in later times, when the doctrine of the earth's daily rotation had been scientifically proposed; it was not unreasonable for Plutarch or Cicero to ask whether this had perhaps been the motion to which Plato had darkly alluded. In fact the question which Plutarch puts to himself is, whether Plato conceived the earth to be stationary or to revolve according to the theory subsequently proposed by Aristarchus and Seleukus[1].

It seems certain that if Aristotle had not made that unlucky reference to the word ἴλλεσθαι and to the *Timæus*, nobody would ever have supposed that Plato taught the rotation of the earth, either in twenty-four hours or in any other period. The main question is therefore: what did Aristotle mean, and did he refer to Plato? First of all we must notice that he does not begin by referring to Plato by name. In the sentence preceding the disputed one he has discussed the Philolaic

[1] Plutarch, *Questiones Platonicæ*, VIII.

system, and he then quotes "some people" (ἔνιοι) as teaching a rotary motion of the earth, ἔνιοι δὲ καὶ κειμένην ἐπὶ τοῦ κέντρου φασὶν αὐτὴν ἴλλεσθαι καὶ κινεῖσθαι περὶ τὸν διὰ παντὸς τεταμένον πόλον, ὥσπερ ἐν τῷ Τιμαίῳ γέγραπται. By "some people" he can only have meant those later Pythagoreans, who, as we have seen, modified the theory of the motion of the earth round the central fire into a theory of its rotation round an axis. Nothing is more likely than that these people may have tried to give their theory weight by representing Plato as having already taught it[1]. Boeckh[2] thinks that Aristotle, when alluding to these Pythagoreans, compressed into one sentence what ought to have been expressed in two sentences like this: "wound round the axis as is written in the *Timæus*, and is moved round it." Having already disposed of the idea that the earth has a planetary motion round the centre of the universe, Aristotle now refers to the idea that it, although occupying the central position round the axis of the celestial sphere, is not without motion but "moves round" the axis. In condensing his words in this manner Aristotle doubtless never dreamed that he might be misunderstood; to him the adherence of Plato to the generally adopted idea of the rotation of the "Aplanes" was so self-evident, that he somewhat carelessly expressed himself as if the words "and is moved" (καὶ κινεῖσθαι) were found in the *Timæus*. Another suggestion has been made by Martin[3] to the effect that the words "as it is written in the *Timæus*" do not refer to ἴλλεσθαι καὶ κινεῖσθαι but only to the words immediately preceding "as it is written," viz. περὶ τὸν διὰ παντὸς τεταμένον πόλον. The "axis stretching through the whole universe" is essentially a Platonic notion, and Martin points out that Aristotle himself never uses πόλος to express "axis" but only as meaning "pole," so that it would be quite natural for Aristotle to quote the *Timæus* when making use of

---

[1] Perhaps Cicero refers to this when he, apropos of Hiketas, says that "hoc etiam Platonem in Timaeo dicere quidam arbitrantur, sed paulo obscurius"; though it is also possible that he alludes to people who (like Alexander of Aphrodisias) thought that Plato had meant that the earth rotates, without themselves sharing this opinion.

[2] *Untersuchungen*, pp. 81–83.

[3] *Mém. de l'Acad. d. inscr.* t. xxx. p. 77.

this peculiar expression, which probably had been adopted by
those Pythagoreans who pretended to find in that dialogue a
support for their own theory. Zeller has come to the same
result after a very full discussion of Aristotle's manner of
making quotations, apparently without knowing of Martin's
paper[1]. And this simple explanation was brought forward
already six hundred years ago by good old Thomas Aquinas, a
fact which neither Zeller nor Martin nor any other modern
writer seems to have noticed. He says: " Quod autem addit,
quemadmodum in Timæo scriptum est, referendum est non ad
id quod dictum est, *revolvi et moveri*, sed ad id quod sequitur,
quod sit super statutum polum[2]." This is certainly much more
likely to be the case than that Aristotle, who himself never uses
the word ἴλλεσθαι when speaking of rotation with or without
displacement (any more than Plato does), should have assumed
that Plato had hidden a new theory in that word in this one
passage. We must in any case acquit Aristotle of having
believed Plato to have taught that the celestial sphere is immovable while the earth rotates, and it is simply impossible to
suppose (with Grote) that Plato taught the daily rotation of the
heavenly sphere and also the rotation of the earth in the same
direction and in the same period.

We have still to consider the words "guardian and artificer
of night and day" which are applied to the earth in the same
passage, and which by Gruppe are supposed to prove that Plato
held the earth to *produce* day and night by its rotation. But
even if this were not distinctly contradicted by Plato's previous
statement that we are acquainted with day and night " from the
revolution of sameness," there is nothing in the words φύλαξ
καὶ δημιουργὸς inconsistent with a stationary earth. If the
earth did not exist, there would be no change of day and night,
consequently there is nothing peculiar in the earth being considered a guardian and producer of time. This brings us to the
last passage of the *Timæus*, which has been made use of by

---

[1] In a paper "Ueber die richtige Auffassung einiger aristotelischen Citate"
(*Sitzungsberichte der K. Preuss. Akad. d. Wiss.* 1888, p. 1339).

[2] *S. Thomæ Aquinatis Opera Omnia*, t. III. p. 205, Rome, 1886 (*Comment. in
libros Aristotelis de Cœlo*, lib. II. lect. xxi).

Gruppe. On p. 42 D Plato speaks of the souls of the beings which live on the various cosmical bodies; "he planted some of them on the earth, others in the moon, and others in the other different instruments of time" (ὄργανα χρόνου). Plutarch devotes the eighth of his "Platonic Questions" to the consideration of this passage, and explains it as being true in the same sense in which we call a gnomon or sundial an instrument of time, because, though it never moves, it marks the movements of the shadow. There is nothing from an astronomical point of view to be said against this explanation, if Plato really has called the earth an instrument of time, but this is denied by Boeckh[1]. Nowhere does Plato mention the earth as generating time in any way, not even on p. 39, where he, after mentioning the measures of the day, month and year, adds that the periods of the five planets are not understood, though we can conceive "how the perfect number of time completes the perfect year whenever the courses of the eight revolutions are completed and return to the same starting point[2]." And there is, besides, no necessity to assume that the expression "other different instruments" refers back to the earth as well as to the moon, and it is quite unreasonable to consider the earth as included among the instruments of time in this sentence only, while Plato otherwise never classes it among them.

Not content with having made the rotation of the earth a leading feature in the *Timæus*, Gruppe claims for his hero still greater laurels by proclaiming him as "the real author of the heliocentric system[3]." The passages on which he endeavours to found this claim occur in the seventh book of the *Laws*, probably the last great work written by Plato, and Gruppe is certainly

---

[1] *Untersuchungen*, p. 71.

[2] The duration of this annus magnus is variously given by the ancients. According to Censorinus (*De die natali*, c. 18) Aristarchus fixed it at 2484 years, Aretes Dyrrachinus at 5552 years, Herakleitus and Linus at 10,800 years, Dion at 10,884 years; others made it very much longer. The author of the dialogue, *De causis corruptæ eloquentiæ* (attributed to Tacitus), gives 12,954 years (ut Cicero in *Hortensio* scribit), and Macrobius (*Somn. Scip.* II. 11) 15,000 years. We have already mentioned that Plato in the *Phædrus* supposes a period of ten thousand years, but of course there was not any astronomical reason for this.

[3] *Die kosm. Systeme der Griechen*, pp. 151—172.

the first to have seen anything remarkable in the astronomical
allusions in that work[1]. Plato here lets an Athenian guest
discuss with a Cretan and a Spartan the question whether
astronomy should form part of the education of youth. We
Greeks, says the Athenian, tell a falsehood respecting those
mighty divinities, the Sun and Moon; we say that they never
proceed in the same path, and with them some others which we
call wandering or erratic (πλανητὰ). The Cretan acknowledges
this to be true, as he has himself often seen the morning and
the evening star and certain other stars, never proceeding in the
same track, but wandering about, and we all know that the sun
and moon do this. Urged to explain himself further, the
Athenian continues thus:

"The doctrine respecting the sun and moon and the other
stars that they are erratic, is not correct, but the contrary is the
case. For each of them perpetually describes the same path,
not many but one, in a circle, but they appear to be moving in
many. But that which is the swiftest of them is not justly
thought to be the slowest and *vice versâ*. Now if such is the
case, but we do not think so, if we had such notions respecting
the horses that run at Olympia, or of men contesting in the long
course, and we called the swiftest the slowest and the slowest
the swiftest, and celebrated the vanquished as the victor, I think
we shall not distribute our praises properly, nor in a manner
agreeable to the racers, being men. But now when we err in
the very same manner respecting the gods, do we not think that
what, when it took place there, would be ridiculous and wrong,
would be the same in the case of the gods?"

Gruppe maintains that we have here the first sign of the
important distinction being made between apparent and real
motion, and that the doctrine of the earth's rotation is distinctly
enunciated. For only if the earth rotates can we say that sun
and moon have only one motion in a circle, viz. the orbital
motion; whereas, if the heaven rotates, the sun and moon
describe spirals, as they rise every day in a different place.
This looks plausible, but the distinction made by Plato is not

---

[1] *Leg.* 821 ff.

between real motion and the apparent motion of a body which
in reality is fixed, but simply between apparent (diurnal)
motion and real orbital motion. Plato neither here nor anywhere else in his writings makes any attempt to solve the
intricate problem of planetary motion, and doubtless had only
the most general idea of the irregularities of the wanderings
of the planets among the stars. In the helices, which he
describes in the *Timæus*, a planet appears to pursue its way
through the motion of sameness, as the obliquity of the motion
of the diverse turns its daily path round the earth into a helix,
but in reality each planet has a separate motion in a circle.
In the above passage in the *Laws* Plato does not speak of
"only one motion in a circle," but he states that a planet
always moves in one and the same orbit, a circle, which is only
accidentally disguised into a helix, and there is in this no
allusion whatever to the rotation of the earth. Neither is
there in the following sentence about the swiftest and the
slowest. Gruppe interprets this as meaning that if the earth is
at rest Saturn must in twenty-four hours describe an enormous
orbit round it, and is therefore the swiftest body, but if the
earth rotates Saturn has only its orbital motion in a period of
29½ years, and is the slowest one. Now this explanation does
not agree with Plato's comparison with the circus runners, and
the passage differs in no way from the similar passage in the
Timæus which we have already considered. The moon, which
has the greatest angular velocity, is, if we only consider the
rotation of the heavens, erroneously called the slowest, because
it comes to the meridian three-quarters of an hour later every
night, while Saturn, which in reality has the smallest angular
velocity, yet completes its daily revolution in twenty-four hours
and a few seconds. But if Plato in this place does not allude
to the rotation of the earth, he still less refers to any orbital
motion of the earth, as Gruppe imagines him to do in the
expression that each planet "perpetually travels the same
path, not many but one, in a circle, but they appear to describe
many." Here again we have nothing but a repeated assertion
that the planets really describe separate orbits round the earth
from west to east and do not merely perform their daily course

from east to west more slowly than the fixed stars, and by an obliqueness of motion describe helices round the earth[1].

Gruppe has been encouraged in his strange attempt to set up Plato as a precursor of Copernicus by a well-known story told by Plutarch in two of his writings, that Plato considerably changed his opinion about the construction of the world in his old age. But this statement does not support Gruppe's theory when more closely examined. In his life of Numa (c. 11) Plutarch describes the Pythagorean doctrine of the motion of the earth round the central fire, and adds : " These ideas are said to have been entertained by Plato also in his old age ; for he too thought that the earth was in a subordinate position, and that the centre of the universe was occupied by some nobler body." Is it possible to state more distinctly that Plato inclined to the doctrine of Philolaus, and is there in the passage the smallest hint about the sun being in the central position ? The other place in which Plutarch refers to Plato's old age is where he discusses the obscure passage in the *Timæus*. After alluding to the rotation of the earth as taught by Aristarchus and Seleukus he proceeds : "Theophrastus states that Plato, when he was old, repented of having given to the earth the central place in the universe which did not belong to it." This statement is less clear as to what Plato really did mean in his old age, but on the other hand it tells us who Plutarch's authority was. We know from Diogenes (v. 50) that Theophrastus, a disciple of Aristotle, wrote a work in six books on the history of astronomy, and it was doubtless this work which Plutarch quoted. But Theophrastus lived long before Aristarchus and Seleukus, so he cannot have mentioned Plato's opinions in connection with them. But neither does Plutarch, for what would have been more natural than to have added that Plato placed the sun in the position hitherto supposed to be occupied by the earth ? The two statements in the life of Numa and in the Platonic questions are in perfect accordance, and are probably founded on the well-known fact that Plato, who had always been something of an eclecticist in his views

---

[1] Gruppe's extraordinary assertions were refuted in the clearest manner by Boeckh in his *Untersuchungen*, pp. 48–57.

of nature, in his later years inclined in more ways than one to
the Pythagoreans. See for instance the concluding chapters
of the *Timæus*, where he entertains the doctrine of the trans-
migration of souls. Even in the *Symposium*, which no doubt
was written much earlier, we find the Philolaic system hinted
at, though it is only in a playful manner[1]. It is highly probable
that some of Plato's successors went further in the same
direction; at any rate we know from Aristotle that some of
his own contemporaries who were not Pythagoreans believed in
the central fire, for he says[2]: "But also many others may share
this opinion that one ought not to place the earth in the
middle, as they do not take their convictions from the pheno
mena but from considerations (ἐκ τῶν λόγων); for they think
that the most excellent ought to have the most excellent place,
but fire is more excellent than earth...." As Aristotle speaks
in the present tense he must refer to people then living,
and the beginning of the sentence shows that they were not
Pythagoreans. As there was a certain degree of coolness
between the Academy and the Aristotelians, nothing is more
likely than that Theophrastus should have mixed up Plato
with those Platonists who more or less adopted the system of
Philolaus, and who perhaps, to justify themselves, had spread
the report that Plato had done the same.

But that Plato really should have ended as an adherent
of Philolaus is an idea which cannot be seriously maintained.

---

[1] *Symposium*, p. 190. The joker of the party, Aristophanes, says that the
human race consisted originally of three sexes, male, female, and men-women;
they were round in shape, having back and sides "as in a circle," with four
hands and feet and so on, and they ran by turning cartwheels. "Now these
three existed on this account, because the male kind was the produce originally
of the sun, the female of the earth, and that which partook of the other two, of
the moon, for the moon partakes of both the others. The bodies thus were
round, and the manner of their running was circular, through their being like
their parents." In other words their "parents" both rotated and had a progressive motion, and that is the case both with the sun and the earth in the
Philolaic system only.

[2] *De Cœlo*, II. 13, p. 293 a. According to Simplicius (p. 513 Heib.) Alexander of Aphrodisias had tried in vain to find anyone earlier than Aristotle to
whom he might have alluded, but as he could find none, he concluded that
Aristotle could only have meant Pythagoreans. But that Aristotle spoke of
contemporaries of his own was pointed out by Boeckh, l. c., p. 148.

We have already mentioned that the *Laws* was the last work which Plato lived to complete, but a sequel to it was published after his death by Philip of Opus, a pupil of his, who was the author of writings on the distances of the sun and moon, eclipses, and other astronomical subjects, of which nothing has been preserved[1]. Though it is not likely that the *Epinomis* was written by Plato[2], this short treatise was at any rate published by his devoted pupil Philip and represents Platonic ideas throughout. And there is not a word in it which is in opposition to the ordinary geocentric system, nor any passage which can be twisted or distorted into an allusion to the rotation of the earth or heliocentric motion or the system of Philolaus. There are eight powers in the heavens, the fourth and fifth (Venus and Mercury) are nearly equal in velocity with the sun, while the eighth orb is one which may most correctly be called the upper world (ἄνω κόσμον), which moves in the opposite direction to all those, and draws the others along with it, " as it would appear to persons who know little on these matters; but what we know sufficiently it is necessary to speak of and we do speak[3]." Here again we have a refutation of the old Ionian notion that the planets had no orbital motion, but merely lagged behind more or less in the daily revolution of the heavens, but we have no sign whatever that Plato had in his old age formed any new opinions about the system of the world[4].

[1] About him see Boeckh, *Ueber die vierjährigen Sonnenkreise der Alten*, p. 34 et seq.

[2] Both Boeckh and Grote are of the opinion that Plato was the author. It is Boeckh who has first drawn attention to the astronomical parts of the Epinomis (*Untersuchungen*, p. 149); Gruppe does not mention it.

[3] *Epinomis*, 987 B.

[4] It is much to be regretted that Schiaparelli, when writing his memoir, *I Precursori di Copernico*, was not acquainted with Boeckh's *Untersuchungen*, as he could not have failed to become convinced of the correctness of Boeckh's interpretation of the passages in the *Laws* and the *Epinomis*, instead of lending his great authority to the view that Plato after all must have inclined to believe in the rotation of the earth during the last few years of his life, but (remembering Anaxagoras and Sokrates) was afraid to say so openly. This has lately been repeated by Gomperz (*Griechische Denker*, II. p. 609: " Ich folge dieser grossen Autorität "). Boeckh's interpretation is from an astronomical point of view absolutely unassailable.

The *Epinomis* also alludes to a subject on which we have hardly touched, how Plato supposed the stars and the celestial space through which they moved to be constituted. Following the lead of Philolaus and Empedokles, Plato assumes four elements, earth, water, air, and fire, the component particles of which are respectively shaped like four of the regular bodies, the cube, the icosahedron, the octahedron, and the tetrahedron[1]. There is no element formed of the fifth body, the dodecahedron, which had been used by God as a model for the universe, the spherical shape of which is an improvement on it[2]. These four elements exist in various stages of coarseness and purity; they are distributed everywhere in the universe, but each one predominates in some parts of it. The stars are principally formed of fire[3] and they move in the ether, which is a specially pure kind of air, extending from the upper limit of the atmosphere throughout the heavenly space, though according to the *Phædo* (p. 109) the uppermost regions of the earth reach up into it. The moon, being nearest to the earth, partakes more of the nature of the latter; it is dark and only borrows its light from the sun. All the heavenly bodies are looked on as divine beings, the first of all living creatures, the perfection of whose mind is reflected in their orderly motions[4].

We have now come to the end of our examination of Plato's astronomical opinions. There is absolutely nothing in his various statements about the construction of the universe tending to show that he had devoted much time to the details of the heavenly motions, as he never goes beyond the simplest and most general facts regarding the revolutions of the planets. Though the conception of the world as Kosmos, the divine work of art, into which the eternal ideas have breathed life, and possessing the most godlike of all souls, is a leading feature in his philosophy, the details of scientific research had probably no great attraction to him, as he considered mathematics

---

[1] *Timæus*, 54–56.

[2] Ibid. 55 c. The author of the *Epinomis* therefore differs from Plato when he asserts distinctly that the ether is a fifth element.

[3] *Tim.* 40 A; *Epin.* 981.

[4] *Tim.* 39 E, 40 A (above pp. 68 and 71); *Leg.* X. 898 D, XII. 966 D.

inferior to pure philosophy in that it assumes certain data
as self-evident, for which reason he classes it as superior
to mere opinion but less clear than real science[1]. But crude as
his astronomical system, the geocentric system pure and simple,
undoubtedly is, there is a charm in the poetical conception of
the "soul of the world" which makes the study of the *Timæus*
peculiarly attractive. In the history of astronomy it does,
however, not play a very important part, although it must
not be forgotten that Plato through his widely read books
helped greatly to spread the Pythagorean doctrines of the
spherical figure of the earth and the orbital motion of the
planets from west to east. But chiefly on account of the various
controversies to which Plato's astronomical system has given
rise, it seemed desirable to subject it to a somewhat lengthy
analysis in the present work.

[1] *Republic*, p. 533.

CHAPTER IV.

THE HOMOCENTRIC SPHERES OF EUDOXUS.

THE examination of the astronomical doctrines of Plato has shown us that philosophers in the first half of the fourth century before the Christian era possessed some knowledge of the motions of the planets. No doubt astronomical instruments, even of the crudest kind, cannot be said to have existed, except the gnomon for following the course of the sun; but all the same the complicated movements of the planets through the constellations must have been traced for many years previously. That the moon, though its motion is not subject to very conspicuous irregularities, does not pursue the same path among the stars from month to month and from year to year must also have been perfectly well known, since Helikon, a disciple of Eudoxus, was able to foretell the solar eclipse of the 12th May, 361, for which he was rewarded by Dionysius II of Syracuse with a present of a talent[1]. But the clearest proof of the not inconsiderable amount of knowledge of the movements of the heavenly bodies, which was available at the time of Plato, is supplied by the important astronomical system of his younger contemporary, Eudoxus of Knidus, which is the first attempt to account for the more conspicuous irregularities of those movements.

Eudoxus was born at Knidus, in Asia Minor, about the year

---

[1] Boeckh, *Ueber die vierjährigen Sonnenkreise der Alten, besonders den Eudoxischen.* Berlin, 1863, p. 153. For an account of the life of Eudoxus see ibid. p. 140, and about the geographical researches attributed to him see Berger's *Erdkunde d. Gr.* II. pp. 68-74.

408 B.C., and died in his fifty-third year, about 355[1]. At the age of twenty-three he went to Athens and attended Plato's lectures for some months, but not content with the knowledge he could attain in Greece, Eudoxus afterwards proceeded to Egypt, furnished with letters of recommendation from the Spartan King Agesilaus to Nectanebis, King of Egypt. He stayed at least a year in Egypt, possibly much longer (about 378 B.C.), and received instruction from a priest of Heliopolis. According to Seneca[2] it was there that he acquired his knowledge of the planetary motions, but although this is not unlikely to have been the case, we have no reason to believe that Eudoxus brought his mathematical theory of these motions home from Egypt, in which country, as far as we know, geometry had made very little progress[3]. Diogenes of Laerte, who does not say a word about the scientific work of Eudoxus, does not omit to mention that the Egyptian Apis licked his garment, after which the priests prophesied that he would be short-lived but very illustrious. If this prophecy was really uttered it was a true one, as Eudoxus stands in the foremost rank of Greek mathematicians. Most, if not the whole, of the fifth book of Euclid is due to him, as well as the so-called method of exhaustion, by means of which the Greeks were able to solve many problems of mensuration without infinitesimals. We are told by Plutarch[4] that Plato, on being consulted about the celebrated Delian problem of the duplication of a cube, said that only two men were capable of solving this problem, Eudoxus and Helikon; and if the story is apocryphal, it shows at any rate the high renown of Eudoxus as a mathematician. In the history of astronomy he is also known as the first proposer of a solar cycle of four years, three of 365 and one of 366 days, which was three hundred years later introduced

---

[1] Strabo (p. 119) mentions the observatory of Eudoxus (at Knidus) as not having been much higher than the houses, but still he was able to see the star Canopus from it.

[2] *Quæst. Nat.* VII. 3.

[3] Cantor, *Gesch. der Math.* chap. 2. Whatever the Egyptians may have known of geometry, there is no doubt that the Greeks had long before the time of Eudoxus outstripped them completely.

[4] *De genio Socratis*, cap. VIII.

by Julius Cæsar. He was therefore fully capable of grappling successfully with the intricate problem of planetary motion, which Plato (according to Simplicius) is said to have suggested to him for solution[1], and his labours produced a most ingenious cosmical system which represented the principal phenomena in the heavens as far as they were known in his time.

This system of concentric spheres, which was accepted and slightly improved by Kalippus, is known to us through a short notice of it in Aristotle's *Metaphysics* (Λ 8), and through a lengthy account given by Simplicius in his commentary to Aristotle's book on the Heavens[2]. The systems of Hipparchus and Ptolemy eventually superseded it, and the beautiful system of Eudoxus was well-nigh forgotten. One historian of astronomy after another, knowing in reality nothing about it, except that it supposed the existence of a great number of spheres, contented himself with a few contemptuous remarks about the absurdity of the whole thing. That the system, mathematically speaking, was exceedingly elegant does not seem to have been observed by anybody, until Ideler in two papers in the Transactions of the Berlin Academy for 1828 and 1830 drew attention to the theory of Eudoxus and explained its principles. The honour of having completely mastered the theory and of having investigated how far it could account for the observed phenomena, belongs, however, altogether to Schiaparelli, who has shown how very undeserved is the neglect and contempt with which the system of concentric spheres has been treated so long, and how much we ought to admire the ingenuity of its author. We shall now give an account of this system as set forth by Schiaparelli[3].

[1] Simpl. *De Cælo*, p. 488 (Heib.).   [2] ii. 12, pp. 493–506 (Heib.).

[3] Schiaparelli: "Le sfere omocentriche di Eudosso, di Callippo e di Aristotele," *Pubblicazioni del R. Osservatorio di Brera in Milano*, No. ix. Milano, 1875. German translation in *Abhandlungen zur Geschichte der Mathematik*, Erstes Heft. Leipzig, 1877. Schiaparelli does not mention a paper by E. F. Apelt: "Die Sphärentheorie des Eudoxus und Aristoteles," in the *Abhandlungen der Fries'schen Schule*, Heft ii. (Leipzig, 1849), which gives a fairly full exposition of the theory. Later than Schiaparelli's paper appeared one by Th. H. Martin in the *Mém. de l'Acad. des Inscr.* t. xxx. 1881. In this objections are raised to Schiaparelli's interpretation of the theories of the sun and moon, but they have been sufficiently refuted by Tannery in the *Mém. de la Soc. des sc. phys. et nat. de Bordeaux*, 2ᵉ Série, t. v. 1883, pp. 129–147.

Although the various cosmical systems suggested by philosophers from the earliest ages to the time of Kepler differ greatly from each other both in general principles and in matters of detail, there is one idea common to them all: that the planets move in circular orbits. This principle was also accepted by Eudoxus, but he added another in order to render his system simple and symmetrical. He assumed that all the spheres which it appeared necessary to introduce were situated one inside the other and all concentric to the earth, for which reason they long afterwards became known as the homocentric spheres. No doubt this added considerably to the difficulty of accounting for the complicated phenomena, but the system gained greatly in symmetry and beauty, while it also became physically far more sensible than any system of excentric circles could possibly be. Every celestial body was supposed to be situated on the equator of a sphere which revolves with uniform speed round its two poles. In order to explain the stations and arcs of retrogression of the planets, as well as their motion in latitude, Eudoxus assumed that the poles of a planetary sphere are not immovable but are carried by a larger sphere, concentric with the first one, which rotates with a different speed round two poles different from those of the first one. As this was not sufficient to represent the phenomena, Eudoxus placed the poles of the second sphere on a third, concentric to and larger than the two first ones and moving round separate poles with a speed peculiar to itself. Those spheres which did not themselves carry a planet were according to Theophrastus called ἀνάστροι, or starless. Eudoxus found that it was possible by a suitable choice of poles and velocities of rotation to represent the motion of the sun and moon by assuming three spheres for each of these bodies, but for the more intricate motions of the five planets four spheres for each became necessary, the moving spheres of each body being quite independent of those of the others. For the fixed stars one sphere was of course sufficient to produce the daily revolution of the heavens. The total number of spheres was therefore twenty-seven. It does not appear that Eudoxus speculated on the cause of all these rotations, nor on the material, thickness, or mutual distances of the spheres. We only know from a statement of Archimedes (in his ψαμμίτης) that Eudoxus estimated the sun to be nine times greater than the moon, from which we may conclude that he assumed the sun to be nine times as far distant as the moon. Whether he merely adopted the spheres as mathematical means of representing the motions of the planets and subjecting them to calculation thereby, or whether he really believed in the physical existence of all these spheres, is uncertain. But as Eudoxus made no attempt to connect the movements of the various groups of spheres with each other, it seems probable that he only regarded them as geometrical constructions suitable for computing the apparent paths of the planets.

Eudoxus explained his system in a book " On velocities," which is lost, together with all his other writings. Aristotle, who was only one generation younger, had his knowledge of the system from Polemarchus, an acquaintance of its author's. Eudemus described it in detail in his lost history of astronomy, and from this work the description was transferred to a work on the spheres written by Sosigenes, a peripatetic philosopher who lived in the second half of the second century after Christ. This work is also lost, but a long extract from it is preserved in the commentary of Simplicius, and we are thus in possession of a detailed account of the system of Eudoxus[1].

While all other ancient and medieval cosmical systems (apart from those which accept the rotation of the earth) account for the diurnal motion of sun, moon, and planets across the sky by assuming that the sphere of the fixed stars during its daily revolution drags all the other spheres along with it, the system of Eudoxus provides a separate machinery for each planet for this purpose, thereby adding in all seven spheres to the number required for other purposes. Thus the motion of the moon was produced by three spheres; the first and outermost of these rotated from east to west in twenty-four hours like the fixed stars; the second turned from west to east round the axis of the zodiac, producing the monthly

---

[1] Simplicius also quotes in the course of his account Alexander of Aphrodisias and Porphyrius, the Neo-Platonic philosopher (p. 503 Heib.).

motion of the moon round the heavens; the third sphere
turned slowly, according to Simplicius, in the same direction
as the first one round an axis inclined to the axis of the zodiac
at an angle equal to the highest latitude reached by the moon,
the latter being placed on what we may call the equator of
this third sphere. The addition of this third sphere was
necessary, says Simplicius, because the moon does not always
seem to reach its highest north and south latitude at the same
points of the zodiac, but at points which travel round the
zodiac in a direction opposite to the order of its twelve signs.
In other words, the third sphere was to account for the retrograde motion of the nodes of the lunar orbit in 18½ years.
But it is easy to see (as was pointed out by Ideler) that Simplicius has made a mistake in his statement, that the innermost
sphere moved very slowly and in the manner described; as the
moon according to that arrangement would only pass once
through each node in the course of 223 lunations, and would
be north of the ecliptic for nine years and then south of it for
nine years. Obviously Eudoxus must have taught that the
innermost sphere (carrying the moon) revolved in 27 days[1]
from west to east round an axis inclined at an angle equal to
the greatest latitude of the moon, to the axis of the second
sphere, which latter revolved along the zodiac in 223 lunations
in a retrograde direction. In this manner the phenomena are
perfectly accounted for; that is, as far as Eudoxus knew them,
for he evidently did not know anything of the moon's changeable velocity in longitude, though we shall see that Kalippus
about B.C. 325 was aware of this. But that the motion of the
lunar node was known forty or fifty years earlier is proved by
the lunar theory of Eudoxus.

With regard to the solar theory, we learn from Aristotle
that it also depended on three spheres, one having the same
daily motion as the sphere of the fixed stars, the second
revolving along the zodiac, and the third along a circle inclined
to the zodiac. Simplicius confirms this statement, and adds
that the third sphere does not, as in the case of the moon,
turn in the direction opposite to that of the second, but in the

[1] More accurately in $27^d\ 5^h\ 5^m\ 36^s$, the draconitic or nodical month.

same direction, that is, in the direction of the zodiacal signs, and very much more slowly than the second sphere. Simplicius has here made the same mistake as in describing the lunar theory, as, according to his description, the sun would for ages have either a north or a south latitude, and in the course of a year would describe a small circle parallel to the ecliptic instead of a great circle. Of course the slow motion must belong to the second sphere and be directed along the zodiac, while the motion of the third sphere must take place in a year[1] along the inclined great circle, which the centre of the sun was supposed to describe. This circle is by the second sphere turned round the axis of the zodiac, and its nodes on the ecliptic are by Eudoxus supposed to have a very slow direct motion instead of a retrograde motion as the lunar nodes have. The annual motion of the sun is supposed to be perfectly uniform, so that Eudoxus must have rejected the remarkable discovery made by Meton and Euktemon some 60 or 70 years earlier, that the sun does not take the same time to describe the four quadrants of its orbit between the equinoxes and solstices[2].

It is very remarkable that although Eudoxus thus ignored the discovery of the variable orbital velocity of the sun, he admitted as real the altogether imaginary idea that the sun did not in the course of the year travel along the ecliptic, but along a circle inclined at a small angle to the latter. According to Simplicius[3], "Eudoxus and those before him" had been led to assume this by observing that the sun at the summer- and winter-solstices did not always rise at the same point of the horizon. Perhaps it did not strike these early observers that

---

[1] Strictly speaking in a period slightly longer than a tropical year, owing to the supposed slow, direct motion of the second sphere.

[2] This agrees with the statement in the so-called Papyrus of Eudoxus, that this astronomer gave the length of the autumn as 92 days, and that of each of the three other seasons as 91 days. This papyrus was written about the year 190 B.C., and seems to have been a student's note-book, perhaps hastily written during or after a series of lectures. See Boeckh, *Ueber die vierjährigen Sonnenkreise der Alten*, p. 196 and foll. It was published by F. Blass (*Eudoxi Ars astronomica*, Kiel, 1887, 25 pp. 4°), and translated by Tannery, *Recherches sur l'Astr. ancienne*, pp. 283–294.

[3] p. 493, l. 15 (Heib.).

neither these rough determinations of the azimuth of the rising
sun nor the observations with the gnomon were sufficiently
accurate; they had without instruments perceived that neither
the moon nor the five planets were confined to move in the
ecliptic (or, as they called it, the circle through the middle of
the zodiac), and why should the sun alone have no motion in
latitude, when all the other wandering stars had a very perceptible one? This imaginary deviation of the sun from the
ecliptic is frequently alluded to by writers of antiquity. Thus
Hipparchus, who denies its existence, quotes the following
passage from a lost book on the circles and constellations of
the sphere, the Enoptron of Eudoxus: "It seems that the sun
also makes its return ($\tau\rho o\pi\grave{a}\varsigma$, solstices) in different places, but
much less conspicuously[1]." How great Eudoxus supposed the
inclination of the solar orbit to be, or how long he supposed
the period of revolution of the nodes to be, is not known, and
he had probably not very precise notions on the subject. Pliny
gives the inclination as 1°, and the point where the maximum
latitude occurs as the 29th degree of Aries[2]. On the other
hand, Theon of Smyrna, who goes more into detail on this
subject, states on the authority of Adrastus (who lived about
A.D. 100) that the inclination is $\frac{1}{2}$°, and that the sun returns to
the same latitude after $365\frac{1}{8}$ days, so as to make the shadow of
the gnomon have the same length, as he says, while the sun
takes $365\frac{1}{4}$ to return to the same equinox or solstice, and $365\frac{1}{2}$
days to return to the same distance from us. This shows that
the solar nodes were supposed to have a retrograde motion
(not a direct one as assumed by Eudoxus) and in a period of
$365\frac{1}{4} : \frac{1}{8} = 2922$ years[3]. Schiaparelli shows that with an inclination of $\frac{1}{2}$° between the axes of the second and third spheres
the solstitial points would oscillate 2° 28′. This of course

[1] That is, the maximum latitude is much less than that of the moon.
Hipparchus adds, that observations with the gnomon show no latitude, and
lunar eclipses calculated without assuming any solar latitude agree with observations within at most two digits. *Hipparchi ad Arati et Eudoxi Phenomena*,
lib. I.; ed. Manitius, pp. 88–92.

[2] *Hist. Nat.* II. 16 (67). He has doubtless misunderstood his source and
taken a range of 1° to mean an inclination of 1°.

[3] *Astronomia*, ed. Th. H. Martin, pp. 91, 108, 175 (cap. XII.), 263 (cap.
XXVII.), 314 (cap. XXXVIII.).

influences the length of the tropical year, and it is very possible
that the whole theory of the sun's latitude originally arose
from the fact that the tropical year had been found to be
different from the sidereal year, the true cause of which is the
precession of the equinoxes. To whom this theory in the first
instance is due is not known. Notwithstanding the great
authority of Hipparchus and Ptolemy the strange illusion is
still upheld by the compiler Martianus Capella in the fifth
century[1], who improves on it by stating that the sun moves in
the ecliptic except in Libra, where it deviates $\frac{1}{2}°$! The meaning
is probably that the latitude of the sun was insensible to the
instruments of the day except in Libra (and in Aries) where it
reached $\frac{1}{2}°$, and consequently the nodes must have been supposed nearly to coincide with the solstices. It is to be noticed
that the precession of the equinoxes is unknown to all these
writers[2].

The solar theory of Eudoxus was therefore practically a
copy of his lunar theory. But the task he had set himself
became vastly more difficult when he took up the theories of
the five other planets, as it now became necessary to account
for the stations and retrograde motions of these bodies. Of
the four spheres given to each planet the first and outermost
produced the daily rotation of the planet round the earth in
twenty-four hours; the second produced the motion along the
zodiac in a period which for the three outer planets was respectively equal to their sidereal period of revolution, while it for
Mercury and Venus was equal to a year. From the fact that
the revolution of this second sphere was in all cases assumed
to be uniform, we see that Eudoxus had no knowledge of the
orbital changes of velocity of the planets which depend on the

---

[1] *De nuptiis Philologiæ et Mercurii*, lib. VIII. 867, on the authority of a
book by Terentius Varro.

[2] Schiaparelli (l. c. p. 17) shows that Theon's theory cannot have been
designed to explain the motion of the equinoxes discovered by Hipparchus.
He also gives a lengthy refutation of the assertion of Lepsius, that the third
solar sphere of Eudoxus proves that Eudoxus knew precession and had received
his knowledge of it from the Egyptians (l. c. pp. 20–23). This had, however,
already been refuted by Martin, "Mémoire sur cette question: La précession
des équinoxes a-t-elle été connue avant Hipparche" (*Mém. par divers savans*,
t. VIII. 1869, pp. 303–522).

excentricity of each orbit, but that he believed the points of the zodiac in which a planet was found at successive oppositions (or conjunctions) to be perfectly equidistant one from the other. Neither did he assume the orbits to be inclined to the ecliptic, but let the second sphere of every planet move along this circle, while the latitudes of the planets were supposed to depend solely on their elongation from the sun and not on their longitude. To represent this motion in latitude, and at the same time the inequality in longitude depending on the elongation from the sun, a third and fourth sphere were introduced for each planet. The third sphere had its poles situated at two opposite points of the zodiac (on the second sphere), and rotated round them in a period equal to the synodic period of the planet, or the interval between two successive oppositions or conjunctions with the sun. These poles were different for the different planets, but Mercury and Venus had the same poles. The direction of the rotation of this third sphere is not given by Simplicius except as being from north to south and from south to north, but it turns out to be immaterial which of the two possible directions we adopt.

On the surface of the third sphere the poles of the fourth were fixed, the axis of the latter having a constant inclination, different for each planet, to the axis of the third sphere. Round the axis of the fourth sphere the rotation of the latter took place in the same period, but in a direction opposite to that of the third sphere. On the equator of the fourth sphere the planet is fixed, and it is thus endowed with four motions,

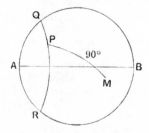

the daily one, the orbital one along the zodiac, and two others in the synodic period. What effect will these two last-mentioned

motions have on the apparent position of the planet in the sky?
In the appended figure a sphere (the third) rotates round the
fixed diameter AB (we may leave the motion of the first, or
daily sphere, altogether out of consideration, and for the
present also neglect that of the second sphere); during this
rotation round AB a certain point P, one of the poles of the
fourth sphere, describes the small circle QPR, while this
fourth sphere in the same period, but in the opposite direction,
completes a rotation round P and its other pole P'. The
planet is at M in the equator of the fourth sphere, so that
$PM = 90°$. The problem is now to determine the path described
by M, projected on the plane of the circle $AQBR$. This is
easy enough with the aid of modern mathematics, but was
Eudoxus able to solve it by means of simple geometrical
reasoning? This question has been admirably investigated by
Schiaparelli, who has shown that the solution of the problem
was well within the range of a geometrician of the acknowledged ability of Eudoxus. The result is that the projected
path is symmetrical to the line AB, that it has a double point
in it, and is nothing but the well-known "figure of eight" or
lemniscate, the equation of which is $r^2 = a^2 \cos 2\theta$, or, strictly
speaking, a figure of this kind lying in the surface of the
celestial sphere, for which reason Schiaparelli calls it a spherical
lemniscate. The longitudinal axis of the curve lies along the

zodiac, and its length is equal to the diameter of the circle
described by P, the pole of the sphere which carries the planet.
The double point is 90° from the two poles of rotation of the
third sphere. The planet describes the curve by moving in
the direction of the arrow, and passes over the arcs 1–2, 2–3,
3–4, 4–5, etc., in equal times,

So far we have only considered the motion of the point M under the influence of the rotations of the third and fourth sphere. But we must now remember that the axis AB revolves round the ecliptic in the sidereal period of the planet. During this motion the longitudinal axis of the lemniscate always coincides with the ecliptic, along which the curve is carried with uniform velocity. We may therefore for the third and fourth sphere substitute the lemniscate, on which the planet moves in the manner described above. The combination of this motion with the motion of the curve along the ecliptic gives the apparent motion of the planet through the constellations. The motion of the planet on the lemniscate consists in an oscillation forward and backward, the period being that of the synodical revolution, and during one half of this period the motion of the planet along the ecliptic becomes accelerated, and during the second half it becomes retarded, when the two motions are in opposite directions. Therefore when on an arc of the lemniscate the backward oscillation is quicker than the simultaneous forward motion of the lemniscate itself, then the planet will for a time have a retrograde motion, before and after which it is stationary for a little while, when the two motions just balance each other. Evidently the greatest acceleration and the greatest retardation occur when the planet passes through the double point of the lemniscate. The motions must therefore be so combined that the planet passes through this point with a forward motion at the time of superior conjunction with the sun, where the apparent velocity of the planet in longitude is greatest, while it must again be in the double point, but moving in a retrograde direction, at the time of opposition or inferior conjunction, when the planet appears to have the most rapid retrograde motion. This combination of motions will of course be accompanied by a certain amount of motion in latitude depending on the breadth of the lemniscate.

This curve was by the Greeks called the hippopede (ἵππου πέδη), because it was a favourite practice in the riding school to let the horse describe this figure in cantering; and Simplicius in his account of the planetary theory of Eudoxus expressly

states that a planet describes the curve called by Eudoxus a
hippopede. This word occurs in several places in the commentary to the first book of Euclid written by Proklus, in which
he describes the plane sections of the solid generated by the
revolution of a circle round a straight line in its plane,
assuming that the line does not cut the circle[1]. A section by
a plane parallel to the line and touching the inner surface of
the "anchor ring" is by Proklus called a hippopede, and it is
therefore proved that Eudoxus and his followers had a clear
idea of the properties of the curve which represents the resultant
motion of the third and fourth sphere. The curve and its
application is thus alluded to by Theon of Smyrna in his account
of the astronomical theory of the Platonist Derkyllides: " He
does not believe that the helicoid lines and those similar to the
Hippika can be considered as causing the erratic motions of
the planets, for these lines are produced by chance[2], but the
first cause of the erratic motion and the helix is the motion
which takes place in the oblique circle of the zodiac." After
this Theon describes the helix apparently traced by a planet in
the manner of Plato in the *Timæus*: but the opinion rejected
by Derkyllides is undoubtedly the motion in the lemniscate
invented by Eudoxus[3].

If we now ask how far this theory could be made to agree
with the actually observed motions in the sky we must first
of all remember that we possess no knowledge as to whether
Eudoxus had made observations to ascertain the extent of the
retrograde motions, or whether he was merely aware of the
fact that such motions existed, without having access to any
numerical data concerning them. To be able to test the
theory we require to know the sidereal period, the synodic
period, and the distance between the poles of the third and
fourth sphere, which Schiaparelli calls the inclination. The
length of this distance adopted by Eudoxus for each planet is

---

[1] Cantor, *Gesch. der Math.* I. pp. 229–30 (2 d.).

[2] Does this allude to the loops y the planets about the time of
opposition, and not to the machiner sed to produce them?

[3] Theon, ed. Martin, p. 328,

not stated either by Aristotle or Simplicius, and the periods are only given by the latter in round numbers as follow[1]:

Star of	Synodic Period		Modern value	Zodiacal Period	Modern value
Hermes	110 days		116 days	1 year	1·0 year
Aphrodite	19 months		584 ,,	1 ,,	1·0 ,,
Ares	8 ,,	20 days	780 ,,	2 years	1·88 years
Zeus	13 ,,		399 ,,	12 ,,	11·86 ,,
Kronos	13 ,,		378 ,,	30 ,,	29·46 ,,

With the exception of Mars these figures show that the revolutions of the planets had been observed with some care, and Eudoxus may even have been in possession of somewhat more accurate figures, as the Papyrus of Eudoxus gives the synodic revolution of Mercury as 116 days, a remarkably accurate value, which he had most probably obtained during his stay in Egypt[2]. If only we knew the inclination on which the dimensions of the hippopede depend, we should be able perfectly to reconstruct each planetary theory of Eudoxus. As the principal object of the system certainly was to account for the retrograde motions, Schiaparelli has for the three outer planets assumed that the values of the inclinations were so chosen as to make the retrograde arcs agree with the observed ones. The retrograde arc of Saturn is about 6°, and with a zodiacal period of 30 years, a synodic period of 13 months, and an inclination of 6° between the axes of the third and fourth sphere the length of the hippopede becomes 12° and half its breadth, i.e. the greatest deviation of the planet from the ecliptic turns out to be 9′, a quantity insensible for the observations of those days. We have therefore simply a retrograde motion in longitude of about 6° between two stationary points. Similarly, assuming for Jupiter an inclination of 13°, the length of the hippopede becomes 26°, and half its breadth 44′, and with periods of respectively 12 years and 13 months this gives a retrograde arc of about 8°. The greatest distance from the ecliptic during the motion on this arc, 44′, was probably hardly

[1] p. 496 (Heib.).

[2] This papyrus gives the zodiacal periods of Mars and Saturn as two years and thirty years, in perfect accordance with Simplicius (Blass, p. 16 ; Tannery, p. 287).

noticeable at that time. For these two planets Eudoxus had thus found an excellent solution of the problem proposed by Plato, even supposing that he knew accurately the lengths of the retrograde arcs.

But this was not the case with Mars, which indeed is not to be wondered at, when we remember that even Kepler for a long time found it hard to make the theory of this planet satisfactory. It is not easy to see how Eudoxus could put the synodic period equal to 8 months and 20 days (or 260 days), whereas it really is 780 days, or exactly three times as long. All editions of Simplicius give the same figures, and Ideler's suggestion that we should for 8 months read 25 months seems therefore unwarranted; besides, it does not help matters in the least. For with a synodic period of 780 days and putting the inclination equal to 90° (the highest value reconcilable with the description of Simplicius), the breadth of the hippopede becomes 60°, so that Mars ought to reach latitudes of 30°. And even so, the retrograde motion of Mars on the hippopede cannot in speed come up to the direct motion of the latter along the zodiac, so that Mars should not become retrograde at all, but should only move very slowly at opposition. To obtain a retrograde motion the inclination would have to be greater than 90°; in other words the third and fourth sphere would have to rotate in the same direction. And even this violation of the rule would be of no use, since Mars in that case would reach latitudes greater than 30°, and Eudoxus was doubtless not willing to accept this. On the other hand, if we adopt his own value of the synodic period, 260 days, the motion of Mars on the hippopede becomes almost three times as great as before, and with an inclination of 34° the retrograde arc becomes 16° long and the greatest latitude nearly 5°. This is in fair accordance with the real facts, but unfortunately this hypothesis gives two retrograde motions outside the oppositions and four additional stationary points, which have no real existence. The theory of Eudoxus fails therefore completely in the case of Mars.

With regard to Mercury and Venus, we have first to note that the mean place of these planets always coincides with the

sun, so that the centre of the hippopede always lies in the
sun.   As this centre is 90° from the poles of rotation of the
third sphere, we see that these poles coincide for the two
planets.   This deduction from the theory is confirmed by the
remark of Aristotle that "according to Eudoxus the poles of
the third sphere are different for some planets, but identical
for Aphrodite and Hermes," and this supplies a valuable proof
of the correctness of Schiaparelli's deductions.   As the greatest
elongation of each of these planets from the sun equals half
the length of the hippopede, i.e. the inclination of the third
and fourth spheres, Eudoxus doubtless determined the inclination by observing the elongations, as he could not make use of
the retrograde motions, which in the case of Venus are hard to
see, and in that of Mercury out of reach.   With a hippopede
for Mercury 46° in length the half breadth or greatest latitude
becomes 2° 14', nearly as great as that observed.   For Venus
we may make the hippopede 92° in length, which gives half its
breadth equal to 8° 54' in good accordance with the observed
greatest latitude.   But, as in the case of Mars, Venus can
never become retrograde, and no different assumption as to
the value of the inclination can do away with this error of the
theory.   And a much worse fault is, that Venus ought to take
the same length of time to pass from the east end of the
hippopede to the west end and *vice versâ*, which is not in
accordance with facts, since Venus in reality takes 440 days to
move from the greatest western to the greatest eastern elongation, and only about 143 days to go from the eastern to the
western elongation, a fact which is very easily ascertained.
The theory is equally unsatisfactory as to latitude, for the
hippopede intersects the ecliptic in four points, at the two
extremities and at the double point; consequently Venus
ought four times during every synodic period to pass the
ecliptic, which is far from being the case.

But with all its imperfections as to detail the system of
homocentric spheres proposed by Eudoxus demands our admiration as the first serious attempt to deal with the apparently
lawless motions of the planets.   For Saturn and Jupiter, and
practically also for Mercury, the system accounted well for the

motion in longitude, while it was unsatisfactory in the case of Venus, and broke down completely only when dealing with the motion of Mars. The limits of motion in latitude were also well represented by the various hippopedes, though the periods of the actual deviations from the ecliptic and their places in the cycles came out quite wrong. But it must be remembered that Eudoxus cannot have had at his command a sufficient series of observations; he had probably in Egypt learned the main facts about the stationary points and retrogressions of the outer planets as well as their periods of revolution, which the Babylonians and Egyptians doubtless knew well, while it may be doubted whether systematic observations had for any length of time been carried on in Greece. And if the old complaint is to be repeated about the system being so terribly complicated, we may well bear in mind, as Schiaparelli remarks, that Eudoxus in his planetary theories only made use of three elements, the epoch of an upper conjunction, the period of sidereal revolution (of which the synodic period is a function), and the inclination of the axis of the third sphere to that of the fourth. For the same purpose we nowadays require six elements!

If, however, the system was founded on an insufficient basis of observations, it seems that some of the adherents of Eudoxus must have compared the movements resulting from the theory with those actually taking place in the sky, since we find Kalippus, of Kyzikus, a pupil of Eudoxus, engaged in improving his master's system some thirty years after its first publication. Kalippus is also otherwise favourably known to us by his improvement of the soli-lunar cycle of Meton, which shows that he must have possessed a remarkably accurate knowledge of the length of the moon's period of revolution. Simplicius states that Kalippus, who studied with Polemarchus, an acquaintance of Eudoxus, went with Polemarchus to Athens in order to discuss the inventions of Eudoxus with Aristotle, and by his help to correct and complete them[1]. This must have happened during the reign of Alexander the Great (336-323), which time Aristotle spent at Athens. From the

[1] Simpl. *De Caelo*, p. 493 (Heib.).

investigations of Kalippus resulted an important improvement of the system of Eudoxus which Aristotle and Simplicius describe; and as the former solely credits Kalippus with it, it does not seem likely that he had any share in it himself, though he cordially approved of it[1]. Kalippus wrote a book about his planetary theory, but it was already lost before the time of Simplicius, who could only refer to the history of astronomy by Eudemus, which contained an account of it.

The principle of the homocentric spheres, as we shall see in the next chapter, fitted in well with the cosmological ideas of Aristotle, and had therefore to be preserved, so that Kalippus was obliged to add more spheres to the system if he wished to improve it. He considered the theories of Jupiter and Saturn to be sufficiently correct and left them untouched, which shows that he had not perceived the elliptic inequality in the motion of either planet, though it can reach the value of five or six degrees. But the very great deficiencies in the theory of Mars he tried to correct by introducing a fifth sphere for this planet in order to produce a retrograde motion without making a grave error in the synodic period. This is only a supposition, as we are not positively told why Kalippus added a sphere each to the theories of Mars, Venus, and Mercury[2], but Schiaparelli has shown how the additional sphere can produce retrogression without unduly adding to the motion in latitude. Let AOB represent the ecliptic, A and B being opposite points in it which make the circuit of the zodiac in the sidereal period of Mars. Let a sphere (the third of Eudoxus) rotate round these points in the synodic period of the planet, and let any point P_1 in the equator of this sphere be the pole of a fourth sphere which rotates twice as fast as the third in the opposite direction carrying the point P_2 with it, which is the pole of a fifth sphere rotating in the same direction and period as the third and carrying the planet at M on its equator. It is easy to see that if at the beginning of motion the points P_1, P_2, and M were situated in the ecliptic

---

[1] *Metaph.* XI. 8, p. 1073 b.

[2] Simplicius merely says that Eudemus has clearly and shortly stated the reasons for this addition (*De Cœlo*, ed. Heiberg, p. 497, l. 22).

in the order AP_2P_1MB, then at any time the angles will be as
marked in the figure, and as $AP_1 = MP_2 = 90°$, the planet M
will in the synodic period describe a figure symmetrical to the

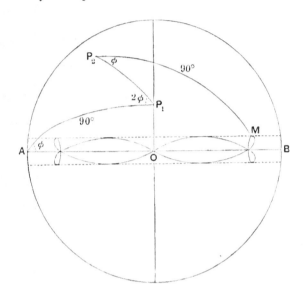

ecliptic which alters its form with the adopted length of the
arc P_1P_2, and, like the hippopede, may produce retrograde
motion.   And it has this advantage over the hippopede, that it
can give the planet in the neighbourhood of O a much greater
direct and retrograde velocity with the same motion in latitude.
It can therefore make the planet retrograde even in the cases
where the hippopede of Eudoxus failed to do so.   Thus, if
P_1P_2 is put equal to $45°$, the curve assumes a figure like that
shown; the greatest digression in latitude is $4° 11'$, the length
of the curve along the ecliptic is $95° 20'$, and it has two triple
points near the ends, $45°$ from the centre.   When the planet
passes O, its velocity is $1·293$ times the velocity of P_1 round
the axis AB, and as the period of the latter rotation is 780
days, the daily motion of P_1 is $360°/780 = 0°·462$, which number
multiplied by $1·293$ gives $0°·597$ as the daily velocity of the
retrograde motion on the curve at O.   But as O has a direct
motion on the ecliptic of $360°/686 = 0°·525$, the resulting daily

retrograde motion of the planet in the heavens is $0°·072$, which is a sufficient approximation to the real motion of Mars at opposition. It must however be remembered that we have no way of knowing what value Kalippus assumed for the distance $P_1 P_2$; but that the introduction of another sphere could really make the theory satisfactory has been proved by Schiaparelli's investigation.

In the same way an additional sphere removed the errors in the theory of Venus. If $P_1 P_2$ is $= 45°$, the greatest elongation becomes $47°40'$, very nearly the true value; and the different velocity of the planet in the four parts of the synodic revolution is also accounted for; as in the curve depicted above the passage from one triple point to the other takes one fourth of the period, the same passage back again another fourth, while the very slow motion through the small loops at the end of the curve occupies the remaining time. In the case of Mercury the theory of Eudoxus was already fairly correct, and no doubt the extra sphere made it better still.

In the solar theory Kalippus introduced two new spheres in order to account for the unequal motion of the sun in longitude which had been discovered about a hundred years previously by Meton and Euktemon through the unequal lengths of the four seasons. The so-called papyrus of Eudoxus, to which we have already referred, gives us the values adopted by Kalippus for the lengths of the seasons (taken from the Parapegma, or meteorological calendar of Geminus), and though only given in whole numbers of days (95, 92, 89, 90, beginning with the vernal equinox), the values are in every case less than a day in error, while the corresponding values determined by Euktemon about B.C. 430 are from $1\frac{1}{4}$ to 2 days wrong[1]. The observations of the sun had therefore made good progress in Greece during the century ending about B.C. 330. By adding two more spheres to the three spheres of Eudoxus, Kalippus had only to follow the same principle on which Eudoxus had represented the synodic inequalities of the planets, and a hippopede $4°$ in length and $2'$ in breadth gives in fact the necessary maximum inequality of $2°$ in a perfectly satisfactory manner.

---

[1] Boeckh, *Vierjährige Sonnenkreise der Alten*, p. 46.

Similarly the number of lunar spheres was increased by two, and though Simplicius is not very explicit, we can hardly doubt that he means us to understand the cause to be similar to that which he has just stated in the case of the sun. In other words, Kalippus must have been aware of the elliptic inequality of the moon. Indeed he can hardly have failed to notice it, even if he merely confined his attention to lunar eclipses without watching the motion of the moon at other times, since the intervals between various eclipses compared with the corresponding longitudes (deduced from those of the sun) at once show how far the moon's motion in longitude is from being uniform. A hippopede 12° in length would only be twice 9′ in breadth, and would therefore not sensibly affect the latitude, while it would produce the mean inequality of 6°. The improved theory was therefore quite as good as any other, as long as the evection had not been discovered.

Such then was the modified theory of homocentric spheres, as developed by Kalippus. Scientific astronomy may really be said to date from Eudoxus and Kalippus, as we here for the first time meet that mutual influence of theory and observation on each other which characterizes the development of astronomy from century to century. Eudoxus is the first to go beyond mere philosophical reasoning about the construction of the universe; he is the first to attempt systematically to account for the planetary motions. When he has done this the next question is how far this theory satisfies the observed phenomena, and Kalippus at once supplies the observational facts required to test the theory and modifies the latter until the theoretical and observed motions agree within the limit of accuracy attainable at the time. Philosophical speculation unsupported by steadily pursued observations is from henceforth abandoned; the science of astronomy has started on its career.

CHAPTER V.

ARISTOTLE.

THE system of homocentric spheres was fully accepted by
Aristotle (384—322), the last great speculative philosopher
who figures in the history of ancient astronomy. Unlike Plato,
he sought the idea in its concrete realisation in the phenomena
of nature, and all results of experience and observation had
therefore claims on his attention. As a consequence of this
tendency of his philosophy to view the universe as a system of
units, each of which is of importance to the conception of the
whole, his writings are of a somewhat encyclopædic character,
embracing all branches of knowledge ; but though they are
considerably more dry and matter-of-fact than the poetical
dialogues of Plato, they were of much greater value in the
development of science, while they also afford us good pictures
of the state of knowledge at the time when Greek intellectual
life was at its height.

It does not come within the scope of this work to consider
the principles of the Aristotelian natural philosophy, which are
contained in the eight books on " Physics," a purely metaphysical work, which deals with the general conditions of natural
being : motion, space, and time. In this work Aristotle in
reality does little but analyse the meaning of every-day expressions and words in order thereby to solve the problems of
nature, instead of attacking these solely by observation and
experiment. The works in which he discusses astronomical
matters are his four books " On the Heavens," and to some
extent also the four books on " Meteorologica," in which also
some astronomical subjects (comets, the Milky Way) are dis-

cussed. The work on the Heavens is, however, not altogether devoted to astronomy, with which in fact only the second book deals; but it must be remembered that Aristotle is probably neither responsible for the form in which his writings have come down to us, nor for the titles by which we know them. The first of the four books is quite metaphysical and discusses such questions as whether the universe is infinite or finite, created, or without beginning, &c. With regard to the former question Aristotle argues that the material universe cannot be infinitely extended, since a line from the earth's centre to an infinitely distant body could not perform a complete circular revolution in a definite time (twenty-four hours); and as there cannot be bodies at an infinite distance, neither can space be infinite, since it is only a receptacle of bodies. The heavens are without beginning and imperishable, since they cannot be the one without being the other, while Plato had supposed that though the world had been created, it would last for ever.

The second book on the heavens discusses the form of the Kosmos, the motions and nature of the stars, and finally the position and form of the earth, which is at rest in the centre of the universe. The third and fourth books contain nothing astronomical, but commence the discussion further carried on in the work "On Generation and Destruction," in which is set forth Aristotle's theory of the two pairs of opposites, hot and cold, wet and dry; the first pair active, the second passive, from the various combinations of which the four elements, fire, air, water, and earth, are produced.

In his general conception of the Kosmos Aristotle is guided by purely metaphysical arguments[1]. The universe is spherical, because the sphere is among bodies, as the circle among plane figures, the most perfect owing to its unique form limited by a single surface, and is the only body which during its revolution continually occupies the same space. This is an unfortunate argument, since the same may be said of any solid of revolution. In this spherical universe that sphere is the best which is endowed with the most perfect motion, and as the quickest is the most perfect, the outermost sphere which has the fastest

[1] *De Cœlo*, II. chapter IV. and foll.

revolution is the most perfect sphere of all, and is the seat of
unchangeable order. It is under the immediate influence of
the primary divine cause of motion, which from the circumference extends its power to the centre, instead of being like
the motive power of the Pythagoreans seated in the centre, or
like the cosmic soul of the *Timæus* omnipresent in the universe.
The motion of the heavens is towards the right (from east to
west[1]) because this is the more honourable direction, and it
proceeds with uniform spéed, since the single parts of it do not
move *inter se,* as we can see from the want of change in the
constellations, while the sphere as a whole is not subject to any
occasional acceleration or retardation, which would be unnatural,
since it would mean that the motive power was sometimes in a
state of weakness and sometimes in a state of vigour. With
regard to the composition of the eternal and divine stars,
Aristotle considers it most reasonable to assume that each star
consists of that material in which it has its motion, and he
shows that while rectilinear motion naturally belongs to the
four elements known to us (fire having an upward motion and
earth a downward motion), the circular motion must be natural
to the primitive and superior element[2]. The spheres and the
stars are composed of this element, and not of fire, and Aristotle
holds that the heat and light of the heavenly bodies are caused
by friction with the ether during the revolutions of the spheres,
but so that the adjacent ether is heated, and not the stars or
the spheres[3].

Turning to the motions of the heavenly bodies Aristotle first
considers whether both the stars and their spheres are in
motion, and he comes to the conclusion that it is unreasonable
to think that each star should travel along with precisely the
same velocity as its sphere if both were detached from each
other, which they would have to do "as it appears that they
and their spheres appear again together at the same point."
Therefore the stars are at rest in their spheres, and only the
latter are in motion. "Furthermore, since the stars are spherical,

[1] Compare above under Plato, p. 64.
[2] *De Cælo,* i. 2, p. 269 a, and ii. 7, p. 289 a.
[3] Ibid. ii. 7, p. 289 a.

as others maintain and we also grant, because we let the stars
be produced from that body, and since there are two motions of
a spherical body, rolling along and whirling, then the stars, if
they had a motion of their own, ought to move in one of these
ways. But it appears that they move in neither of these ways.
For if they whirled (rotated), they would remain at the same
spot and not alter their position, and yet they manifestly do so,
and everybody says they do. It would also be reasonable that
all should be moved in the same motion, and yet among the
stars the sun only seems to do so at its rising or setting, and
even this one not in itself but only owing to the distance of our
sight, as this when turned on a very distant object from weakness becomes shaky. This is perhaps also the reason why the
fixed stars seem to twinkle, while the planets do not twinkle.
For the planets are so near that the eyesight reaches them in
its full power, but when turned to the fixed stars it shakes on
account of the distance, because it is aimed at too distant a
goal ; now its shaking makes the motion seem to belong to the
star, for it makes no difference whether one lets the sight or
the seen object be in motion. But that the stars have not a
rolling motion is evident ; for whatever is rolling must of
necessity be turning, while of the moon only what we call its
face is visible[1]." For these reasons Aristotle concludes that
stars have not any individual motion ; and as they are spherical
in form, as we can see by the phases of the moon, and he holds
this form to be the least suitable for progressive motion, he
first concludes from their form that they have no motion, and
further on argues from their want of independent motion that
they must be spherical in form ! The Pythagorean idea of the
music of the spheres does not find favour with him, he rejects
the notion that we do not hear it because it is always going on,
and remarks that so many and so great bodies, if they made
any noise, ought to make a great one, which could not fail to
be noticeable, since thunder can burst stones and the strongest
bodies asunder. And this is another proof that the planets are
not travelling through an unmoved medium, but are attached

---

[1] *De Cœlo*, II. 8, p. 290 a. In other words, Aristotle holds that the fact
that we only see one side of the moon proves that the moon does not rotate.

to spheres, since they would produce audible sounds if moving freely through that medium[1].

In this place we might have expected that Aristotle would have given an account of the arrangement of the planetary spheres. But (whatever may have been his reason) he merely says that this matter may be sought in books on astronomy (περὶ ἀστρολογίαν), for in them it is sufficiently dealt with. In the eleventh book of that composite work known as Metaphysics (cap. 8) he gives, however, when treating of Pythagorean and Platonic systems of numbers, a short account of the system of spheres of Eudoxus and Kalippus (to which we have already referred), and adds some considerations of his own in order to adapt it to his principle of the motive power working from the outer surface of the Kosmos towards the centre[2]. To Aristotle the spheres are therefore not merely representatives of mathematical formulæ, though he says that the object of the arrangement is to account for the phemonena; the spheres are physically existing parts of a vast machinery by which the celestial bodies are kept in motion by the soul of each. The problem was now how to connect all the groups of spheres and yet to prevent the motion of the outer spheres from being transmitted to the inner ones. After the last and innermost sphere of each planet and before the outermost sphere of the planet next following below it he therefore inserted a number of additional spheres which he merely describes as "unrolling" (ἀνελίττουσαι). Let I, II, III, and IV represent the four spheres of the planet Saturn in the theories of Eudoxus and Kalippus, so that I is the outermost one, immediately inside that of the fixed stars, while the planet is attached to IV. Inside IV Aristotle assumes an extra sphere IVa rotating round the poles of IV with equal but opposite velocity, then the rotations of IV and IVa will neutralize each other, and any point of IVa will move as if it were attached to III. Similarly, he adds an extra sphere IIIa inside IVa, having the same poles as III and equal but opposite velocity, and the rotations of III and IIIa

---

[1] *De Cœlo*, II. 9, p. 291 a.

[2] Further explained by Simplicius (*De Cœlo*, II. 12, p. 497 Heib. and foll.), who gives Sosigenes as his authority.

neutralise each other, so that any point of IIIa will move as if firmly connected with II. Finally, inside IIIa he adds a sphere IIa with the same poles as II and equal and opposite velocity, so that any point of IIa will move as if attached to sphere I. But as the sphere I moves together with the sphere of the fixed stars, IIa will also move in the same manner, and the first sphere of the next planet, Jupiter, will therefore move as if all the spheres of Saturn did not exist.

For the same purpose Aristotle introduced for each of the other planets extra spheres, numbering in each case one less than the actively working spheres of Kalippus, so that he gave three new ones to Jupiter, and four to each of the planets Mars, Mercury, Venus, and sun. The moon did not in his opinion require any, as there was nothing below it which might be disturbed by it. The total number of extra spheres was therefore twenty-two, which added to the thirty-three spheres of Kalippus makes a grand total of fifty-five! This is the number given by Aristotle, and it is no wonder that philosophers after him found his machinery rather cumbersome. Obviously he might have practised economy a little by leaving out six spheres. For as the spheres IIa of Saturn and I of Jupiter are next to each other, and move with the diurnal speed of the fixed stars, we may amalgamate them into one, and similarly five others might be spared, which would bring the number down to forty-nine.

Though Aristotle does not enter into any particulars about the spheres in the work on the heavens, he devotes some space to various general considerations respecting them. It evidently disturbs him a little that the planets do not possess the same number of spheres, or that the number does not (as he thinks we might have expected) gradually increase from the single sphere of the fixed stars and downwards. Instead of this we find, he says[1], that sun and moon are moved in fewer motions than some of the planets, and yet the latter are certainly more distant, as he has himself seen the moon cover Mars, while Egyptians and Babylonians possess many observations of occultations of the other planets. We have seen above that

[1] *De Cœlo*, ii. 12, p. 292 a.

Aristotle simply adopted the five spheres of Kalippus for the moon, while he added four spheres to the five spheres of Kalippus for the sun. As he in this place classes the sun with the moon as having fewer spheres than the planets (although he ought to give Mars, Mercury, Venus, and the sun nine spheres each), it would seem that he had his doubts as to the necessity of adopting the new spheres given to the sun and moon by Kalippus; and this is confirmed by a remark at the end of his account in the *Metaphysics,* where he says that, if we leave out the spheres added (by Kalippus) for sun and moon, the number of spheres becomes forty-seven—an obvious slip for forty-nine[1], which Sosigenes did his best to excuse or account for[2]. The different numbers of spheres Aristotle endeavours to explain in this way. The earth is at rest and is furthest removed from the Divine principle, but the sphere of the fixed stars is under the immediate control of the Divine motor and is only subject to a single motion; the moon and the sun are nearest to the motionless earth, and are therefore less extensively moved than the planets somewhat further out, whose motions are more manifold, while Jupiter and Saturn being nearer to the Divine principle are moved in a rather simpler manner. Equally metaphysical, but more obscure, is the explanation offered of the fact that the primary motion governs an immense number of bodies (the fixed stars), while each planet requires several spheres for itself alone. Aristotle seems to think that this very unequal distribution of matter is more apparent than real, as we have on the one hand many stars partaking of one motion, on the other hand few stars acted on by many motions, so that the variety of these motions may be supposed to make up for the deficiency in the number of stars subject to them. Considerations of this kind, which look strange to a modern reader, are quite in keeping with the speculative tendency of Hellenic science.

---

[1] Four Kallipean and two extra spheres being subtracted from the fifty-five.

[2] Simplicius, ibid., p. 503, l. 10 seq. (Heib.). The simplest explanation would be to suppose that Aristotle meant to abolish the sun's motion in latitude, which would remove two spheres more. But this would be a real step forward which Aristotle probably hardly would have thought of. It has not occurred to any commentator to suggest this explanation.

The questions as to the position, possible motion, and form
of the earth are very fully discussed by Aristotle, with constant
references to the opinions of previous philosophers, and we
have in the preceding chapters frequently made use of these
valuable references. He first mentions the system of the
Pythagorean school, which he rejects as founded on the supposition that the most excellent body ought to occupy the
centre, while his opinion is that the centre is not the origin of
anything but is more like a boundary. "For that which is
to be defined is the middle, but what defines is the extremity;
and more excellent is that which encompasses and is the limit
than that which is completed: the latter is matter, the former
is the essence of the composition[1]." On the other hand, some
place the earth in the centre, wound round the axis extended
through the universe[2], while there has also been much difference of opinion as to its form, as some people have held that
if the earth were a sphere the sun when rising or setting ought
not to be bounded by a straight line but by a curved one; an
opinion which Aristotle refutes by referring to the great
distance of the sun and the greatness of the circle of the
horizon. Intimately connected with the problem of the figure
of the earth has been the question as to what preserves its
stability, and while the Ionian idea, that the earth floats on
water, is contrary to experience, the theory advocated by
Anaximenes, Anaxagoras, and Demokritus, that it fits over the air
in the lower half of the sphere, is also rejected, since it cannot
be assumed that it could only be at rest by being flat. Likewise the idea of Empedokles, that the particles forming the
earth sought the centre in the beginning of time owing to the
whirling of the heavens, is untenable; for why should then at
the present time everything heavy tend towards the earth
when the whirling takes place far from us? And why does
fire move upwards? Heaviness and lightness must have
existed before the whirling began; therefore the condition of
the earth cannot be a consequence of the motion of the

---

[1] *De Cælo*, II. 13, p. 293 b; comp. II. 4, p. 287 a: πρῶτον δὲ σῶμα τὸ ἐν τῇ
ἐσχάτῃ περιφορᾷ.

[2] Ibid. This is the passage discussed at length above, p. 77.

heavens. Anaximander suggested that the earth could not fall in any particular direction because it is placed in the middle and has the same relations to every part of the circumference. But we find that the earth not only rests at the centre but also moves towards the centre; for whither all the parts are moving there the whole is moving, and consequently it is not owing to its relations to the circumference that it remains at the centre. Besides, in that case fire placed at the centre ought also to remain there instead of striving upwards and becoming dissipated in equal proportions over the uppermost parts (τοῦ ἐσχάτου), and this would also happen to the earth if the centre were not by nature its proper place.

Aristotle next considers whether the earth is in motion or not, " as some make it out to be one of the stars, while others place it in the middle and assert that it is packed and moved (ἴλλεσθαι καὶ κινεῖσθαι) round the middle axis[1]." A motion of this kind, he maintains, cannot naturally belong to the earth, since it would in that case also be natural to its single parts, instead of which we see them move in straight lines towards the centre. As the motion would therefore have to be produced by force, it would be contrary to nature and could not be everlasting. "Furthermore it appears that the bodies, which have circular motions in space, all, with the exception of the first sphere, also have a backward motion, and in fact have more motions than one, so that necessarily also the earth, whether it moved round the centre[2] or was moved at the centre itself, would move in two courses; but in this case there would of necessity be passings by and turnings (παρόδους καὶ τροπὰς) of the fixed stars; but we find that none such appear, but that the same stars both rise and set at the same places."

It seems evident that Aristotle in this passage merely refutes the Philolaic system, and even when speaking of a motion of the earth "at the centre itself" does *not* think of a rotation of the earth in twenty-four hours. At least it seems very unlikely that he would have neglected to remark that it was more proper for the most exalted sphere to perform this

---

[1] ii. 14, p. 296 a. Compare above, p. 72.
[2] As in the theory of Philolaus.

rotation, while his whole mode of reasoning shows that he never dreamt that the motion of the earth could be supposed to be a mere copy of the rotation of the first sphere *and intended to replace it*. It seems almost that when he wrote this passage Aristotle cannot have been aware that anybody had proposed to account for the daily motion of sun, moon, and stars by assuming a uniform rotation of the earth. Can anyone doubt that he would have used quite different arguments in order to combat this theory, particularly if it had been supported by the great authority of Plato, and that he would have repeated his assertion that none of the heavenly bodies rotate (στρέφεσθαι)? Instead of which he again points out that the natural motion for parts of the earth and the whole earth is towards the centre of the universe, for which reason it is situated at the centre itself; and if we ask whether that which has weight is moved towards the centre, because it is the centre of the universe or because it is the centre of the earth, he explains that it is the centre of the universe towards which it is moving, just as light things and fire move in the opposite direction towards the limits of the world. That the centre of the earth coincides with that of the universe we may see from the fact that heavy bodies do not move in parallel lines but under equal angles, consequently towards one centre, which is that of the earth. It is also well-known that bodies thrown upwards with great force fall straight down again to the point from which they started. It is thus clear, he thinks, that the earth is neither in motion nor is outside the centre, and as its parts are by nature intended to move from all sides towards the centre it is impossible that any part of it could be moved away from the centre, and therefore also that the whole earth could do so. A further proof of the immovable position of the earth is, as mathematicians say of astronomy, that the actual phenomena are exactly as they ought to be if the earth really were at the centre, if we make allowance for the changes of forms by which the arrangement of the stars is determined.

That the earth is a sphere is first proved by Aristotle by arguing that, when heavy particles are moved uniformly from

all sides towards a centre, a body must be formed, the surface
of which is everywhere equidistant from this centre; and even
if the particles were not equally moved towards the centre the
greater would push the smaller on, until the whole was everywhere uniformly settled round the centre. But in addition to
this more metaphysical reasoning he also produces more solid
arguments appealing to direct observation[1]. First he refers to
lunar eclipses, during the progress of which the edge of the
shadow is always circular without showing any of the changes
to which the line terminating the illuminated part of the moon
is subject in the course of a month, so that the earth which
throws the shadow must be a sphere. Secondly, a very slight
journey north or south is sufficient to change the horizon
sensibly and make different stars be situated overhead; and
some stars are visible in Egypt and the neighbourhood of
Cyprus which are not seen further north, while those which
are circumpolar for northern countries are seen to set when we
go south. This shows that the earth is not a large sphere, as
a small change of position makes so much difference in the
appearance of the heavens. Aristotle adds that those who
believe that the region about the Pillars of Herakles is connected with India, so that in this way the ocean is one, do not
assert anything very improbable, and they point to the fact
that there are elephants both in India and in West Africa[2].
"And those among the mathematicians who attempt to calculate the extent of the circumference maintain that it is about
400,000 stadia, from which it follows that the bulk of the earth
must not only be spherical, but not large in comparison with
the size of the other stars."

This statement (which finishes the astronomical part of
Aristotle's book on the heavens) is the oldest attempt to

---

[1] *De Cœlo*, ii. 14, pp. 297 b—298 a.

[2] In the *Meteorol.* ii. 5 (p. 362 b), Aristotle says that owing to the sea the land
is not continuous between the Pillars of Herakles and India, but in the passage
quoted above he merely insists on the fact that these two regions are not
diametrically opposite parts of a flat earth, but are within a reasonable distance
of each other on the surface of a moderately sized sphere. See Simplicius,
pp. 547-48 (Heib.), and Seneca, *Quest. Nat.* i. præfatio, where he points out
that India is only a few days' sail from Spain if you have a good wind.

estimate the size of the earth. We neither know by whom it was made, nor how it was carried out, but as Eudoxus appears to have been the first scientific astronomer, it is not unlikely that he is responsible for it, and that it was connected with his visit to Egypt. Of course the only manner in which an observer could determine the size of the earth was by observing the meridian altitude of the sun or a star from two stations north and south of each other and estimating the linear distance between the stations, and as neither of these operations could be accurately carried out, a rough approxima-tion was all that could be attained. The result given by Aristotle is equivalent to the diameter of the earth being equal to 12,461 miles[1], and as the real diameter is 7920 miles, it is curious that he should have considered the earth to be rather small. His remark that it is " not large in comparison to the other stars " is not to be taken to mean that it is the smallest of all, and according to Stobæus he held the moon to be smaller than the earth[2], while he in the *Meteorologica* merely says that the earth is smaller than some of the stars[3]. In dealing with the difficult question as to the dimensions of the universe Aristotle was therefore not able to add anything to the vague surmises of previous philosophers.

The Aristotelean system of the world distinguishes sharply between the heavens, the region of unchangeable order and of circular motion[4], and the space below the moon's sphere, the region of the unsettled and changeable and of rectilinear motion. The latter region is occupied by the four elements,

[1] This is on the assumption that the stadium is 157·5 metres, as it resulted from the steps taken when marching a long distance. See below, chapter VIII.

[2] *Ecl. Phys.* I. 26, Diels, p. 357; not mentioned in the *Plac. Phil.*

[3] *Meteorol.* I. 3, p. 339 b. A little further on he points out that it is absurd to think that the stars are really small merely because they look small. There is a valuable edition of the *Meteorology* by J. L. Ideler (2 vols. Leipzig, 1834–36), which includes a Latin version and copious extracts from the commen-taries of Alexander of Aphrodisias, Olympiodorus, and Johannes Philoponus, as well as lengthy notes by the Editor. I quote, however, as usual, the pages of Bekker's edition.

[4] As far back as human tradition reaches there has never been any change in the outermost heavens or any of its parts. Therefore the upper place already in early times received the name of αἰθήρ, because it always runs (ἀεὶ θεῖ). *De Cælo*, I. 3, p. 270 b; the same derivation is given by Plato, *Krat.* 410 n.

of which earth occupies the place nearest to the centre, water next, air higher up, and fire rising highest of all. The several elements are however not separated by definite limits, and Aristotle takes special care to state that there is not a layer of fire above the air. But "fire" does not mean flame, which is only a temporary product of the transformation of the moist and dry elements air and fire, while the element of fire is matter which on the slightest motion being given to it burns like smoke, and there can be no fire without something being consumed[1]. There can therefore be no fire in the celestial space, as it would long ago have consumed everything; besides, as all the earth and all the water is confined within a very small space, the amount of air and fire would be out of all proportion to that of these two other elements, if the immense upper region were filled with air and fire[2]. The fire predominates in the upper part of the atmosphere, the air in the lower, but matter in the celestial region is much purer than our elements, and the circular motion is natural to it. Even this ether is however not uniformly distributed as regards purity, which gradually increases with the distance from the earth[3]. It transmits to the earth through the medium of the air the heat generated by the motion of the sun, which is much greater than that produced by the motion of the moon, notwithstanding the greater proximity of the latter, owing to the greater speed of the sun[4].

The upper part of the atmosphere is an important factor in the Aristotelean system of the world. Here shooting stars and meteors are produced, which are hot and dry products of evaporation which on rising to the upper layer of the atmosphere are dragged along in its rotation and thereby become ignited. The aurora is produced in a similar manner[5]. With regard to

---

[1] *De part. animal.* II. 2, p. 649 a; *Meteorol.* I. 3, p. 340 b, II. 2, p. 355 a.

[2] *Meteorol.* I. 3, 339 b—340 a.   [3] Ibid. 340 b.

[4] Ibid. 341 a. The ether was accepted both by the Stoics and by the Epicureans. It was by later writers called *quinta essentia*, from which our modern word quintessence is derived. The author of the *Epinomis* had already assumed the existence of this fifth element, while Plato had only recognised four.

[5] Ibid. I. 4—5, 341 b—342 b.

comets, Aristotle was obliged to resort to an explanation of the
same kind, on account of his theory of the unchangeableness
of the ethereal regions, and probably partly also because the
nature of the solid celestial spheres made it impossible to accept
the doctrine of the Pythagoreans that they were appearances
of one single planet of some sort which appeared above the
horizon as rarely as did the planet Mercury. Aristotle rejects
this idea, and points to the fitful appearance of comets, and to
their not being confined to the zodiac, as proving that they have
nothing in common with planets. He also disproves easily the
idea of Anaxagoras and Demokritus, that comets are generated
by conjunctions of planets and stars, by referring to repeated
observations of conjunctions of Jupiter with a star in Gemini,
which did not produce any comet. Aristotle's own theory is
that dry and hot exhalations, similar to those which cause
shooting stars and aurora, occasionally are carried up into the
uppermost or fiery part of the atmosphere, which partakes of
the daily rotation of the heavens from east to west, and while
carried along there they take fire under the influence of the sun
and appear to us as comets ; they last as long as there is any
inflammable material left, or as long as this is replenished from
the earth. Most comets are seen outside the zodiac, because the
motion of the sun and planets prevents the aggregation of
material in the neighbourhood of their orbits. A phenomenon
of the same kind is the Milky Way, which is formed constantly
under the influence of the motion of the fixed stars, and therefore always occupies the same position and divides the heavens
as a great circle along the solstitial colure. Aristotle thinks
that this theory explains both why there is in general such a
great number of stars near the Milky Way, as also why there
are so many bright stars in that neighbourhood where the
Milky Way is bifurcated. The permanent accumulation of
ignited vapour is the cause of the scarcity of comets, as the
material from which the latter might be formed goes to keep
up the Milky Way[1].

Although the system of homocentric spheres which Aristotle
had borrowed from Eudoxus and Kalippus, and modified from

[1] *Meteorol.* I. 6—8, pp. 342 b seq.

a mathematical theory into a physical representation of Kosmos, did not maintain its hold over the minds of succeeding philosophers very long, his ideas about the non-celestial character of comets and the Milky Way held sway until the revival of astronomy in the sixteenth century. He was least fortunate in his explanations of these phenomena, but this should not make us blind to the merits of many others of his cosmical ideas. His careful and critical examination of the opinions of previous philosophers makes us regret all the more that his search for the causes of phenomena was often a mere search among words, a series of vague and loose attempts to find what was "according to nature" and what was not; and even though he professed to found his speculations on facts, he failed to free his discussion of these from purely metaphysical and preconceived notions. It is, however, easy to understand the great veneration in which his voluminous writings on natural science were held for so many centuries, for they were the first, and for many ages the only, attempt to systematise the whole amount of knowledge of nature accessible to mankind; while the tendency to seek for the principles of natural philosophy by considering the meaning of the words ordinarily used to describe the phenomena of nature, which to us is his great defect, appealed strongly to the medieval mind, and, unfortunately, finally helped to retard the development of science in the days of Copernicus and Galileo.

CHAPTER VI.

HERAKLEIDES AND ARISTARCHUS.

WHILE Plato and Aristotle, as we have seen, never abandoned the idea of the daily rotation of the heavens from east to west, their contemporary, Herakleides of Pontus, clearly and distinctly taught that it is the earth which turns on its axis from west to east in twenty-four hours.

We know very little about the life of this philosopher. His life seems to have extended over the greater part of the fourth century B.C., since he described the destruction of the town of Helike in Achaia by an earthquake (B.C. 373) as having happened in his own lifetime[1], and lived till after the foundation of Alexandria[2]. He was born at Heraklea in Pontus, but emigrated to Athens, where he became a pupil of Speusippus the Platonist, and afterwards possibly of Plato himself[3], but at the same time he is said to have attended the schools of the Pythagorean philosophers. Finally, he seems also to have received instruction from Aristotle. As a philosopher, he appears to have carried some of Plato's views about the Kosmos rather further than his master did, as he called the world a god and a divine mind, and attributed divinity to the planets[4], but, on the other hand, we are also told that he considered the Kosmos to be infinite, and that he, with the Pythagoreans, considered each planet to be a world

---

[1] Strabo, VIII. pp. 384–85.   [2] Plutarch, *Alexander*, XXVI.

[3] Proklus (*in Tim.* p. 281 E) says expressly that Herakleides had *not* heard Plato, but in another place (p. 28 c) he mentions that Herakleides himself claimed to have been on intimate terms with Plato, and this agrees with what Simplicius (in Arist. *Phys.*, Brandis, *Scholia*, p. 362 a) tells us, that H. like Aristotle and Hestiaeus had taken down Plato's lectures in writing.

[4] Cicero, *De Natura Deorum*, I. 13, § 34.

with an earth-like body and an atmosphere[1]. He was fond of
telling " puerile fables " (as Cicero tells us) and of embellishing
his books with marvellous stories. Diogenes Laertius gives
a list of his writings, which appear to have dealt with a great
variety of subjects, one being " On the things in the Heavens "
(Περὶ τῶν ἐν οὐρανῷ), which, however, possibly did not altogether
deal with astronomical matters, since the next book mentioned
is " On the things in Hades," of which Diogenes says that it was
written in the style of the Tragedians. His writings are unfortunately all lost, so that what we know of the astronomical
doctrines of Herakleides is only derived from the allusions of
later writers, but these are sufficiently detailed to make it quite
certain that he really held some views which were more advanced
than those of his contemporaries. We have no way of knowing
whether these doctrines were published before Aristotle wrote
his book on the heavens, but the absence of any reference to
them in that work seems to point to their having been unknown
to Aristotle at the time when he wrote it. It is much to be
regretted that the opinions of Herakleides are not included
among those discussed in that work, since we cannot, in one or
two instances, be sure that subsequent commentators did not,
in their interpretation of the doctrines of Herakleides, view
them a little too much in the light of knowledge acquired by
astronomers long after his time. But this does not in any way
affect the credibility of the principal doctrine for which he is
famous.

The statement of Diogenes, that Herakleides attended the
Pythagorean schools[2], is of special importance to us, as it is
extremely likely that their influence (which is also perceptible
in his ideas about atoms, which he calls masses, ὄγκοι), tended
to convince him of the truth of the very simple explanation
of the daily motion of the stars proposed by Hiketas and
Ekphantus. We have already seen his name coupled with that
of Ekphantus[3], but the accounts given by Simplicius do not
mention the latter. He first alludes to Herakleides when dis-

---

[1] Plac. II. 13, Diels, p. 343. Comets he held to be clouds at a very great
height, illuminated from above. Plac. III. 2, Diels, p. 366.

[2] Diog. v. 86.   [3] See above, p. 51.

cussing the chapter in which Aristotle considers the motion
of the starry vault[1]. Aristotle first remarks that, taking for
granted that the earth is at rest, the starry sphere (the *aplanes*)
and the planets might either both be at rest, or both be in
motion, or one be at rest and the other in motion. And these
cases he considers (says Simplicius) " on account of there being
some, among whom were Herakleides of Pontus and Aristarchus[2],
who believed they could save the phenomena (account for the
observed facts) by making the heavens and the stars be immovable, but making the earth move round the poles of the
equator (τοῦ ἰσημερινοῦ) from the west, each day one revolution
as near as possible; but 'as near as possible' is added on
account of the [daily] motion of the sun of one part (degree[3]);
so that, if then the earth does not move, which presently he
(Aristotle) is going to show, the hypothesis of both being at
rest cannot possibly save the phenomena." Again, when considering what Aristotle might have held to be Plato's meaning
in the disputed passage about the earth and its axis, Simplicius
winds up the discussion by adding that Aristotle in this chapter
deals with all the various theories as to the earth's position and
figure; " but Herakleides of Pontus, assuming the earth to be
in the middle and to move in a circle, but the heavens to be at
rest, considered the phenomena to be accounted for[4]." Once
more Simplicius, when defending and commenting on Aristotle's
statement that the observed phenomena only agree with the
assumed position of the earth in the centre of the world, refers
to those who gave it a progressive motion, and also alludes to
the doctrine of Herakleides, that the earth " moved in a circle
round its centre, while the heavenly things were at rest[5]."

All this agrees perfectly with the statement which we have
already quoted from Aëtius, that Herakleides and Ekphantus

---

[1] Simplicius to Arist. *De Cælo*, ii. pp. 444—445 (Heib.).

[2] Aristotle cannot have alluded to Aristarchus, who lived much later, so we
may also doubt that he thought of Herakleides in this connection.

[3] Herakleides would therefore seem to have been aware of the difference
between the sidereal and the mean solar day.

[4] Simplicius, p. 519. An anonymous scholiast to Aristotle uses almost the
same words (Brandis, *Scholia*, p. 505 b, 16).

[5] Simplicius, p. 541.

the Pythagorean let the earth move, " not progressively, but in a turning manner, like a wheel fitted with an axis, from west to east round its own centre[1]." Proklus also contrasts Herakleides with Plato, who held the earth to be at rest, and says that " Herakleides of Pontus, not having heard Plato, held this doctrine, that the earth moves in a circle[2]." The doctrine of the earth's rotation was certainly not taught in the Academy, but Herakleides must have been quite capable of discovering it for himself, since he was also, with regard to the motion of the planets, much in advance of Plato. Chalcidius, in his commentary to the *Timæus*[3], tells us that Herakleides let Venus move round the sun instead of round the earth, so that it is sometimes nearer to us and sometimes farther off than the sun. We have seen that there was considerable difference of opinion among philosophers as to whether Mercury and Venus were nearer than the sun or not, and it is indeed strange that this, in connection with the fact that these two planets are never seen at any great distance from the sun, were not sufficient indications of the true nature of their orbits. Chalcidius possibly goes further than Herakleides did, when he states that the sun itself moves on a small circle or epicycle with which the larger orbit of Venus is concentric, though it is very likely indeed that Herakleides, like Kalippus, was aware of the variable velocity of the sun in the course of a year, and we shall presently consider another passage which seems to point to his having possessed

---

[1] Plac. III. 13, Diels, p. 378. Compare above, p. 51.

[2] Proklus *in Tim.* p. 281 E. This has naturally roused the ire of Gruppe, who devotes a chapter in his book (*Die kosm. Systeme der Griechen*, pp. 126 ff.) to an attempt to prove that Herakleides got the idea of the earth's rotation from the *Timæus* and published it as his own, at the same time as he falsified the Pythagorean system by pretending that only the properly initiated knew that the central fire was in the interior of the earth, while the antichthon really was the other hemisphere of the earth or even the moon. Whether the later Pythagoreans, who modified the system of Philolaus, really endeavoured to give their own ideas weight by claiming that these represented the old system correctly, may be doubtful, but this anyhow does not concern Herakleides, as there is nothing to prove that Simplicius alluded to him or to the rotation of the earth when mentioning the modified doctrine (see above p. 52). Compare Boeckh, *Untersuchungen*, pp. 127—133.

[3] Cap. cx—cxii. ed. Wrobel, pp. 176—178, reprinted by Martin, *Theonis Smyrnæi Astr.* p. 423. Chalcidius lived about three hundred years after Theon.

this knowledge. Martin suggests that Chalcidius may have copied this sentence about the solar epicycle from Theon or from the lost commentary to Plato's *Republic* written by Adrastus, from which writers he borrowed all the astronomical parts of his treatise; indeed he copies their mistakes as well, for instance when he gives the greatest elongation of Venus from the sun as 50°, or when he attributes to Plato a knowledge of the theory of epicycles. Chalcidius does not mention Mercury in connection with Herakleides, but he has already described the motions of both the interior planets, and evidently makes no distinction between them. There is a chapter in Theon's *Astronomy* bearing on this question[1]. Theon (or rather Adrastus) first suggests that the sun, Mercury and Venus each moves on its own sphere (or epicycle), the centres of which move with equal velocity on separate spheres (deferents) round the earth, that of the sun being the smallest[2], "but it may also be, that there is one hollow sphere (or deferent) common to the three stars, and that the three solid spheres (or epicycles) have a common centre in this, of which the smallest and truly solid one is the sun, and round it the sphere of Stilbon[3], and surrounding them both, and occupying the whole depth of the hollow and common sphere, the sphere of Phosphorus." After this Theon goes on to describe the advantage of this arrangement, which would only allow Mercury to be at most 20° and Venus at most 50° from the sun. Theon does not mention Herakleides in connection with this theory, and the system first suggested no doubt belongs to some astronomer of Alexandria and has probably nothing to do with Herakleides.

Although the name of Herakleides had not been connected with this planetary system until Martin drew attention to the above-mentioned passage in Chalcidius, the system itself has always been known to have had some adherents among the ancients. Martianus Capella, a writer in the fifth century of

---

[1] Theon, ed. Martin, p. 296.

[2] That is, the sun is always nearer to the earth than Mercury or Venus, and the line from the earth to the sun produced always passes through the centres of the epicycles of the two planets. Chalcidius also mentions this arrangement and attributes it to Plato!

[3] Mercury.

our era and the author of a curious encyclopædic work, "*De nuptiis Philologiæ et Mercurii*," in which he treats of the various free arts, has in his eighth book, on astronomy, set forth this system. He says that although Mercury and Venus rise and set daily, yet their circles do not go round the earth, but round the sun in a wider circuit, so that, when they are beyond the sun, Mercury is nearer to us than Venus, and the reverse when they are on our side of the sun[1]. The Roman author who calls himself Vitruvius, in his celebrated book on architecture, when describing sun-dials, deals with the periods of revolution of the various planets, but without setting forth any particular system of the world. His remarks about Mercury and Venus have been supposed to show that he was an adherent of the heliocentric motion of these planets. But it seems to me that his words (clothed in his usual affected language) can equally well apply to the first of the two above-mentioned proposals of Theon of Smyrna, viz., that of letting the epicycles of Mercury and Venus move round the earth on circles concentric with the orbit of the sun, in such a manner that the centres of the epicycles are always in a line with the sun and the earth[2]. At all events, he says nothing about the orbits or the varying distances from the earth, and in a preceding paragraph, when first mentioning the planets, he distinctly says that they move from west to east, in the order: Moon, Mercury, Venus, Sun, Mars, Jupiter, Saturn[3]. But as this pseudo-Vitruvius was, almost certainly, not what he pretends to be, a contemporary of Augustus, but a rather ignorant compiler who lived much later (perhaps about A.D. 400), his opinion would not be of much value unless we could be sure

---

[1] Martianus Capella, VIII. § 857, ed. Eyssenhardt, p. 317.

[2] *De Architectura*, lib. IX. cap. 4, §§ 18—21. Mercurii autem et Veneris stellæ circum solis radios, solem ipsum uti centrum, itineribus coronantes, regressus retrorsum et retardationes faciunt. Etiam stationibus propter eam circinationem morantur in spatiis signorum. Id autem ita esse, maxime cognoscitur ex Veneris stella, quod ea cum solem sequatur, post occasum ejus apparens in cœlo, clarissimeque lucens, Vesperugo vocitatur; aliis autem temporibus eum antecurrens et oriens ante lucem Lucifer appellatur. Ex eoque nonnumquam plures dies in uno signo commorantur, alias celerius ingrediuntur in alterum signum.—After this he goes on describing the stations and retrogradations of both planets, but there is not a word about the actual orbits.

[3] Ibid. § 15.

where he had borrowed it. It seems probable that in many cases his authority was the encyclopædia (*De novem disciplinis*) of Terentius Varro, and it is probably this author who is responsible for the paragraph in question. But, as already remarked, it does not seem to state that the sun is in the centre of the orbits of Mercury and Venus[1].

For ages this system, in which Mercury and Venus alone move round the sun, has been known as "the Egyptian," on the authority of a somewhat questionable passage in the commentary on Cicero's fragment, *Somnium Scipionis*, by Macrobius, who lived towards the end of the fourth century. Cicero himself, both in his *Somnium Scipionis* and elsewhere, adheres to the geocentric system, placing the planets in the order: Moon, Mercury, Venus, Sun, etc.[2] Macrobius[3] remarks that Cicero in this followed Archimedes and the Chaldeans, while "Plato followed the Egyptians, the parents of all sciences, who placed the sun between the moon and Mercury." He then describes how, according to the same, the sphere of Saturn is the outermost, how those of Jupiter and Mars come next, how Venus is so much inferior to Mars that a year is enough for it to pass round the zodiac, Mercury and the sun coming next, so that these three go round the heavens in about a year. This does not look as if Macrobius believed the Egyptians to have made the two planets go round the sun. But the next passage is different, in which he, after pointing out that the nearness to each other of Venus, Mercury, and the sun confused the order in which they had been placed, continues thus: " But the theory did not escape the skill of the Egyptians, and it is the following: the circle by which the sun moves round is surrounded by the circle of Mercury, but the circle of Venus encloses that again, being above it, and thus it happens that these two stars, when they run through the upper part of their circles, are found to

---

[1] About the authorship of Vitruvius see Ussing's memoir, " Observations sur Vitruve," in the Memoirs of the R. Danish Academy of Science, *Hist. philos. Afd.* VI. Series, Vol. 4, p. 93 (1896).

[2] Cic. *De Nat. Deorum*, II. 20, §§ 52–53, where the places of the sun and moon are not mentioned and Venus is put nearest to the earth. In the *Somnium* and *De Diviuat.* II. 43, § 91, Venus is placed third, between Mercury and the sun.

[3] *In somn. Scip.* lib. I. cap. 19.

be placed above the sun, but when they go through the lower parts of their circles, the sun is considered to be above them."

At first sight it looks as if Macrobius here simply enunciates the geocentric system[1], but if so, there is no sense in the last paragraph. If we, however, assume[2] that he, like Chalcidius, means epicycle when he uses the word *circulus*, his theory is exactly the same as that which Chalcidius ascribes to Herakleides, and this seems indeed to have been what Macrobius meant, as he, in the preceding passage, when explaining the order of the planets according to Plato, used the word sphere for orbit, which Chalcidius expresses by *globus*[3]. All the same, he ascribes both the Platonic arrangement and the heliocentric motion of Mercury and Venus to the Egyptians, so that his testimony is quite worthless as to the origin of the system which has so long borne the name of the Egyptian. There is, indeed, no proof whatever that the astronomers or priests of ancient Egypt, either during the time of the Pharaohs or later, were aware of the fact that Mercury and Venus travel round the sun. On the contrary, Achilles states distinctly that the Egyptians placed the sun fourth in order, "which the Greeks call the sixth[4]."

Chalcidius is thus the only author who ascribes this important discovery to Herakleides; but when we remember the intimate connection between Chalcidius and the previous most reliable writer, Adrastus, and when we also bear in mind what enlightened views Herakleides must have held, since he acknowledged the rotation of the earth, we may consider it as thoroughly proved that he also made this other step forward. Though all his writings are lost, we are not without proofs that he had devoted some attention to other astronomical questions besides the movements of the interior planets. In his commentary to the *Physics* of Aristotle, Simplicius gives us an interesting quotation from a commentary to the *Meteorology* of Posidonius, written by

---

[1] Martin, *Études*, II. pp. 132–133, and Humboldt, *Kosmos*, III. p. 466, are of this opinion.

[2] As already done by Schiaparelli, *I Precursori di Copernico*, p. 27.

[3] It is deserving of notice that Macrobius in the following chapter, when discussing the dimensions of the sun, twice writes orbis instead of sphæra.

[4] Isagoge in Arati Phenom. § 17, Petavius, III. p. 80.

Geminus in the first half of the first century B.C.[1] Dealing
with the difference between physics and astronomy, Geminus
says that to the former science belongs the examination of the
nature, power, quality, birth, and decay of the heavens and the
stars, but astronomy does not attempt any of this, it makes
known the arrangement of the heavenly bodies, it investigates
the figure and size and distance of earth and sun and moon, the
eclipses and conjunctions of stars and the quality and quantity
of their motions; in which numerical investigations astronomy
obtains help from arithmetic and geometry.   But although the
astronomer and the physicist often prosecute the same research,
such as the size of the sun or the sphericity of the earth, still
they do not proceed in the same manner, the latter seeking for
causes and moving forces, while the astronomer finds certain
methods, adopting which the observed phenomena can be accounted for.   " For why do sun, moon, and planets appear to
move unequally?   Because, when we assume their circles to be
excentric or the stars to move on an epicycle, the appearing
anomaly can be accounted for ($\sigma\omega\theta\eta\sigma\epsilon\tau\alpha\iota$ $\dot{\eta}$ $\phi\alpha\iota\nu o\mu\acute{\epsilon}\nu\eta$ $\dot{\alpha}\nu\omega\mu\alpha\lambda\acute{\iota}\alpha$
$\alpha\dot{\upsilon}\tau\hat{\omega}\nu$), and it is necessary to investigate in how many ways the
phenomena can be represented, so that the theory of the wan-
dering stars may be made to agree with the etiology in a
possible manner.   Therefore also a certain Herakleides of Pontus
stood up and said that also when the earth moved in some way
and the sun stood still in some way, could the irregularity
observed relatively to the sun be accounted for.   In general it
is not the astronomer's business to see what by its nature is
immovable and of what kind the moved things are, but framing
hypotheses as to some things being in motion and others being
fixed, he considers which hypotheses are in conformity with the
phenomena in the heavens.   He must accept as his principles
from the physicist, that the motions of the stars are simple,

---

[1] Simplicii in Aristot. phys. ed. Diels, p. 291, reprinted in Gemini *Elem.
Astr.* ed. Manitius, p. 283. Compare Boeckh's *Untersuchungen*, p. 134, and
about the commentary of Geminus see also Boeckh, *Ueber die vierjährigen
Sonnenkreise der Alten*, p. 12. In the passage (*Phys.* II. 1, p. 193 b, 22) commented on by Simplicius, Aristotle explains the difference between the points
of view from which a mathematician and a natural philosopher consider the
phenomena of nature.

uniform, and regular, of which he shows that the revolutions are circular, some along parallels, some along oblique circles."

This paragraph is remarkable in more ways than one. It distinguishes clearly between the physically true causes of observed phenomena and a mere mathematical hypothesis which (whether true or not) is able to "save the phenomena." This expression is rather a favourite one with Simplicius, who doubtless had it from the authors long anterior to himself, from whose works he derived his knowledge. It means that a certain hypothesis is able to account for the apparently irregular phenomena revealed by observation, which at first sight are puzzling and seem to defy all attempts to make them agree with the assumed regularity of all motions, both as to velocity and direction. In this passage Geminus points out that an astronomer's chief duty is to frame a theory which can represent the observed motions and make them subject to calculation, while it is for this purpose quite immaterial whether the theory is physically true or not. The passage in which he among such theories mentions one put forward by Herakleides, presents many difficulties and has been the subject of a great deal of controversy. In the original it runs thus:

διὸ καὶ παρελθών τις φησὶν Ἡρακλείδης ὁ Ποντικός, ὅτι καὶ κινουμένης πως τῆς γῆς, τοῦ δὲ ἡλίου μένοντός πως δύναται ἡ περὶ τὸν ἥλιον φαινομένη ἀνωμαλία σώζεσθαι.

In the first place we must mention that formerly, that is in the Aldine edition of Simplicius and in Brandis' collection of *Scholia* (p. 348 b), the word ἔλεγεν appeared after ὁ Ποντικός, so that the passage could only be translated "therefore somebody stood up, said Herakleides of Pontus, and said that," etc. It appears, however, that this word does not occur in the best codices, which somewhat simplifies matters. "Therefore also one Herakleides of Pontus stood up (came forward) and said, that also when the earth moved in some way and the sun stood still in some way, could the irregularity observed relatively to the sun be accounted for." That a well-known philosopher is mentioned as τις, as if he were an obscure person, looks strange[1], and so does the expression "stood up" (παρελθών), as if Herakleides came forward and spoke in an assembly or a conference, like that at Athens between Kalippus, Polemarchus and Aristotle, at which the theory of Eudoxus was discussed, to which Simplicius alludes elsewhere. But it does not seem unreasonable to assume that the passage has been corrupted, when we bear in mind that it had gone through several hands before Simplicius wrote it down. He says that he took it from the commentator Alexander of Aphrodisias, who had it from an abstract (epitome) of a commentary written by Geminus on the *Meteorology* of Posidonius. The passage taken by itself ought therefore not to be made a foundation for far-reaching theories. And yet that is precisely the use which has recently been made of the second half of the passage[2].

How are we to interpret the expression, that even if the earth had some motion the irregularity connected with the sun could be accounted for? As the words stand they can only refer to the want of uniformity of the annual motion of the sun, which causes the four seasons to be of unequal length[3]. We have seen that Kalippus had accounted for this by adding two other spheres to the solar theory of Eudoxus. What Herakleides did was probably merely to throw out the suggestion, that it might *also* be possible to account for this irregularity by assuming that the earth was not absolutely at rest but moved in some way or other (πως). Every word in the sentence seems to indicate that it was in no way intended to formulate a theory of any kind, but merely to hint in a general manner that there might be more than one way of "saving the phenomena," and that this might *also* be done by abandoning the usual idea of the earth being at rest. The whole argumentation of Geminus about hypotheses, the actual physical truth of

---

[1] Gomperz, *Griechische Denker*, i. p. 433 (2nd ed.), suggests that the words Ἡρακλείδης ὁ Ποντικός have been interpolated by some well-informed reader.

[2] By Schiaparelli in his memoir, "Origine del sistema planetario eliocentrico presso i Greci" (*Mem. del R. Istituto Lombardo*, Vol. xviii. Fasc. 5, 1898).

[3] This obvious explanation has been thought sufficient by Martin (*Mém. de l'Acad. des Inscriptions*, T. xxx. 2ᵉ Partie, p. 34), and Bergk, *Fünf Abhandlungen zur Geschichte der griechischen Philosophie und Astronomie*, Leipzig, 1883, p. 151.

which was of no consequence, so long as they represented the phenomena geometrically, seems to point to the correctness of this view. The vagueness of the expressions renders it rather useless to speculate about what kind of motion Herakleides may have been thinking of[1]. He may merely have thought of some sort of libration or oscillation along a straight line, whereby the distance of the earth from the sun became variable, or he may more likely have thought that the rotation of the earth was not quite uniform, so that the length of the day was not quite the same during the different seasons. No doubt the latter hypothesis would have been contrary to the principle of uniformity of motion, which the Greeks always upheld; still a Greek philosopher may in passing have thrown out a hint of this kind—for it was nothing more than a hint.

Not content with the simple and plain interpretation of the words of this passage, Schiaparelli has endeavoured to found on them a claim for Herakleides as a precursor of Copernicus. He believes the expression ἡ περὶ τὸν ἥλιον ἀνωμαλία to be the same as what Ptolemy afterwards called ἡ πρὸς or ἡ παρὰ τὸν ἥλιον ἀνωμαλία, that is, the irregularities in the motion of the *planets*, of which the period is the synodic revolution, or the time between two successive oppositions to the sun, which therefore show a dependence on the position of the planet relatively to the sun (πρὸς τὸν ἥλιον), and which we now know to be caused by the earth's motion round the sun. Schiaparelli therefore concludes that already at the time of Alexander the Great the idea had been suggested in Greece, that the anomalies of planetary motions might be explained by letting the earth be in motion. But this seems to be an interpretation of the words of Herakleides which does too much honour to that philosopher. In the paragraph in question περὶ can only refer

---

[1] Martin (l. c. p. 35) maintains that Herakleides must have thought of an annual motion of the earth in a small circle, but this would only account for the solar anomaly and not for those of the planets, so that it could only have been as an example of a hypothesis that Herakleides brought it forward. Besides (as remarked by Schiaparelli, p. 91), the period would have had to be six and not twelve months, and then summer and winter would have been equal in length, and spring and autumn likewise. But Martin is all the same right in holding that Herakleides was merely giving an *example* of an hypothesis.

to an anomaly in the motion of the sun itself, and though no
doubt a transcriber might change παρὰ into περὶ, how can we
account for the fact that the authorities to whom we owe our
knowledge of the doctrine of Herakleides concerning the rotation of the earth, do *not* credit him with having taught any
progressive motion of the earth? Simplicius, as we have seen,
says that Herakleides assumes the earth to be in the middle,
and in another place that he lets it move in a circle round
its own centre, while Aëtius most particularly states that
Herakleides lets the earth move *not progressively* (οὐ μήν γε
μεταβατικῶς) but in a turning manner[1]. If Herakleides had
really taught the orbital motion of the earth as an alternative
to the intricate system of Eudoxus, the doxographers could
hardly have been ignorant of it, nor Simplicius, nor Chalcidius
(Adrastus) when describing his theory of Mercury and Venus,
nor the writers who mention Aristarchus as having let the
earth be in motion. There would really be no way of accounting for this conspiracy of silence. For if Herakleides had
seriously believed that he had found the true explanation of
the complicated motions of the planets, he would not have been
afraid to publish his theory, judging by all we know of him.
No doubt it is very tempting to assume that the motion of the
earth, which was taught by Aristarchus in the following century,
had already fifty or sixty years earlier been known to the man
who had made the first step in the right direction by discovering
the heliocentric motion of Mercury and Venus. But to build
this assumption on a vague and probably corrupted passage,
and on the supposition that παρὰ and περὶ are equivalent, or
that παρὰ has been changed into περὶ, and to do this in the
face not only of the absolute silence of all writers, but of the
directly opposing testimony of Aëtius, seems too hazardous.

Before we consider the question what inducement there
might be to abandon the theory of Eudoxus for the theory of
the earth's motion round the sun, we shall set forth the evidence
we possess, showing that this was actually done. The man

---

[1] Compare Simplicius, p. 541 (already quoted), where those who gave the
earth a progressive motion (μεταβατικὴν κίνησιν) are first mentioned, and then
Herakleides who gave it a rotating one only.

who adopted this novel way of "saving the phenomena" was
Aristarchus of Samos, a pupil of Strato (called ὁ φυσικός) the
disciple and successor of Theophrastus. He must have flourished
about the year 281 B.C., as he is said by Ptolemy to have observed the summer solstice of that year[1]. The only book of his
which has been preserved is a treatise "On the dimensions and
distances of the sun and moon[2]," in which we find the results
of the first serious attempt to determine these quantities by
observation. He observed the angular distance between the
sun and the moon at the time when the latter is half illuminated
(when the angle at the moon in the triangle earth-moon-sun is
a right angle) and found it equal to twenty-nine thirtieths of
a right angle or 87°. From this he deduced the result that the
distance of the sun was between eighteen and twenty times as
great as the distance of the moon. Although this result is
exceedingly erroneous[3], we see at any rate that Aristarchus was
more than a mere speculative philosopher, but that he must
have been an observer as well as a mathematician. This treatise
does not contain the slightest allusion to any hypothesis on the
planetary system, so that we have to depend on the statements
of subsequent writers when we endeavour to give Aristarchus
his proper place in the history of cosmical systems. The principal
authority is Archimedes (287—212 B.C.), a younger contemporary of Aristarchus, who, in the curious book (ψαμμίτης) in
which he attempts to find a superior limit to the number of
grains of sand which would fill the universe, incidentally gives
the following account of the ideas of Aristarchus about the
universe[4].

"You know that according to most astronomers the world
(κόσμος) is the sphere, of which the centre is the centre of the
earth, and whose radius is a line from the centre of the earth to

---

[1] Probably in Alexandria (*Almag.* III. 1, p. 206, Heiberg).

[2] "Traité d'Aristarque de Samos sur les grandeurs et les distances du Soleil
et de la Lune, traduit en français par le Comte de Fortia d'Urban." Paris, 1823.
The Greek text was first published by Wallis (Oxford, 1688); there is a modern
edition by E. Nizze, Stralsund, 1856, 4°, 20 pp.

[3] Because the method, though theoretically correct, is not practical, as the
moment when the moon is half illuminated cannot be determined accurately.
The angle of "dichotomy" is in reality 89° 50′ instead of 87°.

[4] Arenarius 4–6 (Heiberg, *Quæstiones Archimedeæ*, Hafniæ, 1879, p. 172).

the centre of the sun. But Aristarchus of Samos has published in outline certain hypotheses[1], from which it follows that the world is many times larger than that. For he supposes ($\dot{v}\pi o \tau \iota \theta \acute{e} \tau a \iota$) that the fixed stars and the sun are immovable, but that the earth is carried round the sun in a circle which is in the middle of the course[2]; but the sphere of the fixed stars, lying with the sun round the same centre, is of such a size that the circle, in which he supposes the earth to move, has the same ratio to the distance of the fixed stars as the centre of the sphere has to the surface. But this is evidently impossible, for as the centre of the sphere has no magnitude, it follows that it has no ratio to the surface. It is therefore to be supposed that Aristarchus meant that as we consider the earth as the centre of the world, then the earth has the same ratio to that which we call the world, as the sphere in which is the circle, described by the earth according to him, has to the sphere of the fixed stars."

In this interesting and important passage we see that Archimedes first defines "the world" as the sphere of which the orbit of the sun is a great circle. Obviously this does not imply that there is nothing outside this orbit, but it either means that Mars, Jupiter, Saturn, and the fixed stars, being all situated at distances as to which no estimates could be made, were assumed to be immediately outside the orbit of the sun; or it is an allusion to the Pythagorean division of the universe into three regions, the Olympos, the Kosmos, and the Uranos,

---

[1] $\dot{v}\pi o\theta \epsilon \sigma \iota \omega v \ \tau \iota v \hat{\omega} v \ \dot{\epsilon} \xi \acute{\epsilon} \delta \omega \kappa \epsilon v \ \gamma \rho a \phi \acute{a}s$. Bergk (*Fünf Abh.* p. 160) explains $\gamma \rho a \phi \acute{a}s$ as "sketched in outline," not worked out or proved.

[2] $\tau \grave{a} v \ \delta \grave{e} \ \gamma \hat{a} v \ \pi \epsilon \rho \iota \phi \epsilon \rho \acute{e} \sigma \theta a \iota \ \pi \epsilon \rho \grave{\iota} \ \tau \grave{o} v \ \ddot{a} \lambda \iota o v \ \kappa a \tau \grave{a} \ \kappa \acute{v} \kappa \lambda o v \ \pi \epsilon \rho \iota \phi \acute{e} \rho \epsilon \iota a v, \ \ddot{o}s \ \dot{\epsilon} \sigma \tau \iota v \ \dot{\epsilon} v \ \mu \acute{e} \sigma \omega \ \tau \hat{\omega} \ \delta \rho \acute{o} \mu \omega \ \kappa \epsilon \acute{\iota} \mu \epsilon v o s$. Barrow translates: Terram ipsam circumferri circa circumferentiam circuli qui est in medio cursu constitutus. The circle which we now call the ecliptic was in those days called the middle circle of the zodiac, the latter being a broad belt inside which the planets moved, and Aristarchus seems to have meant that the earth (and not the sun) moved in this circle in the middle of the belt. This interpretation is not mentioned by Bergk (p. 162) who says: "Die Worte $\ddot{o}s \ \dot{\epsilon} \sigma \tau \iota v$...bezieht man auf $\ddot{a} \lambda \iota o s$, und eine andere Beziehung ist nicht möglich; dann befremdet aber die ungewöhnliche Stellung des Relativsatzes." He therefore proposes to read $o\dot{v} \rho a v \hat{\omega}$ for $\delta \rho \acute{o} \mu \omega$, the circle lying in the middle of the heavens. If $\ddot{o}s$ should refer to the sun, it would show that Aristarchus assumed the sun to be in the centre of the earth's orbit and ignored the difference in length of the four seasons.

the Kosmos being the region of uniform and regular motions. This, then, is the sphere whose capacity of holding grains of sand Archimedes is going to consider, and this leads him to refer to an hypothesis proposed by Aristarchus, that the sun is the centre of the universe. He does not attempt to argue for or against this hypothesis, he merely objects to the unmathe-matical idea of there being a certain ratio between a point, which has no magnitude, and the surface of a sphere. Of course the meaning of Aristarchus is clear enough, that if we suppose the earth to move round the sun in a large orbit, the distance of the fixed stars must be immensely great as compared with that of the sun, as our motion round the latter would otherwise produce apparent displacements among the stars, if they are at different distances from the centre of the world, or at any rate, if they are on the surface of a sphere, make the stars in the neighbourhood of the ecliptic appear to close up or spread out according as the earth is at the part of its orbit farthest from them or nearest to them.

This is indeed a very startling hypothesis to meet with so far back as the third century before our era, and our regret is great that Archimedes did not tell us something more about it. It would almost seem that there was nothing more to say; that Aristarchus had merely thrown out this suggestion or hypothesis without devoting a book or essay to its discussion, and the fact, that his book on the distance of the sun does not contain anything on the subject, tends to confirm this im-pression. We possess only two other very brief references to the hypothesis by other writers.

The first of these is in Plutarch's book *On the face in the disc of the Moon* (§ 6). One of the persons in the dialogue, being called to account for turning the world upside down, says that he is quite content so long as he is not accused of impiety, " like as Kleanthes held that Aristarchus of Samos ought to be accused of impiety for moving the hearth of the world (ὡς κινοῦντα τοῦ κόσμου τὴν ἑστίαν), as the man in order to save the phenomena supposed (ὑποτιθέμενος) that the heavens stand still and the earth moves in an oblique circle at the same time as it turns round its axis."

This is as clear as possible, and shows that the hypothesis of Aristarchus included the rotation of the earth, as might be expected. The other reference is by the doxographers and occurs in the paragraph on the cause of solar eclipses[1]: " Aristarchus places the sun among the fixed stars but lets the earth move along the solar circle, and [says] that it[2] becomes overshadowed according to its inclination." Galenus does not say anything about the motion of the earth but has merely: "Aristarchus [said] that the disc of the sun is obscured by the earth." Evidently the text has in the course of time become greatly corrupted so far as the cause of eclipses is concerned[3], but the passage referring to the motion of the earth in the ecliptic is distinct enough.

We must therefore accept it as an historical fact, that Aristarchus proposed as a way of "saving the phenomena" that the earth performed an annual motion round the sun[4].

The hypothesis does not appear to have attracted much attention, since it is only alluded to in these three passages, while his doctrine of the daily rotation of the earth probably appeared less fanciful and therefore was taken more notice of. Thus an anonymous scholiast to Aristotle[5] states that " it is the opinion of Aristarchus and his followers that the stars and the heavens stand still and that the earth is moved from east to west and reversely (ἀνάπαλιν)." He meant well, no doubt, that scholiast, but his ideas got mixed somehow! The well-known sceptic Sextus Empiricus, who lived in the first half of the third century after Christ, alludes to the diurnal motion of the earth in these words[6]: "Those who do not admit the motion of the world and believe that the earth moves, such as

[1] Aet. II. 24; Stobæus, I. 25 (Diels, p. 355); Galenus, 66 (Diels, p. 627).

[2] In the *Placita*: " the disc."

[3] The meaning is probably that an eclipse takes place or not according as the line from the earth to the moon coincides with that from the earth to the sun or is inclined to it.

[4] We are unable to fix a date for this remarkable proposal, unless we are to assume with Susemihl (*Gesch. d. griech. Litteratur in der Alexandrinerzeit*, I. p. 719) that it must have been after 264/3, in which year Kleanthes became leader of the Stoa.

[5] Brandis, *Scholia*, p. 495 a.   [6] *Adversus mathematicos*, x. 174.

the followers of Aristarchus the mathematician, are not hindered from discerning the time."

We shall hardly be justified in inferring from the words, "the followers of Aristarchus (οἱ περὶ ᾿Αρίσταρχον)," that the proposer of the earth's motion founded a school in which his doctrines were taught. One man there was, however, who took up the doctrine of the diurnal motion (if not that of the annual one), although he cannot have been an immediate disciple of Aristarchus, since he lived more than a hundred years later, about the middle of the second century B.C. Seleukus was a Babylonian, according to Strabo an inhabitant of Seleukia on the Tigris[1], but we know very little about him. Plutarch associates him with Aristarchus in the eighth of his Platonic Questions, in which he considers the question what Plato's meaning was in the disputed passage in the *Timæus*. He asks, whether Plato held that the earth is fixed round its axis and rotates with it, "as Aristarchus and Seleukus have afterwards shown, the one supposing it only (ὑποτιθέμενος μόνον), but Seleukus affirming it as true (ἀποφαινόμενος)[2]." We know from Strabo, that Seleukus was an observer of the tides, and that he also had his own theory as to their origin appears from the following passage of the doxographers[3].

[1] Strabo, p. 739, speaking of the Chaldean astrologers, mentions "Seleukus of Seleukia" as one of them. On p. 174 he tells us (on the authority of Posidonius) that a certain dependence of the tides on the sign in which the moon was situated had been observed by "Seleukus of the Erythrean sea." The latter included the Persian Gulf, so that no doubt the same man is referred to. Seleukus was the originator of the idea, adopted by Hipparchus (Strabo, pp. 5–6), that the Indian Ocean was surrounded by land on all sides. As he was quoted by Hipparchus and wrote against Krates of Mallus, he must have lived in the middle of the second century B.C. About him see a little monograph: "Der Chaldäer Seleukos. Eine kritische Untersuchung aus der Geschichte der Geographie, von Sophus Ruge." Dresden, 1865, 23 pp. 8°.

[2] Immediately after this comes the passage already quoted in Chapter III. that Plato in his old age is said to have regretted that he had placed the earth in the centre of the world. This, however, does not prove that Plutarch here speaks of Aristarchus and Seleukus as having taught the orbital motion of the earth, and as nobody else attributes this doctrine to Seleukus, we shall hardly be justified in concluding from Plutarch that he held it, as Schiaparelli does (*I Precursori*, p. 36).

[3] Aet. III. 17; Stob. I. 38, Diels, p. 383.

"Seleukus the mathematician, writing against Krates[1], and himself letting the earth be in motion, says that the revolution of the moon is opposed to its (i.e. the earth's) rotation, but the air between the two bodies being drawn forward falls upon the Atlantic Ocean, and the sea is disturbed in proportion." Evidently Seleukus supposed the atmosphere to reach to the moon if not further; from another passage we learn that he considered the universe to be infinite[2].

These are the only allusions found in classical writings to the last philosophers of Greek antiquity who taught that the earth had any motion, except an allusion to the doctrine of the earth's rotation by Seneca, who however does not mention any names in connection with it[3]. But scanty though they are, these allusions leave no doubt that Aristarchus taught the annual motion of the earth round the sun, and that both he and Seleukus taught the diurnal rotation of the earth. When we consider that seventeen hundred years were to elapse before the orbital motion of the earth was again taught by anybody, we cannot help wondering how Aristarchus can have been led to so daring and lofty an idea.

After Kalippus and Aristotle the homocentric system never received any further development or improvement, simply because, as Simplicius tells us[4], the great changes in the brightness of the planets, especially Venus and Mars, rendered the idea of each planet being always at the same distance from the earth utterly untenable. According to him, Autolykus of Pitane (about 300 B.C.), writing against Aristotherus (teacher of the poet Aratus and probably an adherent of Kalippus[5]),

---

[1] Krates had also written on the cause of the tides, as stated by Stobæus a few lines above.

[2] Aet. II. 1; Stob. I. 21, Diels, p. 328. Stobæus has: "Seleukus the Erythrean and Herakleides of Pontus say that the world is infinite."

[3] Seneca, *Quæst. Nat.* VII. 2: "Illo quoque pertinebit hoc exenssisse, ut sciamus, utrum mundus terra stante circumeat, an mundo stante terra vertatur. Fuerunt enim qui dicerent, nos esse, quos rerum natura nescientes ferat, nec cœli motu fieri ortus et occasus, sed ipsos oriri et occidere. Digna res est contemplatione, ut sciamus in quo rerum statu simus: pigerrimam sortiti an velocissimam sedem: circa nos Deus omnia, an nos agat."

[4] *De Cœlo*, p. 504 (Heib.).

[5] Susemihl, l. c. I. pp. 286 and 703.

tried in vain to explain this. "For the star called after Aphrodite as well as that called after Ares appears in the midst of the retrograde movement many times brighter, so that that of Aphrodite on moonless nights causes bodies to throw shadows." He adds that it is also easy to see that the moon does not always keep at the same distance from us, as observations show that sometimes a disc of eleven inches and sometimes one of twelve inches, held at the same distance from the eye, is required to hide it. Simplicius further alludes to the fact that central eclipses are sometimes total and sometimes annular, which confirms the observation of the varying distance of the moon; and although he here manifestly follows Sosigenes, who lived towards the end of the second century A.D.[1], still we cannot doubt that the generation after Aristotle found the great variation of distances a very awkward fact to deal with, since Simplicius also refers to Polemarchus of Kyzikus as having known it, though the difficulty is said not to have troubled him much, as he all the same adhered to the homocentric spheres. Simplicius even adds[2] that Aristotle himself in his "Physical Problems" was not perfectly satisfied with the hypotheses of astronomers on account of the variation of brightness; but as this book is lost we have unfortunately no way of testing this statement, which certainly is not confirmed by anything in the writings of Aristotle still existing[3].

The necessity of grappling with this difficulty led gradually to the complete abandonment of the system of Eudoxus, although, as we shall see, the word "sphere" continued from time to time to make its appearance. There can be no doubt that it was this very difficulty which gave rise to new systems; in fact Simplicius states this distinctly a page further on[4],

---

[1] Comp. Proklus, *Hypotyposes*, ed. Halma, p. 111, where he quotes the book of Sosigenes Περὶ τῶν ἀνελιττουσῶν, using almost the same words as Simplicius about solar eclipses.

[2] p. 505.

[3] Though it is evident from the way in which Aristotle expresses himself, *De Cœlo*, II. 12, p. 292 a, about the great distances of the stars and the few starting points we have, that he did not consider the problems of the planets finally solved.

[4] p. 507, l. 14–20.

where he says that "astronomers therefore introduced the hypothesis of excentric circles and epicycles in the place of the homocentric spheres, if the hypothesis of excentrics was not already invented by the Pythagoreans, as some people say, among whom Nikomachus and on his authority Iamblichus." These two late authorities[1] are not to be relied on, since the Neo-Pythagorean school was always ready to attribute every-thing of value to the early Pythagoreans, and their statement has probably been copied by Proklus, who in his account of Ptolemy's planetary theory[2] says that "History teaches us that by the Pythagoreans have the excentrics and epicycles been invented as a sufficient means and the simplest of all." Geminus evidently means the same thing when he says that the Pythagoreans first assumed circular and uniform motion and put this question: how under this assumption the phe-nomena could be accounted for[3]? But Geminus lived in the first century B.C., at the very time when the first attempts were made to palm off as handed down from Pythagoras a system of philosophy which shows strong signs of having been influenced by much later, especially Stoic doctrines[4], and his testimony cannot therefore be depended on. There were no doubt at the time of Aristotle some philosophers who still tried to explain the heavenly phenomena after the manner of the Pythagoreans[5], but the Pythagorean school seems to have be-come finally extinct about the time of the death of Aristotle, and nothing indicates that its last members were distinguished as mathematicians. We must therefore reject as utterly un-supported by reliable testimony the statement of the later worshippers of Pythagoras, that the theory of excentric motion was started at the time when his school was at the point of extinction.

The statement of Nikomachus, that the excentric circles

---

[1] Nikomachus lived about A.D. 100, and Iamblichus early in the fourth century.

[2] *Hypotyp.* ed. Halma, pp. 70–71.

[3] Gemini *Elementa Astron.* ed. Manitius, p. 11.

[4] Alexander Polyhistor in Diog. Laert. VIII. 24 sq.; compare Zeller, *Ph. d. Gr.* III. b, p. 88 (3rd ed.).

[5] *De Cælo*, p. 293 a; comp. above Chapter III. p. 83.

were invented before the epicycles seems, however, correct (as
we shall see in the next chapter), and combined with that
of Sosigenes about the great variation of distances, it indicates
the manner in which the way was cleared for the bold conception of Aristarchus[1]. It was remarked that Mars was always
brightest when it culminated at midnight and therefore was
in opposition to the sun, while the planet gradually became
fainter and fainter as it approached the sun. To the Greek
mind a heavenly body could not possibly move in any other
curve than a circle, and it followed therefore that Mars must
move on a circle which was excentric to the earth, and furthermore that the centre of this circle lay somewhere on the
straight line through the earth and the sun, since Mars
evidently was nearest to the earth when opposite the sun.
Obviously this was not a fixed line but one which turned
round one of its extremities in a year; consequently the centre
of the orbit of Mars described a circle round the earth in a
year. This explained the fact that the oppositions of Mars
do not take place in one particular point of the zodiac, but
may occur at any point of it; and all that was necessary was
to assume that Mars moved round the excentric circle in a
period equal to its synodic revolution (i.e. the interval between
two successive oppositions or two years and fifty days), while
the centre of the excentric moved round the sun in a year.
By a suitable selection of the ratio of the radii of the two
circles it became possible completely to account for the length
of the retrograde motion of Mars at opposition, a problem
which had baffled the ingenuity of Eudoxus and Kalippus.

The complete planetary system according to the excentric
circle theory was therefore as follows. In the centre of the
universe the earth, round which moved the moon in 27 days,
and the sun in a year, probably in concentric circles. Mercury
and Venus moved on circles, the centres of which were always

---

[1] The importance of the movable excentrics in the development of the
system of Aristarchus has been especially insisted on by Schiaparelli; see his
"Vorläüfer des Copernicus im Alterthum," *Altpreuss. Monatsschrift*, XIII. p. 113
(not in the Italian original), and especially his second paper, " Origine del
sistema eliocentrico." Also by Tannery, *Recherches sur l'astronomie ancienne*,
p. 256 ff.

on the straight line from the earth to the sun[1], so that the earth was always *outside* these circles, for which reason the two planets are always within a certain limited angular distance of the sun, from which the ratio of the radius of the excentric to the distance of its centre from the earth could easily be determined for either planet. Similarly the three outer planets moved on excentric circles, the centres of which lay somewhere on the line from the earth to the sun, but these circles were so large as always to surround both the sun and the earth.

This system of "movable excentrics" was known to and employed by Apollonius in the middle of the third century B.C., and as he is not by any writer said to have invented it, it was most probably already known to Aristarchus, who was only about thirty or forty years senior to Apollonius. The part of the system which refers to Mercury and Venus had indeed, as we have seen, been found by Herakleides, who had even gone further and let the centres of the orbits of these two planets coincide with the sun. But is it likely that he or anyone else went another step further and let the centres of the orbits of Mars, Jupiter, and Saturn fall in the sun? This elegant system, which eighteen hundred years later was actually proposed by Tycho Brahe, is not mentioned or hinted at by any writer of antiquity. It is obvious that it presents the same advantages as the system of movable excentrics, of which it is a special and simple case, all the five planetary orbit-centres falling in one point. Very probably Aristarchus may first have been led from the general idea of movable excentrics to this simple case of it, and then his mathematical mind cannot have failed to be struck by the fact, that the phenomena would be precisely the same if he went one step further and made the earth go round the sun instead of the sun round the earth, leaving everything else unaltered. In fact there is no other way in which he can have been led to the heliocentric system, unless the epicyclic theory had already at that time been fully developed. But there is absolutely nothing to indicate that any other thinker either before or after Aristarchus has proposed this "Tychonic

---

[1] This is the first of the two suggestions of Theon referred to above, p. 127. There are many allusions to the movable excentrics in Theon's book.

system," and it is difficult to see why this system if once proposed should have been needlessly complicated by substi- tuting excentrics with different centres for the beautifully simple Tychonic idea of all the orbits having one centre, the sun. All the same, Schiaparelli insists on the necessity of assuming that this system was a stepping-stone from the old geocentric to the heliocentric (or Copernican) system of Aristarchus, not only in the mind of Aristarchus, but separately deduced long before his time. Schiaparelli urges especially, that the idea of any celestial body moving in a circle round a mere mathematical point must at first have been repugnant, and that the conception of motion round a material body must have preceded it. But he would seem to have overlooked the fact that in the first place it was in the eyes of the Greeks perfectly natural for a celestial body to move in a circle and for a terrestrial one to move in a straight line up and down; and secondly, that no philosophical system previous to the time of Aristarchus had supposed a powerful influence to emanate from the centre, with the one exception of the system of Philolaus. The central fire had been created first, and so to say held the world together, although it may be doubted whether the statements of Stobæus and Simplicius to this effect are not strongly influenced by notions of much later date, as the phraseology seems to show[1]. In the system of Plato the soul of the world permeated the whole universe and there was no thought of the centre exerting a ruling power over the celestial motions. According to Aristotle all influences in a sphere or a circle emanated from the circumference and proceeded towards the centre and not *vice versâ*[2], and we shall see in the next chapter that the Stoics practically were of the same opinion. But let us for the sake of argument accept it

---

[1] Stobæus (Diels, p. 332 b) even uses the Stoic expression ἡγεμονικὸν of the central fire, and Simplicius, p. 512, calls it τὴν δημιουγικὴν δύναμιν. Zeller, I. pp. 412 and 416, from a comparison of all references comes, however, to the result that the world owed its origin to the central fire and continued to be ruled by it.

[2] Compare above, Chapter v. pp. 110 and 115. See also Aristotle's *Physics*, p. 267 b, where he shows that the swiftest motion belongs to that which is nearest to the moving cause.

as an indisputable fact, that the adherents of Philolaus had
been in the habit of thinking of the power of the central fire
"not only mathematically and mechanically, but also dyna-
mically" (as Zeller says), and that this might still influence
the mathematicians who designed the system of excentric
circles. The centres of these travelled round a very material
body, the earth, and one would think that this circumstance
would satisfy even a mind filled with the then obsolete
Pythagorean notions as to the seat of power, though it was
only indirectly that the planets moved round the body in the
centre. The conception of motion round a mathematical point
(which in its turn moved round the earth) need therefore not
have had any difficulties to contend with. No doubt a great
mathematician like Apollonius must have noticed how simple
the system of movable excentrics would become by letting all
the centres fall in the sun, but for some reason or other he
must have considered this hypothesis inadmissible; at any rate
there is no sign of his having proposed it. A system in which
the planets moved round the sun would have been just the sort
of thing which later writers would have mentioned among the
"opinions of philosophers"; and in the total absence of any
mention of the Tychonic system by any writer we can only
conclude, that it was never proposed as a way of "saving the
phenomena," though Aristarchus may have been first led to
it, and then immediately afterwards may have been struck
by the still greater simplicity and beauty of the heliocentric
system, which alone he therefore considered worth proposing
publicly.

The very few and scanty references to the system of
Aristarchus by classical authors prove that it can never have
been favourably received. Although the time was long past,
when a philosopher might be judicially called to account for
proposing startling astronomical theories, as Anaxagoras had
once been, and though the accusation of "impiety" (if it was
really brought forward) can hardly have done the theory much
harm, still the novel proposal of Aristarchus was too much
opposed to the general views about the universe of Platonists
and Aristoteleans, as well as to those of the dominant school

of the Stoics, to obtain a hearing. Possibly it seemed to
superficial observers a mere reminiscence of the old and dis-
credited Philolaic system, and we know from Theon[1] that the
old idea (quoted by Plato) of the earth being "the house of the
gods" had not been forgotten, while those who taught that
it moved were by Derkyllides declared worthy of execration
as subverting the principles of divination, an Eastern pseudo-
science which now had commenced to take a firm hold of the
human mind in Europe and to develop into an important
branch of knowledge.

But beyond a doubt the principal reason why the heliocentric idea fell perfectly flat, was the rapid rise of practical
astronomy, which had commenced from the time when the
Alexandrian Museum became a centre of learning in the
Hellenistic world. Aristarchus had no other phenomena to
"save" except the stationary points and retrograde motions
of the planets as well as their change of brilliancy; he may
even have neglected the inequality of the sun's apparent
motion originally discovered by Euktemon and recognized by
Kalippus.   But when similar and much more marked inequalities began to be perceived in the motions of the other
planets, the hopelessness of trying to account for them by the
beautifully simple idea of Aristarchus must have given the
deathblow to his system, which thereby even among mathematicians lost its only claim to acceptance, that of being able
to "save the phenomena." Most likely, as we have already
said, these new inequalities had already more or less dimly
commenced to make themselves felt in the days of Apollonius
(about B.C. 230), and in that case we can understand why he
did not feel disposed to simplify the system of movable excentrics by gathering the reins of all the unruly planetary
steeds into one mighty hand, that of Helios.

[1] *Theon. Smyrn.* ed. Martin, p. 328.

CHAPTER VII.

THE THEORY OF EPICYCLES.

ARISTARCHUS is the last prominent philosopher or astro-nomer of the Greek world who seriously attempted to find the physically true system of the world. After him we find various ingenious mathematical theories which represented more or less closely the observed movements of the planets, but whose authors by degrees came to look on these combinations of circular motion as a mere means of computing the position of each planet at any moment, without insisting on the actual physical truth of the system. Three names stand out clearly among the astronomers of the next four hundred years as the principal, or perhaps we should say the only, promoters of theo-retical astronomy: Apollonius (B.C. 230), Hipparchus (B.C. 130), and Ptolemy (A.D. 140).

These three great men were naturally not the only workers in astronomy during this long period, in the course of which Greek intellectual life spread its light over almost the whole known world; but in the study of the planetary motions they seem to have had the field to themselves. It is, however, a matter of great difficulty to trace the gradual development of the theories of excentric and epicyclic motion. We have men-tioned that Eudemus, one of the immediate disciples of Aristotle, wrote a history of astronomy, and though this precious work is lost, other writers made good use of it, so that we are able to form a tolerably complete idea of the progress of physical astronomy from Thales to Aristotle. But nobody continued the work of Eudemus; Apollonius does not seem to have left any astronomical writings; from the hand of Hipparchus we possess very little indeed, and though the great work of Ptolemy

covers the whole range of astronomy, he rarely gives detailed
historical information about the work of his predecessors.
Fortunately we can to some extent supplement the little he
tells us by means of the works of a few writers of elementary,
we might almost say popular books from the period in question.
The *Elements of Astronomy* (Εἰσαγωγὴ εἰς τὰ φαινόμενα) by
Geminus, probably a native of Rhodes, were written in the first
half of the first century B.C. and deal chiefly with spherical
astronomy[1]. The *Theory of the Heavenly Bodies* (Κυκλικῆς
θεωρίας τῶν μετεώρων βίβλια δύο) of Kleomedes, who seems to
have lived about the same time or possibly somewhat later, is
like the book of Geminus mainly founded on the work of the
Stoic philosopher Posidonius[2]. It tells us almost nothing about
the planets, except that it gives their periods of synodic revolution and mentions how far they move from the ecliptic; but
(as we shall see in the next chapter) it contains a precious
account of the determinations of the size of the earth by
Eratosthenes and Posidonius. But the treatise on astronomy
by Theon of Smyrna, so often quoted in the foregoing pages,
forms a most valuable supplement to the Syntaxis of Ptolemy.
He must have been almost a contemporary of Ptolemy, or
possibly he lived slightly earlier, about A.D. 100; but we know
next to nothing about him except that he is the author of a
book which has come down to us and is entitled *Exposition of
the mathematical subjects which are useful for the study of
Plato*, comprising three treatises on arithmetic, music, and
astronomy. The part dealing with astronomy was quite unknown till it was published by Martin in 1849[3], and it had not
even been suspected that the commentary of Chalcidius to the
Timæus had to a great extent been copied from Theon's book.
Theon chiefly followed the peripatetic philosopher Adrastus and

---

[1] About Geminus and his writings see *Gemini Elementa Astronomiæ*, ed. C.
Manutius. Leipzig, 1898, pp. 237–252. G. says that he had dealt specially
with planetary theories in another book, but this is lost.

[2] The latest edition is by Ziegler. Leipzig, 1891.

[3] *Theonis Smyrnæi liber de astronomia*, ed. Th. H. Martin. Paris, 1849.
8°. The introduction and notes are very valuable. There is a later edition of
the whole work with French translation by J. Dupuis (*Exposition des connais-
sances math.* Paris, 1892), and another of the Greek text only by Hiller.
Leipzig, 1878.

to a lesser extent Derkyllides the Platonist, both of whom were
identified with the revival of philosophical studies which commenced soon after the beginning of our era and eventually
gave rise to the school known as the Neo-Platonic. Like
Geminus, Adrastus seems to have been somewhat behind his
time, and these two writers rather represent the state of
science just before the age of Hipparchus than that of their
own days.

The philosophers of Greece up to the time of Alexander
the Great had meditated on the construction of the universe
without having at their command many facts accurately determined by lengthy series of observations, and their success in
accounting for the phenomena had naturally been a very
limited one. For the purpose of regulating the calendar it
became necessary to follow more closely the courses of the
sun and moon, and this in turn influenced other branches of
astronomy. At the Museum of Alexandria, founded and continued by the munificence of the Ptolemaic dynasty, a school of
observers now arose, who determined the positions of stars and
planets by graduated instruments, and thereby made it possible
for Hipparchus and Ptolemy to make great discoveries in
astronomy, for which the rapid development of pure mathematics supplied an equally necessary foundation. Astronomy
now became a science; vague doctrines and generalisations
were abandoned, while mathematical reasoning founded on
observations took their place. That this change occurred about
the middle of the third century was a circumstance not unconnected with the simultaneous rise of the school of Stoic
philosophy, which may be considered as a natural reaction
against the idealism of Plato and the dogmatic systematising
of Aristotle. Both in abstract philosophy and in science the
wish to get on more solid ground now became universal, and
no science benefited more by this realistic tendency than
astronomy.

Among the great mathematicians of antiquity one of the
foremost places is held by Apollonius of Perge, whose name is
so closely associated with the theory of conic sections. His
life spans over the second half of the third century, most of

which he spent at Alexandria. He seems to have written
nothing except on pure mathematics, or, if he did write on any
astronomical subjects, such writings must have been lost at an
early date, as his astronomical work appears only to have been
known to Ptolemy through the medium of Hipparchus. Still
Apollonius played an important part in the development of
planetary theory, as we learn from the beginning of the twelfth
book of the Syntaxis of Ptolemy, where the theory of the
retrograde motion of the planets is introduced in the following
manner[1].

"Various mathematicians, and among them Apollonius of
Perge, have investigated one of the two inequalities, that one
namely which depends on the position relative to the sun; if
they represented it by means of the epicyclic theory, they made
the epicycle move in longitude in the direct order of the signs
[i.e. from west to east] in a circle concentric with the zodiac,
while the star moves in the epicycle with a velocity equal to
that of the anomaly, and with a direct motion on that part of
the epicycle which is farthest from the earth. If we then draw
a line from our eye which intersects the epicycle in such a

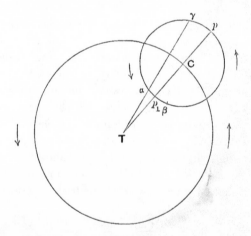

manner that half the chord is to the part of the line from the
eye to the epicycle as the velocity of the centre of the epicycle

[1] Ed. Heiberg, II. p. 450.

is to the velocity of the star on the epicycle, then the nearest
point of intersection of that line with the epicycle divides the
part of the apparent motion which is direct from that which is
retrograde, so that the planet will be stationary when it reaches
that point."

The figure will make this clearer. The earth is at T, the
centre of the circle afterwards known as the deferent, on which
the centre C of the epicycle moves round the earth in the
period in which the planet travels round the entire heavens, its
sidereal revolution (for Mars 687 days, for Jupiter 11·9 years,
for Saturn 29·5 years), while the planet in one sidereal year
travels round the epicycle in the same direction, if we reckon
the period in the modern manner from a radius moving so as to
remain parallel to its original direction. The motion on the
deferent is called the motion in longitude, that on the epicycle
the motion in anomaly[1]. When the planet reaches the point α,
determined by $\frac{1}{2}\alpha\gamma$ being to $T\alpha$ in the ratio of the two linear
velocities, the two angular velocities seen from T will for a
short time be equal and opposite, in consequence of which the
planet becomes stationary; after which its motion becomes
retrograde as seen from T. This lasts until the planet reaches
β, after which the planet again becomes stationary for a while
and then resumes its ordinary direct motion. It is evidently
possible to fix the ratio of the radii of the two circles so that
the observed length of the retrograde arc corresponds exactly
to that given by the theory. When the planet is at p_1, it is
nearest to the earth, and this happens at the moment when the
line CT produced beyond T passes through the sun, that is,
when the planet is in opposition; and when the planet is at p
or farthest from the earth, it is at the same time in conjunction
with the sun, the latter being between T and p_1. For Mercury
and Venus the case is somewhat different; the motion on the
deferent being in the period of a year (the line TC always
pointing towards the sun), while the motion of the planet on
the epicycle (reckoned in the modern way) takes place in what

---

[1] Anomaly means inequality; in the simple excentric circle (without an
epicycle) it means the angle passed over by the radius vector since the passage
through the aphelion.

we call the heliocentric period, 88 days for Mercury and 225 days for Venus. We must, however, not omit to mention that the ancients always reckoned the motion on the epicycle from the point p on the produced radius TC, and for them the period of revolution on the epicycle became therefore for all planets their synodic period, i.e. for the two inner planets the period from one inferior conjunction with the sun to the next one, and for the outer ones the time between two successive oppositions to the sun. The line from the centre of an outer planet's epicycle to the planet is always parallel to the line from the earth to the sun.

It is interesting to notice that it is also possible by means of an epicycle to represent the motion of a body which like the sun and moon moves with a somewhat variable velocity without ever becoming stationary or retrograde. In this case the motion on the epicycle has to be in the opposite direction to that of the motion on the deferent. It is impossible not to be reminded of the singular statement of Plato in the *Timæus*, that Mercury and Venus move in the opposite direction to the sun, and in fact the only way of making sound sense of his statement is to assume that he was acquainted with the epicyclic theory when he wrote the *Timæus* towards the middle of the fourth century B.C., as Theon and Chalcidius assumed that he was. But as Plato nowhere betrays any knowledge of the intricate motions of the planets, this speaks decidedly against his having been acquainted with the elegant mathematical system of epicycles, as does also the encouragement he gave Eudoxus to look out for some new cosmical arrangement. So we must reluctantly give up the tempting idea of acquitting Plato of having made a serious mistake.

After this digression we return to the quotation from Ptolemy. "If (he continues) on the other hand we suppose the anomaly depending on the sun to be produced by means of an excentric, which can only be used for the three planets capable of being at any angular distance from the sun [Mars, Jupiter, Saturn], then the centre of the excentric revolves with a velocity equal to the apparent velocity of the sun round the centre of the zodiac in the direct order of the signs [from west

to east], while the star moves on the excentric in the opposite
direction in a period equal to that of the anomaly; and if we
draw from the centre of the zodiac a chord of the excentric so
that half its length is to the minor segment of the two into
which the observer's eye divides the chord, as the velocity of
the excentric is to that of the star, then the star will, when
arriving at the end of the chord nearest to the earth, become
stationary. We therefore set forth summarily and more con-
veniently the above propositions, using a mixed demonstration
common to both hypotheses, in order to see their resemblance
and agreement."

Ptolemy here for the sake of the historical and mathematical
interest places the two hypotheses in juxtaposition in order to
show how perfectly indifferent it is, whether we employ the
movable excentric or the concentric deferent with an epicycle
to account for the motion of one of the outer planets. The
figure will show this clearly[1], as the apparent motion of the
planet P as seen from the earth T will be exactly the same in
the two cases, provided we make the radii of the excentric and
deferent equal ($cP = TC$) and also those of the concentric and

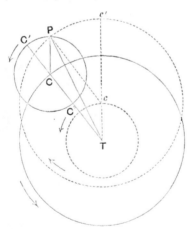

of the epicycle equal ($Tc = CP$). The anomaly is counted from
the point C' or c' in the produced radius TC or Tc (the line

---

[1] The dotted circles represent the movable excentric theory, the others the
deferent and epicycle.

of apsides). But we notice the difference (indicated by the arrows), that the motion on the excentric is retrograde, while that on the epicycle is direct. This is a consequence of the peculiar habit of the ancients, already alluded to, of counting the motion in anomaly from the apogee, that is, from the end of a line (Tc') turning like a material rod round T, instead of, as we should do, from the end of a radius moving parallel to itself, or in other words, from a fixed point on the zodiac. Adopting the apogee as the origin, the mean motion is the difference between the mean heliocentric motion of the planet and the mean motion of the sun. For the three outer planets the latter is greater than the former; the ancients therefore let the motion of the planet be in the opposite direction to that of the centre of the excentric, so that the mean motion is the motion of the sun minus the heliocentric motion of the planet. But in the epicyclic theory the two motions are in the same direction for all the five planets alike.

At the time of Apollonius and for some time afterwards astronomers therefore had their choice between these two hypotheses. It is not difficult to see why the hypothesis of movable excentrics was gradually set aside in favour of the epicycle. The latter had in the first place the undoubted advantage of being much simpler, as the stationary points and retrograde motions of the planets were far more clearly illustrated by it. Secondly, the movable excentrics were only applicable to the three outer planets, while the epicycles had to be used for Mercury and Venus; so that the planetary system became, as it were, split up in two parts requiring different treatment. By adopting the epicycles altogether, on the other hand, the system acquired a simple and homogeneous character which could not but appeal to the mind. We have already, when considering the system of Aristarchus, remarked that the system of movable excentrics might have led mathematicians to notice that there was a special case of it, the simplest of all systems, which made the centres of the three outer planets' excentrics fall in the sun, while the centres of the epicycles of Mercury and Venus likewise coincided with the sun. If the heliocentric system of Aristarchus had been set

forth in a mathematical treatise, the same thing would doubtless
have happened which did happen 1800 years later: Apollonius
would have pointed out that it was possible to avoid offending
old prejudices and yet to have a simple and beautiful system
with all the planetary orbits round the sun, which carried them
round the earth in an annual course. But Aristarchus apparently merely brought forward his theory as a suggestion
without setting forth in a treatise the arguments in its favour;
while mathematicians had probably then already to a great
extent given up the hope of finding the physically true system
of the world and had decided to look for a mathematical theory
which would make it possible to construct tables of the motions
of the planets. If a physically true system was to be thought
of, the only possibility seemed to be some adaptation of the
system of spheres which Aristotle had favoured and which the
Stoics also accepted. Side by side with the new mathematical
development of astronomy went the cosmological speculations
of the philosophical school which now had become the most
prominent one, that of the Stoics; and to writers like Cicero or
Seneca or the doxographic writers the mathematical theories
would appear to have been practically unknown, while they
adopted the opinions of the Stoics about the general constitution
of the world, which on the whole did not differ very much from
those of Aristotle.

The school of the Stoics arose about the end of the fourth
century B.C., and under the successive leadership of its founder
Zeno and of Kleanthes and Chrysippus it became in the course
of a hundred years the leading philosophical school in the
Hellenistic world and maintained this supremacy even later
under the dominion of the Romans, to whose temperament it
was peculiarly suited. Being essentially practical and aiming
solely at acquiring virtue by mental training, the Stoic philosophy concerned itself less with natural science than Aristotle
had done, and with one exception (Posidonius) no prominent
Stoic made a name in the history of science. Still physics
played an important part in their doctrines, owing to their
leading idea of an all-pervading primary substance which is
throughout the world co-extensive with matter and by its

varying degrees of tension causes the different properties of bodies. Kleanthes (about 300—225) held this primary substance to be identical with fire and therefore placed the seat of the world-ruling power in the sun; but this seems to have been rejected by most of the other Stoics, who supposed the ruling power to have its abode in the heavens, the primitive substance being a fiery pneuma or ether, which, though present everywhere in the elements, is in its purity with its tension undiminished only existent in the celestial space[1]. This spiritualized ether is one with the Deity, and the system of the Stoics is therefore a pure pantheism, only Boëthus giving the Deity a local habitation in the sphere of the fixed stars[2].

In the centre of the world the Stoics placed the spherical earth[3], and outside it the planets, each in its own sphere, and farthest out the sphere of the fixed stars[4]. The world is therefore a series of spheres, one inside the other, and it is of finite extent; but outside is a vacuum, if not infinite, at least of sufficient size to allow of the dissolution of the world in the periodically occurring conflagrations in which it is rejuvenated[5]. The Kosmos gives mankind an idea of God by its beauty, by its perfect shape, that of a sphere, by its vastness, the multitude of stars, and the brilliant blue of the heavens[6]. The stars and the

---

[1] Cicero, *Acad. pr.* II. 41, 126; Diog. L. VII. 139; Stobæus, Diels, p. 332 ; Arius Didymus, Diels, p. 471.

[2] Diog. L. VII. 148, but compare ibid. 138, where this seems to be asserted of the Stoics in general. Only one Stoic, Archedemus of Tarsus (in the middle or second half of second century B.C.), placed the ἡγεμονικὸν in the earth (Stob., Diels, p. 332). But according to Simplicius (p. 513, Heib.) quoting Alexander of Aphrodisias, Archedemus agreed with the Pythagoreans in denying that the earth was in the centre of the world. If this is correct, did he believe the central fire to be the seat of the ἡγεμονικὸν? But more likely there is some error in the text of Simplicius as to the name of Archedemus, as this would be too utterly opposed to Stoic ideas in general (Zeller, III. a, p. 46, suggests that Ἀρχέδημος has originally been a badly written [Ἀρίστ]αρχος ὁ Σάμιος). In the pseudo-Aristotelean book Περὶ Κόσμου, written by some eclectic philosopher between B.C. 50 and A.D. 100, the seat of the Deity is placed at the outermost limits of the world, from whence, without moving itself, it produces the manifold motions in the universe.

[3] Achilles, Petav. III. p. 75; Kleomedes, I. 8, ed. Ziegler, p. 75 ff.

[4] Stobæus, Diels, pp. 465–466.

[5] Aet. II. 9, Diels, p. 338.

[6] Aet. I. 6, Diels, p. 293.

sun and the moon are spherical[1], of a fiery nature, but the fire is more or less pure, that of the moon being mixed with earthy matter and air owing to its proximity to the earth. The heavenly bodies are nourished by exhalations from the earth and the ocean (a return to the notions of the earlier philosophers which Aristotle had strongly opposed), and the sun changes the direction of its motion at the solstices in order to be above regions where it can get this nourishment[2]. To the Stoics (at least to the earlier ones) the sun had therefore not any orbital motion, and neither had the planets, which simply moved from east to west round the earth, only not quite as fast as the fixed stars do, and in an oblique direction with sinuous oscillations between certain limits of north and south declination[3]. This was a remarkably great step backward, when we remember how the Pythagoreans, and following them Plato and Aristotle, had agreed in considering the apparent motion of the planets to be made up of the daily rotation of the whole heavens from east to west and the independent motion of the planets at various velocities in inclined orbits from west to east. Without this assumption it is impossible to form any reasonable theory of their motions, and it is no wonder that the Stoics supposed the heavenly bodies to be the highest of rational beings, for mortal man would have found it hard to follow their vagaries, when geometrical explanations were ignored. But the time was past for that kind of hazy talk, and the Stoics remained outside the

---

[1] Aet. II. 14, 22, 27, Diels, pp. 343, 352, 357. Achilles (Petav. III. p. 79) and Aet. II. 14, p. 344, say that Kleanthes alone held the stars (not sun and moon) to be conical, but the moon πιλοειδής, ball-shaped (Stob. p. 467). Perhaps it was not the stars but the world which Kleanthes took to be a cone, comp. Aet. II. 2, p. 329, and Achilles (p. 77), where this opinion is mentioned in connection with some Stoics, while others are said to have taught that it was egg-shaped.

[2] Aet. II. 17, 20, 23, 25, Diels, pp. 346, 349, 353, 356. Pliny (II. 46) says that it is evident that the stars are nourished by terrestrial moisture, because the spots in the moon are earthly dregs drawn up with the moisture and not absorbed.

[3] Aet. II. 16, Diels, p. 345. Only Kleanthes is mentioned here, but II. 23 (p. 353), stating the cause of the solstices, simply mentions "the Stoics." Aet. II. 15, p. 344, says that Xenokrates (Plato's disciple) held that the stars were all on one surface, but that the other Stoics believed some to be farther off than others. Doubtless this refers to the planets and not to the fixed stars.

development of science, Posidonius alone having the courage to prefer mathematical methods to metaphysical arguments.

Thus, the celestial spheres continued to hold their own outside the realm of advanced mathematicians. Theon of Smyrna[1] describes an arrangement of spheres suited to the taste of anyone who might think the motion on imaginary circles too difficult to follow. If in the above figure we describe two circles round T, touching the epicycle at C' and C'', we may assume that these two circles represent two hollow concentric spheres between which a solid sphere, of which the epicycle may be called the equator, can freely revolve. If then the planet is attached at some point in the equator of this solid sphere, its motion round the earth in the centre will be exactly the same as that which we have already described. Therefore it made no difference whatever, whether these spheres were supposed to exist or not, and the adherents of the epicyclic theory doubtless thought to enlist the sympathies both of Aristoteleans and of Stoics by pointing out that it was nothing but a modification of the theory of spheres. The fact, that this notion of solid spheres was accepted by Adrastus, shows that even the later Peripatetics were reluctant to sever themselves from Aristotle in this matter, and in this way the theory of epicycles finally became universally adopted.

In the days of Apollonius, however, the theory of movable excentrics was still able to hold its own among mathematicians, and both it and the rival epicyclic theory were capable of representing the phenomena of planetary motion, as far as they were then known, in a much more satisfactory manner than the theory of homocentric spheres could do. How far Apollonius had succeeded in mastering the theory of the moon's motion is not known, as Ptolemy is silent on the subject; but to some extent he probably had paved the way for the next great astronomer, Hipparchus, who in this as well as in other departments of astronomy advanced science more than any other ancient astronomer before him had done.

Hipparchus was a native of Nicæa in Bithynia, but spent a

---

[1] Theon, ed. Martin, p. 282.

good deal of his life abroad, chiefly at Rhodes, which was still
one of the most flourishing states of the Greek world, renowned
both for the maritime enterprise of her citizens and for the
numerous works of art which adorned her capital.   During the
last century and a half before the Christian era Rhodes was to
a great extent a rival of Alexandria as a centre of literary and
intellectual life, and among the men whose work threw lustre
on the island the foremost place is held by Hipparchus.   Unfortunately nearly all his writings have perished, including the
work in which he laid the foundation of trigonometry, and his
book against Eratosthenes, in which he criticised the geography
of the latter with perhaps undue severity.   These we only know
of through the writings of others.   We only possess one book
written by Hipparchus[1], the date of which is 140 B.C., composed
before he made his great discovery of the precession of the
equinoxes, while the date of his star-catalogue was 129 B.C.[2]
Though nearly all his own writings are lost, his scientific work
has found a most able exponent in Ptolemy, in whose great
work we find the researches of Hipparchus expounded as well
as continued, and in many instances completed; and though
Ptolemy has often neglected to state accurately how much is
due to himself and how much to his great predecessor, there is
in most cases not any great difficulty in dividing the honours
between them.   The Μεγίστη Σύνταξις of Ptolemy, the
Almagest as it is generally called[3], forms a complete compen-

---

[1] "Three books of commentaries to the phenomena of Aratus and Eudoxus,"
included in the *Uranologium* of Petavius and recently edited with a German
translation by Carl Manitius (Leipzig, 1894).   Though most of this book deals
with the risings and settings of stars, it gives at the end of the third book a
very precious list of stars culminating at intervals of exactly an hour, which
supplies us with the only information we possess as to how the water-clocks of
the ancients were corrected.   For a very thorough discussion, see Schjellerup's
paper, "Sur le chronomètre céleste d'Hipparque," in *Copernicus, an internat.
Journal of Astronomy*, I. p. 25.   The result of this investigation is that ancient
astronomers could determine the time during the night to within about a
minute.

[2] According to Ptolemy, whose star-catalogue is for the epoch A.D. 137,
Hipparchus observed 265 years earlier, i.e. B.C. 129.   His earliest observation
known seems to be from the year 161 B.C.

[3] It is hardly necessary to mention that this word is derived from Al-majisti,
an Arabian corruption of μεγίστη (σύνταξις).

dium of ancient astronomy as finally developed at Alexandria.
As we are not here writing the history of theoretical astronomy,
we shall not examine the way in which the available material
of observations was utilised to form theories of the celestial
motions; we shall merely indicate the geometrical constructions
by which these motions were represented and whereby tables
for computing the motions were worked out. There is the less
necessity for a lengthy discussion of the contents of the *Almagest*,
as this has been very thoroughly done by several writers[1].

Hipparchus had at his disposal, in addition to observations
made by himself, the observations made at Alexandria during
the previous 150 years, as well as much earlier Babylonian

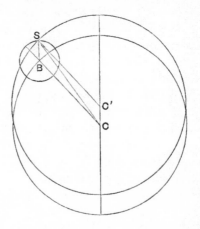

observations of eclipses. From the former he made his brilliant
discovery of the precession of the equinoxes; from Babylonian
and Alexandrian observations combined he worked out the
theories of the sun and moon[2]. In the case of the sun it was a
comparatively easy matter to find an orbit which would satisfy
the observations, since the unequal length of the four seasons

[1] Especially by Delambre, *Hist. de l'astr. anc.* t. II., and recently and better
by Tannery, whose *Recherches sur l'histoire de l'astr. ancienne* are nearly altogether devoted to this subject.

[2] It is very difficult to decide how far Hipparchus was indebted to the Babylonians for the numerical values of the various periods of solar and lunar
motion which he adopted. The recent examination of cuneiform inscriptions
by Kugler has shown that the Babylonian astronomers before the middle of

was the only inequality to be accounted for. Hipparchus showed that the two following hypotheses led to the same result[1].

1. The sun describes in the tropical year a circle with radius r, the earth being at a distance from the centre equal to er, a certain fraction of the radius.

2. The sun moves during the tropical year through an epicycle of radius er in the direction east to west, while the centre of this epicycle in the same period but in the opposite direction describes a circle of radius r round the earth as centre.

Either of these hypotheses was quite sufficient to represent the apparent motion of the sun with an error of less than a minute of arc, a quantity which was utterly insensible not only then but for 1700 years after. The length of the radius r is of course immaterial, but the excentricity e and the longitude of the apogee A must be chosen so as to bring out the observed differences in the length of the seasons. The values fixed by Hipparchus were $e = 0{\cdot}04166$ and $A = 65° 30'$, both of them fairly correct, A being about 35' too small, while the error in determining the excentricity at most could introduce an error of 22' in the equation of the centre.

The great amount of attention which he had paid to the motion of the sun enabled Hipparchus to reject the curious idea that the sun moved in an orbit inclined to the ecliptic, to which we have already alluded in the chapter on Eudoxus. The fact that the doctrine of the sun's motion in latitude is

the second century B.C. made use of almost exactly the same values, but unfortunately we do not know when these had first come into use, and it is very much to be hoped that further discoveries of tablets may solve this problem. The Chaldeans certainly did not get these numerical quantities from the Rhodian astronomer, while there is nothing more probable than that Hipparchus should have obtained these values as well as the observations of eclipses from Babylon, perhaps from Seleukus or second-hand from Diogenes of Babylon. See Kugler's *Die Babylonische Mondrechnung.* Freiburg, 1900.

[1] Theon, ed. Martin, p. 246 ; *Almag.* III. 3, ed. Heiberg, I. p. 216. The figure above (ibid. Heib. p. 225) shows that the sun occupies the same place S, whether we suppose it to describe the epicycle which moves along the circle of which the centre C is the earth, or whether it describes the excentric circle round C'. SB is equal and parallel to $C'C$.

given in a modified form by the later uncritical compilers Pliny, Theon, and Martianus Capella (who apparently all wrote in blissful ignorance of the labours of Hipparchus), renders it exceedingly probable that their accounts represent an attempt made by some early Alexandrian astronomer to account for the difference between the length of the tropical and the sidereal year, made before the discovery of the annual precession had been announced by Hipparchus or before it had obtained credence[1].

The motion of the moon being much more irregular than that of the sun, its theory was found to be much more difficult, but it is still possible to represent the first inequality either by an excentric or by an epicycle. As Ptolemy eventually adopted the epicycle (reserving the excentric for a different use) we shall only consider the epicyclic theory here. Hipparchus assumed first a circle inclined at 5° to the ecliptic, rotating in the retrograde direction round the axis of the latter, so that the nodes perform a complete revolution in 18⅔ years. On this deferent circle moves directly[2] (i.e. from west to east) the centre of an epicycle, while the moon revolves on the circumference of the latter in the retrograde direction. Owing to the direct motion of the line of apsides round the heavens in nearly nine years, the periods of revolution on deferent and epicycle are not quite equal, the motion on the deferent corresponding to change of longitude, that on the epicycle to the change of anomaly, the former being about 3° in excess of the latter[3]. The ratio of the radii of epicycle and deferent was found by the greatest difference between the apparent and the mean place of the moon, which Hipparchus fixed at 5° 1′, the sine of which angle is the ratio sought, or $5\frac{1}{4} : 60 = 0.0875$. This accounted for the

---

[1] Ptolemy estimated the amount of precession at 1° in 100 years or 36″ a year, but Hipparchus appears to have known that the true value is considerably greater. We shall return to this subject in Chapter IX.

[2] In medieval language "in consequentia," i.e. in the order of the signs of the zodiac, the opposite direction being "in antecedentia."

[3] In the simple excentric theory the moon has a direct motion on the excentric circle, the velocity being equal to the mean change of longitude, while the centre of the excentric revolves round the earth with a velocity equal to the excess of the motion in longitude over that in anomaly, i.e. 3° per revolution of the moon.

so-called first inequality of the moon's motion, the equation of
the centre, which in reality is caused by the elliptic form of the
lunar orbit. Hipparchus founded his theory on Babylonian
and Alexandrian observations of lunar eclipses, and the theory
therefore represented the motion of the moon at new and full
moon sufficiently well. He was naturally not content with this,
but examined whether the moon at other points of its orbit
conformed to his calculation, and therefore he observed it at
the quadratures, the time of first and last quarter. He found
that sometimes the observed place of the moon agreed with the
theory, while at other times it did not; but though it was thus
manifest that there must be some other inequality depending
on the relative positions of the sun and moon, Hipparchus had
to leave his successors to investigate its nature properly.

With regard to the five remaining planets Hipparchus did
not succeed in forming a satisfactory theory. Theon of Smyrna
tells us[1] that Hipparchus favoured the theory of epicycles
(which he even claimed as his own) in preference to that of
movable excentrics, saying that it seemed more credible that
the whole system of celestial things was arranged symmetrically
with regard to the centre of the world; "and though he was
not versed in natural science and did not perceive accurately
which motion of the wandering stars was in accordance with
nature and true, and which they performed by accident and
merely apparently, still he supposed that the epicycle of each
moved on a concentric circle, and the planet on the circumference of the epicycle." In other words, Hipparchus merely
considered the motions of the planets from the mathematical
point of view without troubling himself about the physical
truth of his combinations of circles. But Ptolemy gives us
more interesting information[2]. After referring to the difficulties
of following the courses of the planets, he continues : "It was,
I believe, for these reasons and especially because he had not
received from his predecessors as many accurate observations as
he has left to us, that Hipparchus, who loved truth above
everything, only investigated the hypotheses of the sun and

[1] Ed. Martin, p. 300.
[2] *Syntaxis*, ix. 2, ed. Heiberg, ii. p. 210.

moon, proving that it was possible to account perfectly for their revolutions by combinations of circular and uniform motions, while for the five planets, at least in the writings which he has left, he has not even commenced the theory, and has contented himself with collecting systematically the observations and showing that they did not agree with the hypotheses of the mathematicians of his time. He explained in fact not only that each planet has two kinds of inequalities but also that the retrogradations of each are variable in extent, while the other mathematicians had only demonstrated geometrically a single inequality and a single arc of retrograde motion; and he believed that these phenomena could not be represented by excentric circles nor by epicycles carried on concentric circles, but that, by Jove, it would be necessary to combine the two hypotheses."

This important passage gives us not only the historical fact that Hipparchus gave up the hope of forming a complete planetary theory, but also supplies us with his reasons by throwing light on the state of theoretical astronomy at that time. We see that the predecessors of Hipparchus had only had one object in view, that of explaining the annual irregularities which occur about the time when a planet is in opposition to the sun; while they either were not aware of, or had assumed to be negligible quantities, the irregularities in a planet's motion as it passes round the zodiac in a number of years, whereby the arc of retrograde motion varies in length, and which we now know to be caused by the elliptic form of the planet's orbit and the resulting change in its orbital velocity and distance. Obviously the study of these latter phenomena require not only more carefully made observations but also much more prolonged ones than the irregularities depending on the sun do, not only because those phenomena are not nearly as conspicuous, but also because Saturn takes nearly thirty and Jupiter nearly twelve years to go right round the heavens, so that a series of observations had to be properly organised and long persevered in before the law governing the phenomena could be detected. The varying velocity of a planet, independent of the annually recurring apparent disturbance of its motion but influencing the magnitude of this disturbance, had probably more or less dimly been noticed for a long time, but theorists had not attempted to explain it. Hipparchus perceived that without doing this a theory would be extremely inadequate, and he therefore took the first step by sifting and collecting observations which eventually, with those made in the course of the next three hundred years, enabled Ptolemy to produce a satisfactory planetary theory.

In the second century B.C. all that could be said about the five planets was, therefore, that the disturbance in their motion, which seemed connected with their angular distance from the sun, could be accounted for by the epicyclic or the movable excentric theory, but that certain other irregularities had as yet to be left unaccounted for. To us it may look strange that the peculiar way, in which the sun was mixed up with the theory of each planet, did not lead somebody to look for a totally different cause of that extraordinary fact. But in reality it does not seem to have disturbed the mind of anyone interested in such matters. Though the moon did not, like the five planets, stop and retrace its steps for a while, when opposite the sun, yet Hipparchus had found that its velocity at the quadratures was variable, so that the speed of the moon was also to some extent depending on its distance from the sun. Thus it seemed that every wandering star was in some way or other connected with the sun. And it must be remembered that though mathematical astronomy had made considerable progress since the days of Eudoxus, it was a science by itself, a simple means of computing the places of the planets, but not a science which influenced the generally prevailing opinions on the constitution of the world. As regards these, metaphysical arguments were not yet out of date, nor were they indeed destined to be set aside for many centuries. We have mentioned that Kleanthes the Stoic, who had so worthily continued the work of Zeno, the founder of the school, had fixed on fire as the real primary substance. Though the school did not follow him in this, nor in tracing all life to its original source in the sun as the ruler of the world, still we have strong evidence that the idea of the all-pervading influence of the sun in the heavens, as well as on the earth, remained current long after the time of Kleanthes; and no doubt it played an important part in reconciling men to the "solar anomaly" in the planetary theories.   Thus Theon, after mentioning that Mercury and Venus may after all be moving round the sun, delivers himself as follows[1] : "One may conjecture that this position and this order is the more true, as the sun, essentially hot, is the place of animation of the world as being a world and an animal, and so to say the heart of the universe, owing to its motion, its volume and the common course of the stars which are around it (περὶ αὐτόν).   For in animated bodies the centre of the body or animal is different from the centre of magnitude.   For instance, for us who are, as we have said, men and animals, the centre of the animated creature is in the heart, always in motion and always warm, and therefore source of all the faculties of the soul, source of desire, of imagination and intelligence ; but the centre of our volume is elsewhere, about the navel.   Similarly, if we judge the greatest, most worthy of honour and divine things in the same manner as the smallest, accidental and mortal things, the mathematical centre of the universe is where the earth is, cold and immovable; but the centre of the world, as being a world and an animal, is in the sun, which is so to say the heart of the universe, and whence we say that the soul of the world takes its rise to penetrate and extend to its extremities."

Somewhat similar ideas are set forth by Plutarch[2] and Macrobius[3] and are undoubtedly of Stoic origin.   These speculations perhaps helped to fix finally the order in which the orbits of the planets were supposed to be situated.   We have already mentioned more than once that Anaxagoras, the Pythagoreans, Plato, Eudoxus and Aristotle placed them in this order:

Moon, Sun, Venus, Mercury, Mars, Jupiter, Saturn.

This arrangement was also at first adopted by the Stoics[4],

[1] Ed. Martin, p. 296.

[2] *De facie in orbe lunæ*, xv. : "The sun which takes the place of the heart, spreads from itself light and heat like blood and life, to all sides."

[3] *Somn. Scip.* i. 20.

[4] Stobæus, Diels, p. 466; Pseud-Arist. *De Mundo*, p. 392 a.

but afterwards they abandoned it for the following arrangement:

Moon, Mercury, Venus, Sun, Mars, Jupiter, Saturn,

which fell in well with their notions about the dominant position of the sun, as it placed the orbit of this body half-way between the earth and the fixed stars, with three orbits of planets on either side. Cicero[1] states that the Stoic philosopher Diogenes of Babylon (about 160 B.C.) taught this arrangement, and it is not unlikely that it was he who first introduced it (as well as many numerical data used by Hipparchus) into the Greek world from Babylonia, where the planets had been grouped in this order from very early times, as the names of the days of the week testify[2]. Ptolemy attributes this order to "ancient mathematicians." It had probably already been adopted by Hipparchus, it was accepted by all subsequent writers, Geminus, Kleomedes, Pliny[3], Pseudo-Vitruvius, the Emperor Julian[4], as well as by Ptolemy, and up to the time of Copernicus this arrangement was in fact universally adopted. Ptolemy remarks that there is really no way of proving which order is correct, since none of the planets have a sensible

---

[1] *De Divinatione*, II. 43, 91; comp. ibid. 42, 88.

[2] Every hour of the day was dominated by one of the planets, beginning with Saturn and ending with the moon. Saturday was called Dies Saturni because its first hour was ruled by Saturn, and so were the 8th, 15th, 22nd hours. The 23rd was ruled by Jupiter, the 24th by Mars, and the first hour of the following day by the sun, hence its name Dies Solis. And so on. The mythological names of the planets are also Babylonian, and were not adopted by the Greeks before the time of Plato. In earlier times descriptive names were used—Stilbon for Mercury, Hesperos or Phosphoros for Venus, Pyroeis for Mars, Phaethon for Jupiter, Phainon for Saturn.

[3] The only thing remarkable in Pliny's account of the planets is that he gives the period of Venus as 348 days, and that of Mercury as nine days less (II. 38–39), although he says immediately afterwards that they never recede more than 46° and 23° from the sun.

[4] *Oratio* IV. 146 D. There is nothing strange in the statement that the planets are dancing round the sun (περὶ αὐτὸν χορεύοντες, also 135 B), for though this expression is used in the *Timaeus*, 40 C, and Stobaeus, Diels, p. 337 b, of orbital motion, the whole context seems to agree only with the supposition that the sun's orbit has three orbits on either side. After writing this I was surprised to find that Martin interprets this as meaning that Julian (like Tycho Brahe) let all the planets move round the sun (Article "Astronomie grecque et romaine," in Daremberg et Saglio, *Dictionnaire des Antiquités grecques et romaines*, reprint, p. 8). Surely this is a very far-fetched idea.

parallax, but that the ancient arrangement seems probable, as the sun according to it more naturally separates those planets which can pass right round the heavens from those which only reach a limited elongation[1].

[1] *Syntaxis*, IX. 1, ed. Heib. II. p. 207.

CHAPTER VIII.

THE DIMENSIONS OF THE WORLD.

FROM Hipparchus we have to pass over two centuries and a half before we come to the next astronomer whose ideas about the order and arrangement of the world we have to consider. This will give us a convenient opportunity to place together the opinions of the ancients as to the size of the earth and the distances of the planets, to which we have only made passing allusions from time to time.

From the age of Plato the spherical form of the earth was not disputed by any philosophers except by the followers of Epikurus (B.C. 300). This school scarcely deserves to be mentioned in the history of science, since its founder did not take any interest in natural phenomena, and practically left everybody free to form his own conclusions about their causes. To the question how great the sun is, he answered: "as great as it seems to be"; and the flat earth was to him the sediment produced by the fall of the atoms, which he believed to take place vertically in opposition to the circular vortex motion advanced by Demokritus. The Roman expounder of the doctrines of Epikurus, Lucretius (who lived in the first half of the first century B.C.), appears singularly behind the time when talking with contempt of the idea of antipodes, and leaving it an open question whether sun and moon continue their courses under the earth or whether they are daily renewed in the east. On the other hand he appears to more advantage when maintaining that the world is infinite and that we cannot therefore say that the earth is in its centre[1]. When a prominent

---

[1] *De rer. nat.* I. 1070, v. 531, 564, 648; Diog. L. x. 91.

172 of The Dimensions of the World

philosopher could believe the earth to be flat, we need not
wonder that a credulous writer like Ktesias at the beginning of
the fourth century could report that from certain mountains in
India the sun appeared ten times as large as in Greece, or that
Posidonius should condescend to refute the popular idea among
dwellers on the Atlantic, that the sun set in the ocean with a
hissing noise, as well as the statement of Artemidorus of
Ephesus that at Gades (where he had been himself) the sun
when setting appeared an hundred times its usual size[1].
Neither is it strange that the poet Lucanus lets the Atlantic
at the west coast of Libya be heated by the descent of the sun
into it, while the clouds driven by east winds across Spain are
arrested by the solid vault of heaven. Even Vergil, when
describing the sun's course in the zodiac, is uncertain whether
around the opposite hemisphere the silence of night reigns for
ever and the darkness thickens under the pall of night, or
whether Aurora comes to us from thence at the same time as
the reddening vesper lights his late fires there[2]. But though
popular superstitions or poetical vagaries such as these are
hardly sufficient to prove that the arguments of philosophers
and scientists had not yet permeated the minds of non-
scientific writers and become the property of the nations at
large, we have other evidence to the same effect. We have
seen that Aristotle had proved the spherical form of the earth
quite sufficiently; but later writers did not consider it needless
to give as many proofs as possible, though they neglected a
weighty one made use of by Aristotle, viz. the invariable circular
form of the earth's shadow when projected on the moon. Thus
Posidonius[3] shows how the various forms suggested by early
philosophers, flat, like a dinner plate, cubical or pyramidal, are
untenable on account of the change of the part of the starry
heavens visible above the horizon as we proceed north or south,
while the Persians see the sun rise four hours earlier than the
Iberians do. The argument of Aristotle, that the surface of
the ocean must everywhere be equidistant from the earth's

[1] Strabo, III. p. 138.
[2] *Georg.* I. 247–251.
[3] Cleom. I. 8, ed. Ziegler, p. 75 ff.

centre, which can only be the case if the earth is a sphere, is also regularly served out by every writer[1].

With regard to the size of the spherical earth, the earliest attempt to estimate it is that of Aristotle[2], who gives the circumference as 400,000 stadia, without stating his authority for this. There is absolutely nothing to indicate that the Chaldeans or the Egyptians ever attempted to determine the size of the earth, and Aristotle's rough approximation must therefore have been derived from a Greek source, not unlikely from an observation by Eudoxus of the different altitude of stars in Egypt and Greece, as he had lived in both countries. The next estimate is that of Archimedes[3], who gives 300,000 stadia, also without quoting his authority, but he probably followed the only geographer of that time, Dikæarchus of Messana, who died about 285 B.C., and whose estimates of distances in the Mediterranean are referred to by Strabo. His writings are lost except a few fragments, but we know that he taught the spherical form of the earth and stated that the height of Pelion, 1250 passus, was insignificant in comparison to the size of the earth[4]. Posidonius probably refers to this estimate[5] when he (on the occasion of proving that the earth is not flat) mentions that the head of the constellation Draco passes through the zenith of Lysimachia in Thrace, while at

[1] Pliny, II. 165 ; Theon, ed. Martin, p. 146. In Cantor's *Math. Beiträge zum Kulturleben der Völker*, p. 170 and p. 398, I find a quotation from Cassiodorus, according to which M. Terentius Varro (a contemporary of Cicero) in his lost book on geometry stated that the earth was egg-shaped. "Mundi quoque figuram curiosissimus Varro longæ rotunditati in geometriæ volumine comparavit, formam ipsius ad ovi similitudinem trahens, quod in latitudine quidem rotundum sed in longitudine probatur oblongum." But this probably refers to the figure which Empedokles and some Stoics attributed to the *world* and not to the earth. See above, p. 159, note 1.

[2] Horace (*Carm.* I. 28) calls Archytas of Tarentum a measurer of the earth and sea, and in the *Nubes* of Aristophanes certain geometrical instruments are said to serve for measuring the whole earth (Berger, *Gesch. d. wiss. Erdkunde der Griechen*, I. p. 139), so that it is just possible that some attempts in this direction had been made before the time of Aristotle.

[3] *Arenar.* I. 8.

[4] Pliny, II. 162; comp. Martianus Capella, VI. 590; *Dicæarchi Fragm.* ed. Fuhr, p. 117. The suggestion that Dikæarchus is the author of this estimate is made by Berger, l. c. II. p. 93; III. p. 44.

[5] Cleom. I. 8, ed. Ziegler, p. 78.

Syene in Upper Egypt Cancer passes through the zenith. The difference of declination of these stars being 24° and the distance between the places being 20,000 stadia, he points out that the diameter of the world would be only 100,000 stadia, and the circumference of it only 300,000 stadia, from which it follows that the earth cannot be flat, since it is but a point in comparison with the celestial sphere, which his calculation shows would itself be very limited in size. Obviously the earth, if spherical, would be exactly of the size computed. Lysimachia was founded in 309 B.C. by Lysimachus, afterwards king of Thrace; the computation must therefore be later than that date, and yet it must be earlier than the time of Archimedes (who died in 212) since the author makes the circumference exactly equal to three times the diameter. It seems therefore very probable that Dikæarchus was the author of this estimate of the size of the earth, which, however, is but a rough one, since the latitude of Lysimachia was about 40° 33', while the declination of γ Draconis, the brightest and most southerly star in the head of the dragon, was in the year 300 B.C. $= +53°$ 11'.

The next and most celebrated determination is that of Eratosthenes of Alexandria (276 to 194 B.C.), librarian of the great museum in that city. He was a native of Cyrene and studied at Alexandria and Athens, so that he had already acquired a name for learning, when he (about 235) was called to Alexandria, where he spent the rest of his life. He was a man of unusually varied attainments, but it is chiefly as a geographer that he is known to us, though only through the (often hostile) references to him in the works of Strabo and others[1]. He seems in addition to his great work on geography to have written a special book on his determination of the size of the earth, which, however, is lost[2]. He stated that at Syene a gnomon threw no shadow on the day of summer solstice, while the meridian zenith-distance of the sun at Alexandria was $\frac{1}{50}$

[1] All the references to Eratosthenes are put together by Berger, *Die geographischen Fragmente des E.* Leipzig, 1880.

[2] Macrobius, *Somn. Scip.* I. 20, mentions it as Libri dimensionum. The Καταστερισμοί or description of the constellations which goes under the name of Eratosthenes, is a forgery of the second or third century A.D. (Susemihl, *Gr. Litt. in der Alexandrinerzeit*, I. p. 420).

of the circumference of the heavens, which are therefore represented the difference of latitude; while the linear distance of
these two places, which he assumed to be on the same meridian,
was 5000 stadia[1]. Consequently the circumference of the earth
was 250,000 stadia, for which value either Eratosthenes himself
or some successor of his afterwards substituted 252,000 stadia[2],
evidently in order to get a round number, 700 stadia, for the
length of a degree. This value was adopted by Strabo and
Pliny[3].

The question now arises : what was the length of the stadium
adopted by Eratosthenes? The answer to this is given in the
statement of Pliny, that Eratosthenes put a schœnus equal to
40 stadia. Now an Egyptian σχοῖνος was 12,000 royal cubits
of 0·525 meter[4], therefore the stade was 300 such cubits or
157·5m. = 516·73 feet, and 252,000 times this is 24,662 miles,
which corresponds to a diameter of 7850 miles, only 50 miles
less than the true value of the polar diameter of the earth. To
a great extent this close agreement is no doubt due to the
chapter of accidents, though on the other hand it must be
remembered that we only possess the merest outline of the
proceeding of Eratosthenes, but are quite ignorant whether
he took any precautions to guard against error, particularly
in observing the zenith-distance of the sun at Alexandria.
Kleomedes adds, that observations of the shadow of a gnomon
at the winter solstice at Syene and Alexandria gave the same

[1] Cleom. i. 10, ed. Ziegler, p. 100. Strabo, xvii. p. 786, gives the distance
from the small cataract at Syene to the sea=5300 stadia on the authority of
Eratosthenes. The 5000 is therefore only a round number.

[2] Kleomedes seems in fact to be the only author who gives the 250,000,
except Arrianus, quoted by Joh. Philoponus in his commentary to Aristotle's
Meteorology (Ideler, i. p. 138, where the editor even suggests to add καὶ
δισχιλίους !).

[3] Pliny, ii. 247, states that Hipparchus added 26,000 to the 252,000. But
either his text is corrupt (Hultsch, *Metrol.* p. 63, note 6, suggests to read
2600), or Pliny has misunderstood his source (Berger, *Erat.* p. 130, suggests
that Pliny's source referred to the table of latitudes for the quadrant of 63,000
stadia prepared by Hipparchus, and the breadth of the inhabited part of the
earth 38,000 or 37,600 stadia according to Eratosthenes). No other ancient
writer mentions any correction, and Strabo, ii. p. 113 and p. 132, says expressly
that Hipparchus adopted the 252,000 st.

[4] Hultsch, *Griech. und Röm. Metrologie* (Berlin, 1882), p. 364. One schœnus = 6300 meter.

result, $\frac{1}{50}$, but he gives no details. The latitude of Syene is
24° 5'·0¹, that of the Museum of Alexandria about 31° 11'·7
(Ptolemy assumed 30° 58'), the difference is 7° 6'·7, which
happens to be close to the 7° 12' of Eratosthenes. But the
tropic of Cancer did not pass through Syene in the days of
Eratosthenes, as the obliquity of the ecliptic about the year
225 was 23° 43' 20", while Eratosthenes found 23° 51' 20".
Before his time it had been assumed $= 24°$², so that he came
nearer to the truth.

The stade used by Eratosthenes was a shorter one than the
Olympic one of 185m. (400 cubits of 0·462m.) or the Ptolemaic
or Royal Egyptian stade of 210m. (400 cubits of 0·525m.³). It
was an itinerary measure used to express distances, which had
been measured by pacing them, and it has always been known
to have been smaller than the Olympic stade⁴. According to
Martianus Capella⁵, Eratosthenes found the distance between
Syene and Meroe "per mensores regios Ptolemæi," i.e. by the
professional pacers or βηματισταί (itinerum mensores⁶), and it
was therefore natural that he should use the itinerary measure
employed by them. This was apparently also used in the next
attempt to determine the size of the earth, which was made by
Posidonius and is likewise recorded by Kleomedes⁷. Posidonius
was born about 135 B.C. and was a native of Apameia in Syria;
he spent many years in extensive travels (even as far as Spain),
after which he settled at Rhodes and obtained great renown as
a teacher of Stoic philosophy. He was the author of about
twenty works, of which only fragments are left. He died at the

¹ From maps of the First Cataract and information kindly supplied by Capt.
Lyons, Director-General of the Egyptian Survey Department, combined with the
statement of Strabo (XVII. p. 817) that Elephantine was half a stade north of
Syene.

² Eucl. IV. 16 (inscribing a regular quindecagon in a circle) was doubtless
originally worked out for this reason. About the obliquity of 24°, see Theon,
ed. Martin, p. 324, on the authority of Eudemus.

³ Hultsch, l. c. pp. 67 and 355.

⁴ Rennell made it = 505·5 feet (*Geogr. of Herod.* 2nd ed. I. p. 42).

⁵ VI. 598. Martianus imagines that E. made use of this distance to determine the size of the earth.

⁶ Simplicius (*De Cœlo*, p. 549, 8, Heib.) says that the length of a degree had
been found διὰ ὁδομέτρου.

⁷ Cleom. I. 10, ed. Ziegler, p. 94.

age of eighty-four about the year 50 B.C. According to him the bright star Canopus culminated just on the horizon at Rhodes[1], while its meridian altitude at Alexandria was " a quarter of a sign, that is one forty-eighth part of the zodiac." The difference of latitude was therefore 7° 30', and the distance being 5000 stadia[2], the circumference of the earth came out = 240,000 stadia. This result is entitled to much less credit than that of Eratosthenes. Of course it is impossible to see a star when it is exactly on the horizon, but Germinus[3] tells us that Canopus is difficult to see at Rhodes or only visible from high places, while Hipparchus merely says that it can be seen from Rhodes[4]. In reality, the true meridian altitude of Canopus at Rhodes was at that time nearly a degree, or, allowing for refraction, 1° 16', which goes far towards accounting for the error of 2° 15' in the difference of latitude of the two stations, which is only 5° 15'[5]. It was perhaps the recognized difficulty of observing sharply the end of a shadow, which made Posidonius think it desirable to test the result of Eratosthenes by a different method, not using the sun ; but he can hardly have intended his own value to be preferred. He does not seem to have been an habitual observer, and probably he had no proper instrument at Rhodes and therefore found it convenient to use a star, the altitude of which was nearly zero. Possibly he merely used the whole thing as a lecture illustration and did not claim any scientific value for it.

In his *Geography* Ptolemy gives the length of a degree

[1] Proklus *In Timæum*, 277 E : "it grazes the horizon."

[2] Strabo, II. p. 125, says that Eratosthenes had found 3750 stadia, but as this is ¾ of 5000, the unit is obviously the royal stade of 210 m. On p. 95 Strabo makes use of the same unit, as he gives the result of Posidonius for the circumference of the earth = 180,000 stadia.

[3] *Elem. Astr.* c. III. ed. Manitius, p. 42.

[4] *Comment. in Aratum*, cap. XI. ed. Manitius, p. 114.

[5] The latitude of the port of Rhodes is 36° 26'·6. For the year-100 the declination of Canopus was 52° 40'·2, therefore its true meridian altitude was 0° 53', and corrected for mean refraction 1° 16'. At Alexandria its altitude corrected for refraction was 6° 16'. The expression "a quarter of a sign" shows that only a very rough estimate was made. Hipparchus (l. c.) gives the s.P.D. of Canopus as about 38½°, and the latitude of Rhodes as about 36°. Ptolemy in his *Geography* gives the difference of latitude = 5°.

equal to 500 stadia, or the circumference = 180,000 stadia. As this is to the value of Posidonius exactly as the itinerary is to the Egyptian stade, Ptolemy evidently used the official Egyptian stade of 210 meter, which is practically $\frac{1}{7}$ of a Roman mile (barely four feet less) and was therefore a convenient unit for a subject of the Roman Empire and an inhabitant of Egypt to use. He simply adopted the value of Posidonius and expressed it in terms of a different unit[1].

Thus Greek astronomers had a very fair idea of the size of the earth. We shall now examine their opinions as to the distances of the heavenly bodies.

The first philosopher who speculated on these distances was Anaximander, who put the sun's distance equal to twenty-seven and that of the moon nineteen times the radius of the earth, or possibly the distance of the sun twenty-seven times that of the moon[2]. The next who has been credited with speculations of this kind is Pythagoras, and his alleged ideas on the subject are coupled with the theory of the "harmony of the spheres." The only accounts we possess of these speculations are given by very late authors, but most philosophers after the fifth century were more or less imbued with the beautiful fancy of the whole universe being ruled by harmony, which until the end of the Middle Ages continued to captivate the human mind.

> "There's not the smallest orb which thou behold'st
> But in his motion like an angel sings,
> Still quiring to the young-eyed cherubins:
> Such harmony is in immortal souls;
> But, whilst this muddy vesture of decay
> Doth grossly close it in, we cannot hear it[3]."

We have seen how this was the leading idea in Plato's doctrine of the soul of the world, and how it led him to assume that the

---

[1] Hultsch, l. c. p. 64. Simplicius (p. 549, Heib.) says that astronomers took two stars with a difference of declination exactly equal to one degree and then measured the distance between two places on the earth, on the same meridian, through the zeniths of which these stars passed, and found it equal to 500 stadia. Of course nobody ever made an observation of this kind.

[2] See above, Chapter I. p. 15, note 1.

[3] *Merchant of Venice*, v. 1.

radii of the planetary orbits were proportional to the numbers
1, 2, 3, 4, 8, 9, 27, though the last of these numbers does not
correspond to any note in Greek music. The musical scales
ascribed to Pythagoras by various writers are not altogether
identical. According to Pliny[1], the planetary intervals were:

> Earth—moon a tone.
> Moon—Mercury a semitone (dimidium ejus spati).
> Mercury—Venus a semitone (fere tantundem)[2].
> Venus—sun a minor third (sescuplum).
> Sun—Mars a tone.
> Mars—Jupiter a semitone (dimidium).
> Jupiter—Saturn a semitone (dimidium).
> Saturn—fixed stars a minor third (sescuplum).

The scale corresponds to C, D, ♭E, E, G, A, ♭B, B, D. Pliny
makes a sad blunder by adding that this makes "seven tones,
which they call the diapason harmony," whereas it consists of
two parts, each of which comprises a fifth. But as the earth is
at rest it cannot be supposed to emit a sound, and if we leave
it out, the intervals from the moon to the fixed stars form an
octave of the Dorian mode. The last interval is by Censorinus
given as a semitone (limma), while he otherwise agrees perfectly
with Pliny[3]. The fixed stars then correspond to C instead of
D, and the whole forms an octave (diapason), but it does not
agree with the musical system of Pythagoras, in which from the
middle to the highest note is only a fifth. Censorinus also
forgets that he started by saying that only the seven planets
made the music. Exactly the same scale is given by Theon[4] on
the authority of Alexander the poet[5], and a rather similar one
by Achilles[6], i.e. C, D, E, F, G, A, ♭B, B, C, the intervals being

---

[1] *Hist. nat.* II. 84.

[2] Boeckh interprets "fere tantundem" as a hemitonium majus or apotome,
the other semitones being limmas (minor semitones). See "Ueber die Weltseele im Timæus des Platon" (*Kleine Schriften*, Vol. III.), which contains a full
account of Greek harmony.

[3] Even as far as using the expression "fere tantundem." *De die natali*,
XIII. On the other hand Mart. Capella, II. 170-199, gives the interval Sun-
Mars $= \frac{1}{2}$ tone, otherwise the same as Pliny.

[4] Ed. Martin, p. 187, comp. pp. 66 and 359.

[5] Probably Alexander of Ephesus, first century B.C.

[6] Petavius (1703), III. p. 80.

1, 1, $\frac{1}{2}$, 1, 1, $\frac{1}{2}$, $\frac{1}{2}$, $\frac{1}{2}$, and the sun taking the place next after the moon. But it is a very significant circumstance, that all these scales only occur in the writings of very late authors. From the allusions made by Plato and Aristotle we know that the general idea of the harmony of the spheres goes back at least to the beginning of the fourth century B.C. No doubt it may have arisen earlier, though it must have been ignored by Philolaus, with whose system of ten planets it is incompatible. The next allusion to it is in the so-called papyrus of Eudoxus, where it is stated that the sun is as much greater than the moon (and consequently, as their angular diameters are equal, the distance of the sun as much greater than that of the moon) as the diapente is greater than the diatessaron, that is to say : the distance of the sun is to that of the moon as a fifth is to a tone or as 9 to 1[1]. This is the same ratio as that given by Pliny and the other late writers, but otherwise the whole aspect of the case is altered by them, as the distances are made proportional to the intervals and not to the numbers representing these, so that the distance of the sun becomes $3\frac{1}{2}$ times as great as that of the moon instead of nine times as great, as Eudoxus had made it. The more recent way of arranging the planets is also introduced[2], and the whole system becomes one mass of arbitrary assumptions, so that even the idea that the fixed stars gave the highest note (νήτη) and the moon the lowest (ὑπάτη)[3] is reversed by Nikomachus the Neo-Pythagorean[4]. His idea is evidently that the moon as the nearest body should correspond to the shortest string and Saturn to the longest of the seven strings of the lyre. Other speculators assumed, in correspondence with the five tetrachords of the so-called perfect system, five equal intervals in the heavens, one from the moon to the sun (with Mercury and Venus), the second from thence to Mars, the third from Mars to Jupiter, the fourth from

---

[1] *Ars Eudoxi*, ed. Blass, col. xx. 14–15.   Sun : Moon :: 9 : (9 – 8).

[2] Except in the scale recorded by Achilles, in which the sun comes next after the moon.

[3] Cicero, *Somn. Scip.* 18: "Summus ille cœli stellifer cursus...acuto et excitato movetur sono, gravissimo autem hic lunaris."

[4] Boeckh, l. c. pp. 169 and 175, thinks that this system is older than those of Pliny, &c., as it is so simple.   The earth has no place in it.

Jupiter to Saturn, the fifth from Saturn to the sphere of the fixed stars[1]. And while the harmony of the universe was thus blindly sought in the distances of the planets, others looked for it in the dryness, heat, humidity, or solidity of the stars; or, with Ptolemy, they compared the angular distances of the planets in the sky with the musical intervals, so that an octave corresponded to 180° (opposition), a fifth to 120° (trigonus), a fourth to 90° (quadrature), a second to 60° (aspectus sextilis)[2]. In reality therefore we ought hardly to take the planetary intervals, as determined by the sphere-harmony, seriously; the whole doctrine is quite analogous to that of astrology, but is vastly more exalted in its conception than the latter, and it deserves honourable mention in the history of human progress.

Pliny and Censorinus commence their accounts of these scales by stating that Pythagoras believed the moon's distance to be 126,000 stadia, and Pliny adds that the distance from the moon to the sun was twice as great, and the distance from thence to the fixed stars three times as great, " of which opinion our Gallus Sulpicius was also." This information, the details of which do not agree with his musical scale, Pliny has evidently taken from a different authority, either the book on eclipses of C. Sulpicius Gallus or the encyclopædia of Terentius Varro, both of which he mentions among his authorities for the second book of his natural history. The distance of the moon, 126,000 stadia, is exactly half of Eratosthenes' value of the circumference of the earth[3], a sure sign that Pythagoras cannot have had anything to do with this estimate. Although the remarkably small distance of the moon (only $1\frac{1}{2}$ times the diameter of the earth) has an archaic look, it is more likely merely a product of the ignorance of some late writer, to whom the researches of scientific men were unknown or of small moment, unless some numerical error had crept into the copy of some book or other used by Pliny. Equally wild are the estimates which Martianus Capella gives in another place[4].

[1] Boeckh, l. c. p. 173.

[2] Compare F. Boll: *Studien über Claudius Ptolemaus*, p. 163.

[3] Compare Berger, l. c. II. p. 34.

[4] VIII. 856–861.

The ratio of the distances must be simply that of the periods, so that the distance of the moon is $\frac{1}{12}$ that of the sun, while Mars, Jupiter, and Saturn are respectively twice, twelve, and twenty-eight times as far off as the sun. The writer appears to have taken his materials from Varro, and his eighth book therefore merely represents the knowledge current in Rome in the first century B.C.[1] The moon's distance he makes out to be 100 earth-radii, which is supposed to result from the apparent diameter of the moon, 36′, found by the time the moon takes to cross the horizon, and the diameter of the lunar shadow during a total solar eclipse, which he says is $\frac{1}{18}$ of the earth's circumference, as found from observations extending from Borysthenes to Meroe. Evidently he copied from sources which he only partly understood, and his information about the dimensions of the earth is equally faulty, as he gives the circumference = 406,010 stadia on the authority of Eratosthenes and Archimedes.

We shall now turn to undoubted results of serious work in this direction, which only refers to the sun and moon. We learn from Aristotle[2] that already in his day the problem of the size and distance of the sun and moon had commenced to attract attention. His contemporary, Philip of Opus, is credited by Suidas[3] with the authorship of books on the distance of the sun and moon, on the size of the sun, moon, and earth, on lunar eclipses, on the planets. Eudoxus, as we have already seen, considered the diameter of the sun to be nine times as great as that of the moon, and the distance of the former consequently nine times as great as the latter, since they both appear of the same size. Archimedes, who mentions this, adds that his own father[4] Phidias found this ratio to be 12 : 1. Archimedes himself adopted 30 : 1. How these estimates were formed we have no way of knowing, but it is not unlikely that the methods practised by Aristarchus in the following century

---

[1] See Eyssenhardt's preface, p. LVI.

[2] *Meteor.* I. 8, p. 345 b.

[3] Article Φιλόσοφος.

[4] The text (*Aren.* I. 9) has Φειδία δὲ τοῦ ᾿Ακούπατρος, but Blass has pointed out that this meaningless word must be a corruption of ἀμοῦ πατρός (*Astron. Nach.* 2488).

are really due to Eudoxus[1]. We have already mentioned that
Aristarchus made an attempt to determine the relative distances
of the sun and moon and found that the former was between
18 and 20 times as far off as the latter. In the book "On the
sizes and distances of the sun and moon" he also indicates the
method of determining the distance of the sun, which according
to Ptolemy was adopted by Hipparchus, and which for about
1600 years was followed by astronomers. It is based on observations of the breadth of the earth's shadow at the average
distance at which the moon crosses it during lunar eclipses.
In the figure, ρ is half this angular breadth, found by measuring

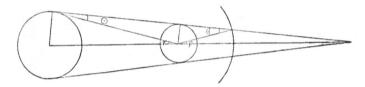

the time taken by the moon to cross the shadow, while r is the
angular radius of the sun and ☉ and ☾ the parallaxes of sun
and moon. It is evident that

$$☉ + ☾ = r + \rho.$$

Hipparchus gave $r = 16'\ 36''\ 55'''$ and ρ $2\frac{1}{2}$ times as much[2], which,
if we with Aristarchus assume ☾ $= 19$ ☉, gives ☉ $= 2'\ 54''$.

But Hipparchus was not content with this method of finding
the sum of the parallaxes of sun and moon, which can only
serve to determine that of the moon; on the contrary, he must
have been aware that the solar parallax was a quantity which
it would be impossible for him to find in this way, owing to
unavoidable errors of observation. From what Ptolemy says[3],
it appears that Hipparchus attempted to find limits within
which the parallax of the sun must lie in order to get observation and calculation to agree in the case of a solar eclipse,
where the difference between the parallaxes of sun and moon
might be expected to reveal itself, while the sum of the

[1] As suggested by Tannery, *Aristarque de Samos, Mém. de la soc. des sc. de
Bordeaux*, 2ᵉ série, v. p. 237.
    [2] *Syntaxis*, IV. 8 (p. 327, Heib.).
    [3] Ibid. v. 11 (p. 402, Heib.).

parallaxes was found by lunar eclipses. Naturally this proceeding did not lead to any result, so that Hipparchus was not even sure whether the solar parallax was at all appreciable, or whether one might assume it = 0. In the latter case the values given above for r and ρ give $\mathbb{C} = 58' 9'' 14'''$ which corresponds to a distance of the moon = 59·1 semidiameters of the earth and $3\frac{1}{2}$ times as many semidiameters of the moon. The upper limit found by Hipparchus is not stated by Ptolemy, but probably Kleomedes had it in his mind when he said[1] that Hipparchus had proved the sun to be 1050 times the size of the earth. As this refers to the cubic contents, it would make the semidiameter of the sun = $10\frac{1}{6}$ times that of the earth, that of the moon $1 : 3\frac{2}{5}$ times the same (which is the figure adopted by Ptolemy), and the distances of moon and sun respectively $60\frac{5}{6}$ and 2103 earth-radii. A medium between these limiting values is given by Theon[2], who states that Hipparchus made the sun 1880 times as great as the earth, i.e. its radius $12\frac{1}{3}$ as great, its distance 2550 earth-radii, the moon's radius $\frac{12}{41} = 0·29$, and its distance 60·5 earth-radii.

As the moon's distance is $60\frac{1}{3}$ times the equatorial radius of the earth, and its semidiameter $1 : 3\frac{2}{3}$ or 0·273 of that of the earth, we see that Hipparchus had a very correct idea of the distance and size of our nearest celestial neighbour[3]. The problem of finding the distance of the sun (and thereby its actual size) was altogether beyond the instrumental means of astronomers until the invention of the telescope, but it does great credit to Hipparchus that, although he attacked the problem in various ways, he saw that it had to be left unsolved. On the other hand Ptolemy, after determining the lunar parallax by comparing observed zenith distances with those resulting from his theory, made a distinctly retrograde step, when he deduced from lunar eclipses a distance of the sun, which is practically the same as that resulting from the

---

[1] II. 1 ; ed. Ziegler, p. 152.   [2] Ed. Martin, p. 320.

[3] And yet Pliny 200 years later says that the moon is larger than the earth, otherwise the moon could not entirely hide the sun from the earth (II. 49). As to the sun, he only says it is very much larger than the earth. Probably his source said that the moon is larger than the eclipsed area of the earth.

proportion found by Aristarchus; and for 1500 years his solar
parallax, 2′ 51″, was accepted without question by astronomers.
The figures adopted by him were[1]:

Mean distance of the moon   59   earth-radii,

,,   ,,   sun   1210[2]   ,,

Radius of the moon   1 : 3·4   ,,

,,   sun   5·5[3]   ,,

It is, however, interesting to notice that Hipparchus is not
the only astronomer of his age who perceived that the sun is
very much more than twenty times as far off as the moon. A
remarkable attempt to determine the actual size of the sun,
founded on a boñí hypothesis as to its distance, was made by
Posidonius and has been handed down to us by Kleomedes and
Pliny[4]. Posidonius knew on the authority of Eratosthenes
that at Syene, under the tropic of Cancer, at the time when
the sun is in the constellation of Cancer, no shadows are seen
at noon within an area 300 stadia in diameter. Every point
within this area was therefore struck by the rays of the sun in
a direction normal to the surface of the earth, and as every
normal to the surface of a sphere passes through its centre,
Posidonius concluded that a cone, having its apex at the centre
of the earth, and as its base the apparent disc of the sun,
would cut off from the surface of the earth a circular area
300 stadia in diameter. Assuming then that the orbit of the
sun is 10,000 times as great as the circumference of the earth,
it followed that the sun's diameter was 10,000 times as great
as the diameter of the shadowless area. If we now ask how
great Posidonius supposed the radius of the earth to be, the
answer is supplied by the following figures given by Pliny,
probably taken from his usual authority, Terentius Varro.
According to him Posidonius supposed the distance from the
surface of the earth to the region of clouds and winds to be
40 stadia, from thence (*inde*) there is pure and liquid air of

[1] *Synt.* v. 15–16, pp. 425–426, Heib.

[2] In reality over 23,000.   [3] In reality 109.

[4] Kleomedes, II. 1; ed. Ziegler, 144, 22 to 146, 16. Pliny, II. 85. This
subject has been dealt with in detail in a paper by F. Hultsch: "Poseidonios
über die Grösse und Entfernung der Sonne"; *Abhandlungen d. K. Gesellsch. d.
W. zu Göttingen*, N. F. I. no. 5, 1897.

uninterrupted light, but from the clouded region to the moon there is a space of 2,000,000 stadia, and from thence (*inde*) to the sun 500,000,000 stadia, " in consequence of which distance the sun, notwithstanding its immense magnitude, does not burn the earth." Combining this with the statement of Kleomedes that the diameter of the sun is 10,000 × 300 stadia, a simple calculation gives the radius of the earth 50,205·02 stadia. But, as remarked by Hultsch[1], the account of Pliny has certainly been corrupted, as there is no sense in supposing the distance from the earth's surface to the sun = 502,000,040 stadia, when evidently only round numbers are employed. Doubtless the second "*inde*" should be left out, and Posidonius simply made the distance of the sun from the centre of the earth = 500 million stadia and the earth's radius = 1 : 10,000 part of that, or 50,000 stadia. In round numbers the circumference would then be 300,000 stadia, or exactly 1000 times the diameter of the shadowless area. The fact that Posidonius did not adopt the value of the circumference of the earth, 240,000 stadia, found by himself, but used the much older value of 300,000 stadia, is another proof that he only cared to operate with round numbers. Similarly the apparent diameter of the sun became = 360° : 1000, although he knew that 360° : 720 was very close to the truth. He followed in fact in the footsteps of Archimedes, the idea of whose "Arenarius" was to form the largest conceivable sphere, to fill it with grains of sand, to count these, and then to prove that a still greater number was always conceivable. For this purpose Archimedes arbitrarily supposed the circumference of the earth equal to three million stadia, and the circumference of the solar orbit 10,000 times as great. This latter figure Posidonius adopted, showing thereby that he did not believe in the large parallax and consequent small distance of the sun adopted by every astronomer of antiquity except by Hipparchus.

As the hypothesis of Posidonius had absolutely no foundation in any observed fact, Ptolemy does not mention it; nor does he take any notice of the exceedingly careless conclusion of Kleomedes[2], that since (according to Aristarchus, whom he

---

[1] l. c. pp. 31–32.   [2] II. 1, ed. Ziegler, p. 146, 17.

does not mention) the earth's shadow at the distance of the
moon is equal to twice the lunar diameter, therefore the latter
must be equal to half the earth's diameter, or 40,000 stadia.
This gives him the moon's distance equal to five million stadia
(more than twice the real distance), and on the assumption
that the actual velocities of the planets in their respective
orbits are equal, the sun's distance equal to thirteen times as
much. No wonder that Ptolemy did not take the trouble to
refute statements like these.

For the sake of completeness we shall shortly mention that
Macrobius in the fifth century of our era makes a curious
statement as to the size of the sun[1]. Without the slightest
attempt at a proof he announces that the shadow of the earth
reaches just to the solar orbit and is equal to sixty diameters
of the earth. On this assumption it follows from the length of
the shadow being equal to the radius of the sun's orbit, that
the sun's diameter is twice as great as that of the earth[2].
Macrobius simply gives "the Egyptians" as his authority,
and everything about his story is crude enough to have come
from a pre-Alexandrian source, though it is more likely that
he merely copied from some ignorant encyclopædic writer,
who had remained a stranger to the rapid rise of scientific
astronomy.

We have thus seen that the extremely imperfect ideas of
earlier philosophers with regard to the extent of the known
part of the universe had gradually given way to more correct
notions as to the size of the earth and the size and distance of
the moon. At the same time the fact that the other members
of the planetary system are at vastly greater distances from us
was pretty clearly perceived by the astronomers of the Alexandrian school and their contemporaries, though their ideas as
to these distances fell far short of reality. Cicero calls the

---

[1] *In Somn. Scip.* I. 20.

[2] This does not quite agree with the immediately following observation (a
marvellously rough one!) that the sun's diameter at the equinox takes ⅛ of an
hour to cross the horizon, so that the solar orbit is 216 (that is, 9×24) times as
great as the sun's diameter. This gives Macrobius its diameter $= \frac{1}{216}$ of
$80,000 \times 120 \times 3\frac{1}{2}$ or $30,170,000 : 216 = 140,000$, which he says is nearly twice the
80,000 stadia of the earth's diameter.

distances which separate the orbits of the three outer planets
from each other and from the starry heavens "infinite and
immense," and adds that the heavens themselves are the
extreme end of the world[1]. Seneca, in his chapter on the
nature of comets[2], points out that the planets are separated
by great intervals, even when two of them appear to be close
together, for which reason the opinion that comets are produced
by the coming together of planets, has no foundation. As to
the fixed stars, the opinion that they are situated on the surface
of a sphere of immense though limited extent continued to
be almost universally accepted[3].

The epicyclic system, which received its last development
by Claudius Ptolemy in the second century, could not give any
clue to the distances of the planets. It could only for every
planet give the ratio between the radii of the deferent and of
the epicycle, which resulted from the observed length of the
retrograde arc. But later writers were not content to be left
in the dark with regard to the dimensions of the planetary
orbits. The distances of the moon and the sun were supposed
to be known, and as the space between them was occupied by
the orbits of Mercury and Venus, the idea was suggested by
"some people" ($\tau\iota\nu\epsilon\varsigma$), as Proklus tells us, that there is no
vacant space in the world, and that all intervals are filled by
intermediate spheres. The greatest distance of the moon is
$64\frac{1}{6}$, and this is also the smallest distance of Mercury, while
the greatest distance of Mercury, computed by Ptolemy's ratio,
$177\frac{33}{60}$, is equal to the smallest distance of Venus, the greatest
distance of which, similarly computed, is 1150, which is nearly
the same as 1160, the perigee distance of the sun[4]. A century
later we find Simplicius laying down the same principle
(without giving any figures) and coolly giving as his authority
the *Syntaxis* of Ptolemy, in which there is not a word about

---

[1] *De Divinatione*, II. 43, 91.

[2] *Quæst. Nat.* VII. 12.

[3] Even Aristarchus believed in the existence of the sphere of the fixed stars;
only Herakleides and Seleukus seem to have discarded it. To say that Epikurus
and his followers were equally enlightened would be to do them too much
honour. See above, p. 171.

[4] *Hypotyposes*, ed. Halma, p. 145.

this idea[1]. But throughout the Middle Ages this notion was universally accepted, and we shall see that down to the time of Copernicus the dimensions of all the planetary orbits, calculated on this principle, were supposed to be well known.

While knowledge of the dimensions of the universe had gradually advanced, philosophers found it more difficult to agree with regard to the physical constitution of the heavenly bodies, though all acknowledged that they were of a fiery nature, the Stoics in particular supposing them to be composed of the very pure fire or ether, which pervaded all the upper regions of space. Naturally the peculiar appearance of the "face of the moon" pointed to its being of a very different constitution, and already Anaxagoras and Demokritus had recognized that it was a solid mass having mountains and plains, while Plato held it to be chiefly composed of earthlike matter. All that the most enlightened minds of antiquity could make out with regard to the constitution of the moon is contained in a most delightful dialogue by Plutarch "On the face in the disc of the moon." In this book the opinion of the Stoics is refuted, that the moon is a mixture of air and gentle fire, since the moon ought not to be invisible at new moon if it did not borrow all its light from the sun; and this also proves that it is not formed of a substance like glass or crystal, since solar eclipses would then be impossible. The manner in which the sunlight is reflected from the moon, and the absence of a bright, reflected image of the sun and of the earth, prove that the substance of the moon is not polished but is like that of our earth. The correct explanation of the fact that the moon remains faintly visible during a lunar eclipse is given. It is of great interest to notice that Plutarch, in order to combat the idea that the moon cannot be like the earth since it is not in the lowest place, boldly asserts that it is not proved that the earth is in the centre of the universe, as space is infinite and therefore has no centre; besides, if everything heavy and earthy were crowded together in one place, we should expect all the fiery bodies to have been likewise brought together.

---

[1] Simplicius, *De Cœlo*, ed. Heiberg, p. 471.

And yet the sun is countless millions of stadia from the upper sphere, while the other planets are not only below the fixed stars, but have their orbits at vast distances from each other; but the moon is so far below the stars that its distance from them cannot be estimated, while it is very near the earth, which it closely resembles in structure.

While the true nature of the moon was clearly perceived by Plutarch, another class of heavenly bodies, the comets, formed the subject of an equally sensible discussion by Seneca in the seventh book of his *Quæstiones Naturales*. He points out how contrary to observed facts is the opinion of Panætius the Stoic, that comets are not ordinary stars but merely false images of stars[1]. He shows that they cannot possibly be mere atmospheric phenomena, since they are not affected by wind or storm but move with perfect regularity. The argument against their being celestial bodies, that they are not confined to the zodiac but roam all over the heavens, he dismisses by saying (with Artemidorus of Parion) that we have no reason to think that the few planets which we know (and which move in the zodiac) are the only ones existing, whereas there may be others which are generally invisible, because their circles are so placed that they can only be seen when they pass through one extremity of them[2]. Of a similar opinion was also Apollonius of Myndus[3], who held them to be a separate class of celestial bodies, which generally travel through the upper regions of space and are only seen by us when they pass through the lower parts of their orbits[4].

[1] *Quæst. Nat.* VII. 30.
[2] Ibid. VII. 13.
[3] A contemporary of Alexander the Great.
[4] Ibid. VII. 17.

CHAPTER IX.

THE PTOLEMAIC SYSTEM.

GREEK astronomy found its last important cultivator and expounder in Claudius Ptolemæus of Alexandria in the second century of our era. We know next to nothing of his life, and neither when he was born nor when he died; we can only fix the time of his life by the fact that the earliest observation, which he mentions as having been made by himself, is from the eleventh year of Hadrian (A.D. 127), while his latest observation was made in A.D. 150. As epoch of his catalogue of stars he adopted A.D. 137 (first year of Antonine). As already mentioned, his principal work is the *Syntaxis*, commonly known as the *Almagest*, but summaries of the numerical data in the *Syntaxis* (with some deviations) are also given in his little book *Hypotheses of Planets* (to which Proklus wrote a commentary) and in an inscription dedicated to the "Saviour God" (Ptolemy Soter) and dated the tenth year of Antonine[1].

In the two hundred and sixty years between Hipparchus and Ptolemy astronomy does not seem to have made any progress, if we except the labours of Posidonius. The only observations recorded by Ptolemy as made during that long span of time, are an occultation of the Pleiades in A.D. 92, observed by Agrippa in Bithynia, and two occultations of Spica and β Scorpii observed by Menelaus at Rome in A.D. 98, but of course this does not prove that he did not make use of others, or at least that others were not made during that period. Possibly he was more indebted to Menelaus for observations of

[1] Edited by Halma, Paris, 1820 ("Hypothèses et époques des planètes de Cl. Ptolémée et Hypotyposes de Proclus Diadochus").

fixed stars than he has acknowledged[1]. But in the astronomy there was absolutely nothing done after Hipparcher until Ptolemy undertook to complete his work and to presᵈ to posterity the first complete treatise embracing the entire range of astronomical science.

In the beginning of his first book Ptolemy shortly recapitulates the fundamental assumptions of astronomy. The heavens is a sphere, turning round a fixed axis, as may be proved by the circular motion of the circumpolar stars and by the fact that other stars always rise and set at the same points of the horizon. • The earth is a sphere, situated in the centre of the heavens; if it were not, one side of the heavens would appear nearer to us than the other and the stars would be larger there; if it were on the celestial axis but nearer to one pole, the horizon would not bisect the equator but one of its parallel circles; if the earth were outside the axis, the ecliptic would be divided unequally by the horizon. The earth is but as a point in comparison to the heavens, because the stars appear of the same magnitude and at the same distances *inter se*, no matter where the observer goes on the earth. It has no motion of translation, first, because there must be some fixed point to which the motions of the others may be referred, secondly, because heavy bodies descend to the centre of the heavens which is the centre of the earth. And if there was a motion, it would be proportionate to the great mass of the earth and would leave behind animals and objects thrown into the air. This also disproves the suggestion made by some, that the earth, while immovable in space, turns round its own axis, which Ptolemy acknowledges would simplify matters very much. In his general conceptions Ptolemy therefore did not differ from his predecessors in any way.

As regards the motion of the sun, Ptolemy contented himself with the theory of Hipparchus. In this he made a great mistake, since in the course of nearly three hundred years

---

[1] It seems to have been commonly believed by the Arabs that Ptolemy had borrowed his whole star catalogue from Menelaus, adding 25′ (41 years' precession at 36″) to the longitudes. See Delambre, *Hist. de l'astr. du Moyen Age*, p. 380, and Al-Sûfi, *Description des étoiles fixes*, p. 42.

precession and the displacement of the line of apsides (of which he was ignorant) had increased the error of 35′ made by Hipparchus to about 5½°. The tropical year had been made too long by Hipparchus; consequently the mean motion was too small, the error in 300 years (from 147 B.C.) amounting to 76½′, to which may be added a maximum error of 22′ in the equation of the centre due to the error in the value of the excentricity adopted by Hipparchus. Thus the error in the position of the sun taken from Ptolemy's tables might amount to nearly 100′[1]. It is indeed very strange that Ptolemy did not make any attempt to improve the accuracy of the solar theory; possibly it did not interest him sufficiently owing to the absence of any but the one inequality in the motion; but no doubt the difficulty of measuring the absolute longitude of the sun with any degree of precision was supposed to throw too many obstacles in the way of finding more accurate values of the numerical quantities of the solar theory.

But when we turn to the theory of the moon, we find that Ptolemy made very substantial improvements in the work of his predecessor. Hipparchus had simply employed an epicycle moving on a deferent concentric to the earth. Ptolemy found that the outstanding errors of this theory, already vaguely noticed by Hipparchus, reached a maximum at the time of quadrature and disappeared altogether at syzygy; but a further difficulty was that the error did not return at every quadrature, sometimes disappearing altogether and sometimes amounting to as much as 2° 39′, its greatest value. Eventually it turned out that when the moon happened to be in quadrature and at the same time in the perigee or apogee of the epicycle, so that the equation of the centre was zero, the moon's place agreed perfectly with the theory of Hipparchus, while the error was greatest whenever the equation of the centre reached its maximum at the time of quadrature. The effect of the second inequality was therefore always to increase the absolute value of the first one, particularly in the quadratures. The obvious inference was, that the radius of the epicycle appeared to be of variable length, greater in quadrature than in syzygy. As the

[1] Tannery, *Recherches*, pp. 169–171.

D.   13

radius could not be supposed really to vary in length, its
distance from the earth had to vary, in order that it might
appear under different angles at different times; in other
words, the centre of the epicycle must move on an excentric
circle, but so that the angular velocity is uniform, not with
regard to the centre of the circle, but with regard to the earth.
But at the same time, the line through the centre and apogee
of the excentric is assumed to rotate in the retrograde direction
round the earth, so that the angle which it forms with the line
from the earth to the centre of the epicycle, the angle ATB, is

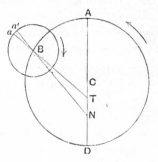

equal to twice the elongation of the moon from the sun, being
180° at the first and last quarter[1]. The distance of B from the
earth T will therefore be greatest at syzygy (the same in fact
as it would be according to the theory of Hipparchus) and
smallest at quadrature. The second inequality is caused by
the epicycle not being in the position in which it would have
been if moving on a concentric circle, and it is equal to the
angle between the lines from the earth to the two places of the
moon according to the two theories. This angle will be nought
at syzygy, because the centres of the epicycle and of the
excentric (B and C) are in a line with the earth and on the
same side of it, while the epicycle lies exactly where the theory
of Hipparchus would place it. At quadrature the centres are
on opposite sides of the earth, and therefore the epicycles

---

[1] In other words, C moves backward with a velocity equal to twice the
elongation minus the argument of latitude (or $24° 22' 53'' - 13° 13' 46'' = 11° 9' 7''$
per day), so that C, the centre of distances, in a synodic month describes a
small circle round the earth from east to west.

according to the two theories are furthest apart. If, however, at this time the moon is in the perigee or apogee of the epicycle, it will be on the line CD, and the angle representing the second inequality will still be nought; while it attains its greatest value ($2°39'$) if the line joining the moon to B is at right angles to that line, i.e. when the moon's anomaly is $90°$ or $270°$. The maximum value of the sum of the two inequalities, $7°40'$, by a simple calculation gave $CT : TA = 49·7 : 10·3$.

Ptolemy had thus made a great step forward by discovering the second inequality, now-a-days called the evection, by fixing its amount at $1°19'30''$, very near the true value, and by adapting the theory of Hipparchus to it. But continued observations showed that the theory was not yet quite sufficiently developed, as there was still some outstanding error. Undauntedly he attacked the problem again, but he did not succeed in discovering the third inequality (variation) but only made the theory still more complicated than it was already. The anomaly, as we have seen, was always counted from the line of apsides of the epicycle, passing through the earth. Ptolemy now supposes that it does not pass through the earth but is always directed to a point N situated at the same distance from T as C is ($TC = TN$). The mean apogee is therefore a', while the apparent apogee a oscillates a little on both sides of the position of the mean one, so that they coincide in the syzygies and quadratures. The correction to the anomaly necessitated by this arrangement ($\pi\rho\acute{o}\sigma\nu\epsilon\nu\sigma\iota\varsigma$ $\tau o\hat{\nu}$ $\dot{\epsilon}\pi\iota\kappa\acute{\nu}\kappa\lambda o\nu$) must be applied to the equation of the centre before finding the correction for evection.

The apparent place of the moon at the time of syzygy and quadrature could be determined according to Ptolemy's theory with an accuracy which was practically sufficient for his time, as he and his contemporaries only possessed crude instruments not capable of fixing the position of any celestial body without an error of perhaps $10'$. But though his theory was thus nearly sufficient for purposes of calculation, it could not claim to give the actual place of the moon in space, since it very grossly exaggerated the variation of the distance of the moon from the earth. From Chaldean observations of two lunar

eclipses occurring near the apogee of the epicycle, Ptolemy
finds the apparent diameter of the moon at apogee of epicycle
and excentric $= 31' 20''$ (only about $2'$ too much). From the
numerical data we have given, it is easy to see that at the
smallest distance from the earth the diameter of the moon
would be nearly a degree. But though Ptolemy cannot have
failed to perceive this, he takes no notice of it. It had now
become a recognized fact, that the epicyclic theory was merely
a means of calculating the apparent places of the planets
without pretending to represent the true system of the world,
and it certainly fulfilled its object satisfactorily, and, from a
mathematical point of view, in a very elegant manner. To the
Greek mind, the theory must have had a grave defect: the
principle of rigorously uniform motion had been violated, both
by introducing a point outside the centre of the deferent, with
regard to which the angular motion was uniform, and by the
prosneusis. This was perfectly indefensible from a physical point
of view but was of course mathematically quite admissible. It
was a stepping-stone in the direction of the discovery of elliptic
motion, but many centuries were to elapse before the work of
Ptolemy was continued.

The man who was capable of advancing the lunar theory so
much was naturally not disposed to leave the theories of the
five other planets in the unsatisfactory state in which he found
them. As if to distinguish these bodies with their far more
conspicuously irregular movements from the sun and moon he
always speaks of them as "the five wandering stars" (οἱ πέντε
πλανώμενοι), although it was more usual among the ancients
to speak of seven planets. He refers their motions to the
plane of the ecliptic[1], to which the plane of the deferent circle
of each planet is inclined at a certain small angle. But the
deferent is not (as in the theory of Apollonius) concentric to
the earth, it is excentric to it, in order to account for the
zodiacal inequality, which in reality is caused by the elliptic

---

[1] Or rather to a plane intersecting the celestial sphere in the ecliptic and
turning with the sphere of the fixed stars round the poles of the ecliptic, so as
to participate in the precession of the equinoxes, which the ancients looked on
as a motion of the sphere and not of the earth's axis.

form of the orbit. The epicycle, on which the planet moves with uniform velocity, accounts for the anomaly or second inequality (stations and retrogradations). The radius from the centre of the epicycle to the planet is (for Mars, Jupiter and Saturn) parallel to the line pointing to the mean place of the sun, while for Mercury and Venus the centre of the epicycle lies on this line. As in the system of Apollonius, the periods of revolution are:

	Of the centre of the epicycle on the excentric (motion in longitude)	Of the planet on the epicycle (motion in anomaly)
For the two inner planets	A sidereal year	Synodic Period[1]
For the three outer planets	Zodiacal period of planet	Synodic Period

But even thus the theory of Apollonius was not sufficient; Ptolemy found it necessary to add a complication somewhat similar to that by which he had put the final touch to his lunar theory. The greatest difference between the mean place and the observed place, in other words the angle which the radius of the epicycle subtended at the earth, turned out to be greater at the apogee and smaller at the perigee than the excentric motion could account for, so that the centre of distances must be nearer to the earth than the centre of uniform motion. He therefore introduced a punctum æquans, situated on the line of apsides, so that the order was: earth (T)—centre of deferent (C)—equant (E), and he found that the observations were best represented by making $TC = CE$. The point E had in the planetary theory nothing directly to do with the motion on the epicycle; but the line from the equant to the centre of the epicycle moved so that it described equal angles in equal times. E was therefore the centre of equal motion, while C was the centre of equal distances. Even this was not enough in the case of the planet Mercury, the centre of whose motion (E) was between the earth and the centre of the deferent, its distance

---

[1] That is, according to the notation of the ancients; according to our modern way of counting the anomaly from a fixed point in the zodiac the period would be, for the inner planets their heliocentric period, for the outer ones a sidereal year. See above, p. 154.

from the former being $\frac{1}{20}$ of the radius of the deferent, but the centre of the deferent, instead of being fixed, describes a small circle with radius $\frac{1}{21}$ in the direction from east to west round a point (C) distant $\frac{1}{21}$ beyond E, in the same period in which the centre of the epicycle moves round the deferent[1].

The inclinations of the planetary orbits to the ecliptic are so small that Ptolemy in his theories of the motion in longi-tude thought it permissible to neglect the deviations from the ecliptic. But the latitudes themselves gave him enough to do, and this part of the work he evidently found very difficult to arrange satisfactorily[2]. For the three outer planets the deferent was assumed to be inclined to the ecliptic at angles of 1° for Mars, 1° 30' for Jupiter, and 2° 30' for Saturn. For Mars the line of apsides of the deferent was perpendicular to the line of nodes, so that it coincided with the line joining the points of greatest north and south latitude; for Jupiter it was 20° west and for Saturn 50° east of the line of greatest latitude. The apogees were in all three cases north of the ecliptic. But the epicycles in their turn were inclined at the same angles to the plane of the deferents, so that their planes were always parallel to the ecliptic. Ptolemy was led to this assumption by remarking that at the apogee and perigee of the deferent the latitude (respectively south and north) was greatest when the planet happened to be at the perigee of its epicycle. As the epicycle of an outer planet was nothing but the earth's annual orbit round the sun transferred to the planet in question, it was of course quite right that the epicycle should be parallel to the ecliptic. In thus remaining parallel to a certain plane the epicycles did what the ancients considered an unusual thing, as they would have thought it natural that the plane of the epicycle should keep at the same angle to the radius joining the centre of the deferent to the centre of the epicycle. The hypothesis therefore demanded the introduction of a small

[1] *Hypotheses*, Halma, p. 48. Except for Mercury, the distance TE, ex-pressed in parts of the radius of the deferent, is practically equal to twice the excentricity in the elliptic theory. In the case of Venus it is too large, $\frac{1}{24} = 0.0417$ instead of 0.0137.

[2] *Syntaxis*, lib. XIII. caps. 1–6. Delambre's account is erroneous in many particulars, and that of Tannery does not deal with the latitudes.

auxiliary circle, the plane of which was perpendicular to the plane of the deferent, the centre of which was in the latter plane and which revolved in the zodiacal period of the planet[1]. If we imagine a stud on the circumference of this circle and let it slide in a slot in the epicycle, we see how the latter was kept parallel to the ecliptic. The varying greatest latitudes could thus be accounted for more or less, but it appears that the agreement was not considered good enough, since Ptolemy afterwards seems to have found it necessary to alter the inclinations of the epicycles to respectively $2° 15'$, $2° 30'$, and $4° 30'$[2], the diameter of the epicycle perpendicular to the line perigee-apogee being always parallel to the plane of the ecliptic.

Mercury and Venus had to be treated quite differently. In the figure, A is the apogee and P the perigee of the deferent; NN' is the line of nodes or the line of intersection of the planes of the deferent and the ecliptic. The angle between these was very small, $10'$ for Venus and $45'$ for Mercury[3], and the plane

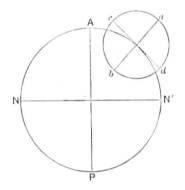

of the deferent oscillated within that limit to both sides of the ecliptic, coinciding with the latter when the centre of the epicycle was at N or N'. As to the epicycle, its line of apsides ab will fall in the plane of the deferent when at A and P, while the diameter at right angles to it, cd, is inclined to that plane at an angle called the λόξωσις (obliquatio). At the

---

[1] *Syntaxis*, xiii. 2, ed. Heiberg, ii. p. 529.

[2] These at least are the values given in the "Inscription" (Halma, p. 59).

[3] The *Hypotheses* give $10'$ for both planets.

nodes NN' the diameter cd falls in the ecliptic and ab is inclined to the deferent. This tilting of the epicycle is called ἔγκλισις, and is zero at A and P. In the case of Mercury, the planet at apogee is south of the ecliptic, in that of Venus (and all the other planets) it is north of it. Therefore, when the epicycle of Venus is at N' (the ascending node) the deferent lies in the ecliptic; as the epicycle advances, the point A rises north of the ecliptic and continues to rise till the epicycle reaches A. After that the latitude decreases until it becomes zero at N, but after that the part NPN' rises north of the ecliptic carrying the epicycle with it, so that the centre of the latter is always in north latitude except at N and N'. Simultaneously the double rocking of the epicycle is going on, like a ship pitching and rolling at the same time. For Mercury everything is reversed, north and south, otherwise the theory is similar.

That Ptolemy found the latitudes of the planets extremely troublesome is not strange, when we remember that in reality their lines of nodes pass through the sun, while Ptolemy had to assume that they passed through the earth. As the inner planets are quite surrounded by the earth's orbit, it was also natural that their motions in latitude should appear more intricate. In no other part of planetary theory did the fundamental error of the Ptolemaic system cause so much difficulty as in accounting for the latitudes, and these remained the chief stumblingblock up to the time of Kepler[1].

That the system as a whole deserves our admiration as a ready means of constructing tables of the movements of sun, moon, and planets cannot be denied. Nearly in every detail (except the variation of distance of the moon) it represented geometrically these movements almost as closely as the simple instruments then in use enabled observers to follow them, and

---

[1] How little the latitude theory satisfied Ptolemy himself appears from the fact that in his treatise "Hypotheses of the planets" he omits all references to the double oscillation of the epicycle and does not speak of the obliquity of the diameter cd. But in the "Inscription," in which only numerical values without explanations are given, he has the following values: For Venus, inclination of deferent 10′, of epicycle 2° 30′, obliquity (λόξωσις) 2° 30′; for Mercury 45′, 6° 15′ and 2° 30′ (Halma, p. 59).

it is a lasting monument to the great mathematical minds
by whom it was gradually developed. It appears from many
statements, not only of Ptolemy himself[1], but also of his commentators, that they merely considered the numerous circles as
a convenient means of calculating the positions of the planets,
and in reality the system is quite analogous to a development in a series of sines or cosines of multiples of the mean
anomaly. Ptolemy generally begins the theory of a particular
part of a planet's motion by saying " let us imagine (νοείσθω)
...a circle," and in the introduction to his *Hypotheses* he
says: " I do not profess to be able thus to account for all the
motions at the same time; but I shall show that each by itself
is well explained by its proper hypothesis[2]." And Proklus at
the end of his commentary[3] states distinctly that the epicycles
and excentrics are merely designed as the simplest way of
accounting for the motions, and in order to show the harmony
which exists among them. The fact (which cannot possibly
have escaped Hipparchus and Ptolemy) that their lunar theory
demanded excessive variations of the moon's distance, and
thereby of its apparent diameter, which never happened in
reality, shows that they did not look upon their work as a real
system of the world, but merely as an aid to computation.
Owing to the state of algebra at that time this had to be
done geometrically, just as Euclid had to adopt a geometrical
representation when dealing with irrational quantities or the
theory of proportion.

To the modern mind, accustomed to the heliocentric idea,
it is difficult to understand why it did not occur to a mathematician like Ptolemy to deprive all the outer planets of their
epicycles, which were nothing but reproductions of the earth's
annual orbit transferred to each of these planets, and also to
deprive Mercury and Venus of their deferents and place the
centres of their epicycles in the sun, as Herakleides had done.
It is in fact possible to reproduce Ptolemy's values of the ratio
of the radii of epicycle and deferent from the semi-axis major

---

[1] See for inst. XIII. 2 (Heiberg, II. p. 532).
[2] Halma, pp. 41–42.
[3] *Hypotyposes*, p. 151, Halma.

of each planet expressed in units of that of the earth, as
shown in the following table:

	Semi-axis major (a)	$\frac{1}{a}$	Ptolemy's ratio.
Mercury	0·3871		0·3708
Venus	0·7233		0·7194
Mars	1·5237	0·6563	0·6583
Jupiter	5·2028	0·1922	0·1917
Saturn	9·5388	0·1048	0·1083

Obviously the heliocentric idea of Aristarchus might just
as well have sprung out of the epicyclic theory as from that of
movable excentrics, and but for the fact that we have some
evidence of the latter being the older of the two, it would be
impossible to say which had formed the starting-point for
Aristarchus. But with regard to the curious dependence of
each planet on the sun in the Ptolemaic system, we have
already mentioned that the zodiacal inequality of the planets
showed that in any case a simple circular motion would not
"save the phenomena"; while the discovery of a strongly
marked inequality of the moon, depending on its position with
regard to the sun, confirmed the notion that the sun was
mixed up in the theories of all the celestial bodies alike.

Although the system of Ptolemy was a mere geometrical
representation of celestial motions, and did not profess to give
a correct picture of the actual system of the world, it would
have been impossible to omit a description of it from our review
of the cosmical systems, chiefly on account of its enormous
historical importance. For more than fourteen hundred years
it remained the Alpha and Omega of theoretical astronomy,
and whatever views were held as to the constitution of the
world, Ptolemy's system was almost universally accepted as the
foundation of astronomical science.

In addition to a complete theory of the planetary motions
the great work of Ptolemy also contains a catalogue of stars,
which, however, is nothing but the catalogue of Hipparchus
brought down to his own time with an erroneous value of the
constant of precession. The precession of the equinoxes was

discovered by Hipparchus[1] by comparing his own determinations of the longitudes of certain stars with those of Timocharis about 150 years earlier. Already in an earlier work, on the length of the year, he had stated that this displacement must amount to *at least* one degree in a hundred years[2], but in his later work "On the displacement of the solstitial and equinoctial signs" he mentions that he had found Spica to be six degrees from the autumnal equinox, while Timocharis had found the distance to be eight degrees. Now, this astronomer observed Spica in B.C. 294 and 283[3], while Hipparchus observed in B.C. 129, so that the change amounts to 45″ or 46″ a year[4]. The values adopted by Hipparchus for the tropical and sidereal year also point to his having adopted this value[5]. Ptolemy, however, by comparing longitudes of four stars found by Timocharis and Hipparchus, with those found by Agrippa and Menelaus in A.D. 93 and 98, found 36″ per annum or one degree in a hundred years, and he adopted this convenient and round number. It is very remarkable that so important a discovery should not have become universally known; and yet we find that precession is never alluded to by Geminus, Kleomedes, Theon of Smyrna, Manilius, Pliny, Censorinus, Achilles, Chalcidius, Macrobius, Martianus Capella! The only writers except Ptolemy who allude to it are Proklus, who flatly denies its existence[6], and Theon of Alexandria, who accepts the Ptolemaic

[1] At least we possess no positive proof that it was known to the Babylonians, though they seem to have been aware that earlier determinations of the equinox required some correction, as three tablets give different positions of the equinox, 10°, 8° 15′, and 8° 0′ 30″ of Aries. Kugler, *Mondrechnung*, p. 103.

[2] *Syntaxis*, VII. 2, Heib. II. pp. 15–16.

[3] VII. 3, pp. 28–29.

[4] Curious, that Ptolemy himself (ibid. p. 30) from the two conjunctions of Spica and the moon, observed by Timocharis, makes out that the longitude of the star had changed 10′ in the interval of (nearly) twelve years, i.e. 50″ a year.

[5] Tannery, *Recherches*, p. 195.

[6] *Hypotyp.* ed. Halma, p. 150, at top, also *In Timæum*, pp. 277 D—278 A. The authority of Julian the Neo-Platonist and similar people is to him far greater than that of Hipparchus and Ptolemy. He imagines that precession has been introduced by the latter to explain the sidereal revolutions of the planets, and he maintains that those who do not use it, such as the Chaldeans, account far better for the phenomena. He also objects that precession ought to have caused circumpolar stars to pass below the horizon before his time.

value of one degree in a hundred years, but who tells the following strange story about it[1]: "According to certain opinions ancient astrologers believe that from a certain epoch the solstitial signs have a motion of 8° in the order of the signs, after which they go back the same amount; but Ptolemy is not of this opinion, for without letting this motion enter into the calculations, these when made by the tables are always in accord with the observed places. Therefore we also advise not to use this correction; still we shall explain it. Assuming that 128 years before the reign of Augustus the greatest movement, which is 8°, having taken place forward, the stars began to move back; to the 128 years elapsed before Augustus we add 313 years to Diocletian and 77 years since his time, and of the sum (518) we take the eightieth part, because in 80 years the motion amounts to 1°. The quotient (6° 28′ 30″) subtracted from 8° will give the quantity by which the solstitial points will be more advanced than by the tables."

The only other ancient writer who alludes to this theory is Proklus, who merely says that the tropical points according to some move, not in a whole circle, but some degrees to and fro[2].

The idea of these people was, therefore, that the longitude of a star increased for 640 years (1° in 80 years), and that it then suddenly began to decrease and went on doing that for 640 years, and then equally suddenly took to increasing again. These good people must have lived before Ptolemy, since he did not agree with them (and since Theon calls them παλαιοί), and they must have been later than Hipparchus, since they knew something about precession and adopted his value of 45″ a year. But what made them think of this extraordinary idea? And why did they fix on the year 158 B.C. as one in which a change of direction occurred; and why was the length of the arc 8°? These questions are difficult to solve. The year is probably that in which they supposed the astronomical work of Hipparchus to have commenced, and the only way in

[1] *Table manuelle*, Halma, p. 53; Delambre, *Astr. anc.* II. p. 625. Not mentioned in Theon's *Commentary*.
[2] *Hypotyposes*, ed. Halma, p. 88.

which a change of 8° in the equinoxes and solstices can be
connected with Hipparchus, is his having finally placed the
beginning of the signs of Aries, Cancer, Libra (claws of
Scorpion), and Capricornus at the equinoxes and solstices, as
Aratus had done, while Eudoxus and others placed these points
in the middle of those signs or at the eighth degree[1]. In
doing so they followed the lead of the Babylonians, whose
ecliptic was a fixed one, determined by the constellations.
Kugler has found that their equinox was at 8° 15′ of Aries, while
their longitudes of new moon are on an average 3° 14′ too
great, so that the beginning of the signs on the Babylonian
ecliptic correspond to about the fifth degree of the signs of our
movable ecliptic[2]. Long afterwards Pliny still gives the eighth
degree as coinciding with the equinoxes and tropics[3], while
Manilius and Achilles say that some writers place them at the
beginning of the signs, others at the eighth degree, others
at the tenth or twelfth[4]. Possibly some ignorant writer by a
misunderstanding concluded from this discrepancy that the
equinoxes oscillated backwards and forwards, and thus started
the theory of the variability of precession which, owing to
the low state, or rather non-existence of practical astronomy
for many centuries after Ptolemy, took firm root, spread to
India and among the Arabs, and was not finally swept aside
until Tycho Brahe appeared on the scene.

Ptolemy not only does not countenance this error, he never
alludes to it anywhere. To him precession is simply a slow
rotation of the sphere of the fixed stars from west to east
around the poles of the zodiac, completed in 36,000 years.
We can hardly doubt that Hipparchus was of the same opinion.
It is true that he called his treatise, not " on the displacement
of the fixed stars " but " on the displacement ($\mu\epsilon\tau\acute{a}\pi\tau\omega\sigma\iota\varsigma$) of
the solstitial and equinoctial signs," and that Ptolemy[5] quotes

[1] Hipparchus, ed. Manitius, pp. 128, 130. Eudoxus let the heliacal rising
of Sirius coincide with the entry of the sun into Leo, the solstice being at the
eighth degree of Cancer. Boeckh, *Sonnenkreise*, p. 190.

[2] Kugler, *Mondrechnung*, p. 103 sq.; Ginzel, *Astr. Kenntnisse der Babylonier*,
p. 204.

[3] *Hist. nat.* ii. 19, § 81, xviii. 25, § 221 and 28, § 264 ; Mart. Capella, viii. 829.

[4] Manilius, iii. 676; Achilles, c. 23.   [5] *Syntaxis*, vii. 2, Heib. ii. p. 15.

him as having said in his book on the length of the year that the tropics and equinoxes "for this cause are moved towards the west"; but he does not say that Hipparchus differed from himself, and in two other places[1] he says distinctly that Hipparchus conjectured that the sphere has a slow rotation in the same direction in which the planets are moving. It is also well known that Hipparchus at first only attributed the eastward drift to the few zodiacal stars, the longitudes of which he had found to have increased, though he soon perceived that the drift was common to all stars.

Ptolemy was the last great astronomer of the Alexandrian school. Several mathematicians of great eminence followed, such as Pappus and Diophantus, but they added nothing to the stock of astronomical knowledge. Ptolemy's works continued to be lectured on in the schools; Theon of Alexandria wrote a valuable commentary to it, but he was probably the last scientific man who could make use of the celebrated library, as it was destroyed in his lifetime by the savage Christian mob of the city (A.D. 389). His renowned daughter Hypatia, who was justly considered a personification of the highest Greek culture and thought, was as such barbarously murdered some years later (415), and the curtain went down for ever on the great stage where Greek science had played its part so well and so long. In Greece the Neo-Platonic school was still alive, and even produced a last philosopher of distinction, Proklus, after whose death it continued to drag on a feeble existence for another half century until the Emperor Justinian suppressed it in the year 529. In company with six other philosophers, Simplicius (who afterwards wrote an elaborate commentary to Aristotle which we have often referred to) sought a refuge in Persia, under the mistaken idea that they would there find unprejudiced rulers and freedom to teach; but they were disappointed and came home in a few years, having found that nowhere in the known parts of the world was the wisdom of the past held in any repute. The long dark night of the Middle Ages had set in.

[1] *Syntaxis*, III. 1, and VII. 1, Heib. I. p. 192, and II. p. 3.

CHAPTER X.

MEDIEVAL COSMOLOGY.

THE Roman Empire was destroyed in the course of the hundred years following the memorable year (375) when the Huns invaded Europe through the natural gateway between the Caspian Sea and the Ural mountains, and drove Gothic and Germanic races headlong before them over most of the provinces of the Roman Empire. In 476 the last nominal Emperor of the West was deposed by a barbarian chieftain; that part of Europe which had formed the Western Empire had been partitioned among the conquerors; ruin and devastation reigned everywhere. There seemed to be an end of all civilisation, as the conquerors were utterly untouched either by the ancient culture of Asia or by anything they might have learned from their new subjects. To some extent their savage state was doubtless softened by the Christian religion, which they gradually adopted; but most of their teachers were unfortunately devoid of sympathy for anything that emanated from the heathen Greek and Roman world; and it was left to the dying Neo-Platonic school and to pagan commentators like Macrobius and Simplicius and the encyclopædic writer Martianus Capella to keep alive for a while the traditions of the past.

But even before the days when enemies from outside had begun to assail the Roman Empire, a fierce onslaught had commenced on the results of Greek thought. A narrow-minded literal interpretation of every syllable in the Scriptures was insisted on by the leaders of the Church, and anything which could not be reconciled therewith was rejected with horror and scorn. In this way some of the Fathers of the Church lent a hand to the barbarians who wrenched back the hand of time about a thousand years, and centuries were to elapse before their work was to some extent undone and human thought began to free itself from the fetters imposed on it in the days when the ancient world was crumbling to decay. In no branch of knowledge was the desire to sweep away all the results of Greek learning as conspicuous as with regard to the figure of the earth and the motion of the planets. When we turn over the pages of some of these Fathers, we might imagine that we were reading the opinions of some Babylonian priest written down some thousands of years before the Christian era; the ideas are exactly the same, the only difference being that the old Babylonian priest had no way of knowing better, and would not have rejected truth when shown to result from astronomical observations.

At first there was no enmity to science exhibited by the followers of the Apostles. Clemens Romanus, in his epistle to the Corinthians[1], written about A.D. 96, alludes in passing to the Antipodes as dwelling in a part of the earth to which none of our people can approach, and from which no one can cross over to us; and in the beginning of the same chapter he uses an expression often found in classical writings, that " the sun and moon and the dancing stars (ἀστέρων τε χοροί) according to God's appointment circle in harmony within the bounds assigned to them without any swerving aside." In Alexandria, where the leaders of the Christians were familiar with the philosophical speculations of Philo and the Neo-Platonists, it was natural that they should feel no desire to place themselves in opposition to science. Clement of Alexandria (about A.D. 200), who had commenced life as a heathen, is indeed the first to view the Tabernacle and its furniture as representing allegorically the whole world; but he is not thereby led astray into sweeping aside the knowledge gained by the Greeks. The lamp was placed to the south of the table, and by it were represented the motions of the seven planets, as the lamp with three branches on either side signifies the sun set in the midst of the planets. The golden figures, each with six wings, represent either the

[1] C. 20. Lightfoot's ed. I. p. 282.

two Bears, or more likely (he thinks) the two hemispheres, while
he believes the ark to signify the eighth region and the world
of thought or God[1]. This desire to find allegories in Scripture
was carried to excess by Origen (185–254), who was likewise
associated with Alexandrian thought, and he managed thereby
to get rid of anything which could not be harmonised with
pagan learning, such as the separation of the waters above the
firmament from those below it, mentioned in Genesis, which he
takes to mean that we should separate our spirits from the
darkness of the abyss, where the Adversary and his angels
dwell[2].

But this kind of teaching was not to the taste of those who
would have nothing to do with anything that came from the
pre-Christian world, and to whom even "the virtues of the
heathen were but splendid vices." A typical representative of
these men was Lactantius, the first and the worst of the adversaries of the rotundity of the earth, whose seven books on
Divine Institutions seem to have been written between A.D. 302
and 323. In the third book, *On the false wisdom of the philosophers*, the 24th chapter is devoted to heaping ridicule on the
doctrine of the spherical figure of the earth and the existence of
antipodes[3]. It is unnecessary to enter into particulars as to his
remarks about the absurdity of believing that there are people
whose feet are above their heads, and places where rain and hail
and snow fall upwards, while the wonder of the hanging gardens
dwindles into nothing when compared with the fields, seas,
towns, and mountains, supposed by philosophers to be hanging
without support. He brushes aside the argument of philosophers that heavy bodies seek the centre of the earth as
unworthy of serious notice; and he adds that he could easily
prove by many arguments that it is impossible for the heavens
to be lower than the earth, but he refrains because he has
nearly come to the end of his book, and it is sufficient to have
counted up some errors, from which the quality of the rest may
be imagined.

---

[1] *Stromata*, l. v. cap. 6; *Ante-Nicene Chr. Libr.* XII. pp. 240–242.
[2] *In Genesim Homiliæ*, *Opera*, ed. Delarue, Paris, 1733, t. II. p. 53.
[3] Ed. Dufresnoy, Paris, 1748, t. I. p. 251.

More moderate in his views was Basil, called the Great, who wrote a lengthy essay on the six days of creation about the year 360[1]. He does not rave against the opinions of philosophers as Lactantius did; he is evidently acquainted with the writings of Aristotle, and generally expresses himself with a certain degree of moderation and caution. Thus he is aware of the fact that there are stars about the south pole of the heaven invisible to us, and he understands perfectly well how summer and winter depend on the motion of the sun through the northern and southern halves of the zodiac[2]. When speaking of the two "great lights," he says that they are really of an immense size, since they are seen equally large from all parts of the earth; no one is nearer to the sun or farther from it, whether it is rising or on the meridian or setting; besides which the whole earth is illumined by the sun, while all the other stars give only a feeble light[3]. But though he is aware of the annual motion of the sun, he does not uphold the spherical form of the heavens or deny that there is more than one heaven; the words of Genesis about the upper waters are too distinct for that; and he sets forth the idea, common among patristic writers, that these waters were placed above the firmament to keep it cool and prevent the world from being consumed by the celestial fire[4]. As to the figure of the earth, he says that many have disputed whether the earth is a sphere or a cylinder or a disc, or whether it is hollow in the middle; but Moses says nothing about this, nor about the circumference of the earth being 180,000 stadia, nor about anything which it is not necessary for us to know[5]. Basil evidently was too sensible to deny the results of scientific investigation, but also too timid to advocate them openly, so that he at most merely mentions them without comment, or endeavours to show that a Christian may accept them without danger to his faith. But for his acceptance of the upper waters, he might seem to have been a comparatively unprejudiced thinker.

The ruthlessly literal interpretation of Scripture was especially

---

[1] *Homiliæ novem in Hexaemeron, Op. omn.*, ed. Garnier, Paris, 1721, t. I.
[2] I. 4, l. c. p. 4 E.   [3] VI. 9–10, l.c. p. 58 D–60 A.
[4] III. 3, l. c. p. 23 E–24 E.   [5] IX. 1, l.c. p. 80.

insisted on by the leaders of the Syrian Church, who would hear
of no cosmogony or system of the world but that of Genesis.
A contemporary of Basil, Cyril of Jerusalem, lays great stress
on the necessity of accepting as real the supercelestial waters[1],
while a younger contemporary of Basil, Severianus, Bishop of
Gabala, speaks out even more strongly and in more detail in
his *Six Orations on the Creation of the World*[2], in which the
cosmical system sketched in the first chapter of Genesis is
explained. On the first day God made the heaven, not the one
we see, but the one above that, the whole forming a house of
two storeys with a roof in the middle and the waters above that.
As an angel is spirit without body, so the upper heaven is fire
without matter, while the lower one is fire with matter, and
only by the special arrangement of providence sends its light
and heat down to us, instead of upwards as other fires do[3]. The
lower heaven was made on the second day; it is crystalline,
congealed water, intended to be able to resist the flame of sun
and moon and the infinite number of stars, to be full of fire and
yet not dissolve nor burn, for which reason there is water on the
outside. This water will also come in handy on the last day,
when it will be used for putting out the fire of the sun, moon
and stars[4]. The heaven is not a sphere, but a tent or tabernacle; "it is He...that stretcheth out the heavens as a curtain
and spreadeth them out as a tent to dwell in[5]": the Scripture
says that it has a top, which a sphere has not, and it is also
written: "The sun was risen upon the earth when Lot came
unto Zoar[6]." The earth is flat and the sun does not pass under
it in the night, but travels through the northern parts "as if
hidden by a wall," and he quotes: "The sun goeth down and
hasteth to his place where he ariseth[7]." When the sun goes

[1] *Catechesis*, IX., *Opera*, Oxford, 1703, p. 116.

[2] Joh. Chrysostomi *Opera*, ed. Montfaucon, t. VII. (Paris, 1724), p. 436 sqq.
Compare also the extracts given by Kosmas, pp. 320–325.

[3] I. 4.   [4] II. 3–4.

[5] Isaiah xl. 22.

[6] Gen. xix. 23. The above is from the Revised Version, but Severianus (III. 4)
has: "Sol egressus est super terram, et Lot ingressus est in Segor. Quare liquet,
Scriptura teste, egressum esse Solem, non ascendisse."

[7] Eccles. i. 5.

more to the south, the days are shorter and we have winter, as the sun takes all the longer to perform his nightly journey[1].

The tabernacle shape of the universe was from that time generally accepted by patristic writers; thus by Diodorus, Bishop of Tarsus (died 394), who in his book *Against Fatalism*[2] declaims against those atheists who believe in the geocentric system; and he shows how Scripture tells us that there are two heavens created, one which subsists with the earth, and one above that again, the latter taking the place of a roof, the former being to the earth a roof but to the upper heaven a floor. Heaven is not a sphere but a tent or a vault[3]. This was also the opinion of Theodore, Bishop of Mopsuestia in Cilicia (d. about 428), but his work is lost, and we only know from the sneers of a later writer, Philoponus, that he taught the tabernacle theory and let all the stars be kept in motion by angels. About the same time St Jerome wrote with great violence against those who followed "the stupid wisdom of the philosophers" and had imagined the Cherubim to represent the two hemispheres, our-selves and the antipodes, and he explained that Jerusalem was the navel of the earth[4].

Somewhat more sensible opinions seem to have prevailed at that time in the Western Church. Ambrose of Milan (d. 397) says that it is of no use to us to know anything about the quality or position of the earth, or whether heaven is made of the four elements or of a fifth[5]; but still he mentions the heaven repeatedly as a sphere[6]. Driven into a corner by the question how there can be water outside the sphere, he some-what feebly suggests that a house may be round inside and square outside[7], or he asks why water should not be suspended

---

[1] iii. 5.

[2] This book is lost, but Photius gives a *résumé* of it, *Bibl.*, Codex 223.

[3] Compare Chrysostom in his comment. on Hebrews viii. 1: " Where are those who say that the heaven is in motion? Where are those who think it spherical? For both these opinions are here swept away." Quoted by Kosmas p. 328.

[4] Comment. on Ezekiel, chs. i. and v.; *Opera*, Benedict. ed., Paris, 1704, ii. pp. 702 and 726.

[5] *Hexaem.* i. 6; Benedict. ed., Paris, 1686, t. i. col. 11, also ii. 2, col. 25.

[6] Ibid., also i. 3, col. 4.

[7] ii. 3, col. 26.

in space just as well as the heavy earth, while its use is
obviously to keep the upper regions from being burned by the
fiery ether[1]. It was natural that Augustine (354–430), who
may be considered a disciple of Ambrose, should express himself with similar moderation, as befitted a man who had been
a student of Plato as well as of St Paul in his younger days.
With regard to antipodes, he says that there is no historical
evidence of their existence, but people merely conclude that the
opposite side of the earth, which is suspended in the convexity
of heaven, cannot be devoid of inhabitants. But even if the
earth is a sphere, it does not follow that that part is above
water, or, even if this be the case, that it is inhabited; and
it is too absurd to imagine that people from our parts could
have navigated over the immense ocean to the other side, or
that people over there could have sprung from Adam[2]. With
regard to the heavens, Augustine was, like his predecessors,
bound hand and foot by the unfortunate water above the
firmament. He says that those who defend the existence of
this water point to Saturn being the coolest planet, though we
might expect it to be much hotter than the sun, because it
travels every day through a much greater orbit; but it is kept
cool by the water above it. The water may be in a state of
vapour, but in any case we must not doubt that it is there, for
the authority of Scripture is greater than the capacity of the
human mind[3]. He devotes a special chapter[4] to the figure of
the heaven, but does not commit himself in any way, though he
seems to think that the allusions in Scripture to the heaven
above us cannot be explained away by those who believe the
world to be spherical. But anyhow Augustine did not, like
Lactantius, treat Greek science with ignorant contempt; he
appears to have had a wish to yield to it whenever Scripture

---

[1] *Hexaem.* II. col. 29.

[2] *De Civitate Dei*, lib. XVI. cap. 9; Benedict. ed., t. VII. cols. 423–24.

[3] *De Genesi ad litteram*, lib. II. cap. 5 (t. III. pp. 134–135). Compare *De Civ.
Dei*, XI. 34 (t. VII. col. 299), where he says that some have thought that heavy
water could not be above heaven, and have therefore interpreted it as meaning
angels; but they should remember that pituita, which the Greeks call φλέγμα,
is placed in the head of man! A beautiful comparison.

[4] *De Genesi*, II. 9; t. III. cols. 138–39.

did not pull him the other way, and in times of bigotry and ignorance this is deserving of credit.

We have thus seen that the Fathers of the Church did not all go equally far in their condemnation of Greek astronomy, and that none of them took the trouble to work out in detail a system to take the place of the detested doctrines of the pagan philosophers. This work was undertaken by one who did not hold high office in the Church, but who had travelled a great deal by land and by sea, and might therefore have been expected to be more liberal-minded in his views than a church-man who had not had that advantage. He is known by the name of Kosmas, surnamed Indicopleustes, or the Indian navi-gator. His book, the *Christian Topography*[1], contains some passages which throw light on his history and enable us to fix the date at which he wrote. He was probably a native of Alexandria, and during the earlier part of his life he was a merchant. He tells us himself (somewhat needlessly) that he was "deficient in the school learning of the pagans[2]," though on the other hand he alludes to the theory of epicycles, and thereby deprives himself of the excuse for his silly notions which total ignorance of Alexandrian learning might have supplied. He travelled in the Mediterranean, the Red Sea, and the Persian Gulf, and on one occasion he even dared to sail on the dreaded Ocean, which "cannot be navigated on account of the great number of its currents and the dense fogs which it sends up, obscuring the rays of the sun; and because of the vastness of its extent[3]." One of the most interesting parts of his book is that which describes his travels in Abyssinia and adjoining countries. As he must have reached places within ten degrees of the equator, it is very remarkable that he could be blind to the fact that the earth is a sphere. His work on *Christian Topography* consisted originally of five books, to which seven others were subsequently added in order to further elucidate

---

[1] "The Christian Topography of Cosmas, an Egyptian monk." Translated from the Greek and edited with notes and introduction by J. W. McCrindle, London, printed for the Hakluyt Society, 1897. The pages quoted are those of Montfaucon's edition in Vol. II. of his *Nova Collectio Patrum*.

[2] p. 124.   [3] p. 132.

various points; it must have been written between the years
535 and 547, as events which happened in these years are
alluded to in the text as occurring while the author was writing.

The first book is *Against those who, while wishing to profess
Christianity, think and imagine like the pagans that the heaven
is spherical.* The daily revolution of the heaven he thinks he
has swept away by saying that the appearance of the Milky
Way shows that the heaven must be constituted of more than
one element, and that it must therefore either have a motion
upwards or downwards, but nothing of the kind has ever been
perceived by anybody. He next asks why the planets stand
still, and even make retrogressions. "They will, perhaps, in
reply assign as the cause those invisible epicycles which they
have assumed as vehicles on which, as they will insist, the
planets are borne along. But they will be in no better case
from this invention, for we shall ask: Why have they need of
vehicles? Is it because they are incapable of motion? Then
if so, why should you assert them to be animated, and that too
even with souls more than usually divine? Or is it that they
are capable? The very idea is, methinks, ridiculous. And why
have not the moon and the sun their epicycles? Is it that they
are not worthy on account of their inferiority? But this could
not be said by men in their sober senses. Was it, then, from
the scarcity of suitable material the Creator could not construct
vehicles for them? On your own head let the blasphemy of
such a thought recoil[1]."

The alleged position of the earth in the centre of the
universe is also in the eyes of Kosmas utterly absurd, as the
earth is so unspeakably heavy that it can only find rest at
the bottom of the universe. The usual cheap arguments against
the existence of antipodes are next served out, but he does not
think these "old wives' fables" worthy of many words. In a
note appended to his fourth book[2] he asks, into which of the
eight or nine heavens (spheres) of the pagans Christ has
ascended, and into which one Christians hope themselves to
ascend? "If the sphere which has motion forces the others

<hr />

[1] p. 119.   [2] pp. 189-191.

to revolve along with it from east to west, whence is produced the motion in the contrary direction of the seven planets? Is it the spheres that have the contrary motion, or the stars themselves? If the spheres, how can they at one and the same time move both westward and eastward? And if the stars, how do the stars cut their way through the heavenly bodies?" After which Kosmas quotes various passages of Scripture in order utterly to crush those Christians who wished to listen to the Greek philosophers ("no man can serve two masters"), and winds up by enquiring how a spherical earth situated in the middle of the world could have emerged from the waters on the third day of creation, or how it could have been swamped by the deluge in the days of Noah?

Kosmas' own idea is, that the figure of the universe can only be learned by studying the design of the Tabernacle, which Moses constructed in the wilderness. We have seen how Severianus and others had already assumed that the earth was like a tabernacle, but their suggestion was now worked out in detail by Kosmas, who points out[1] that Moses had pronounced the outer tabernacle to be a pattern of the visible world, while the Epistle to the Hebrews, in explaining the inner tabernacle, or that which was within the veil, declared that it was a pattern of the kingdom of heaven, the veil being the firmament which divides the universe into two parts, an upper and a lower. The table of shew-bread with its wavy border signified the earth surrounded by the ocean, and another border outside the first one represented another earth beyond the ocean, while every other article in the tabernacle similarly had a cosmographical meaning. As the table was placed lengthwise from east to west, we learn that the earth is a rectangular plane, twice as long as it is broad, and its longer dimension extending from east to west. The ocean which encompasses our earth is in its turn surrounded by another earth, which had been the seat of Paradise and the dwelling-place of man until the deluge, when the Ark carried Noah and his family and animals over to this earth, while the old one has since been inaccessible owing to the

[1] p. 134.

unnavigable state of the ocean. The walls of heaven are four perpendicular planes joined to the edges of the transoceanic earth, and the roof is shaped like a half cylinder resting on the north and south walls, the whole thing being by Kosmas likened to the vaulted roof of a bathroom, while to the modern mind it looks more like a travelling trunk with a curved lid. The whole structure is divided into two storeys by the firmament, which forms a floor for the upper and a ceiling for the lower storey, the latter being the abode of angels and men until the day of judgment, the upper storey being the future dwelling of the blest.

The earth, the footstool of the Lord, is at the bottom of the structure, while the sun, moon and stars are not attached to its sides or roof, but are carried along in their courses below the firmament by angels, who have to carry on this work until the last day. The rising and setting of the sun required a special explanation. Of course it could not possibly go under the earth, and it was necessary (with Severianus) to assume that it was hidden by the northern part of the earth during the night. Quoting the same passage from the book of Ecclesiastes he states[1] that the earth is much higher in the north and west than in the south and east, and that it is well known that ships sailing to the north and west are called lingerers, because they are climbing up and therefore sail more slowly, while in returning they descend from high places to low, and thus sail fast. The Tigris and Euphrates flowing south have far more rapid currents than the Nile, which is " running, as one may say, up[2]." In the north there is a huge conical mountain, behind which the sun passes in the night, and according as the sun during this passage is more or less close to the mountain, it appears to us as if it passes nearer to the top or nearer to the base of the mountain; in the former case the night is short and we have summer, in the latter it is long and we have winter. All the other heavenly bodies are likewise moved in their orbits by angels and pass behind the northern, elevated part of the earth, while eclipses

---

[1] p. 133.

[2] It was absurd to believe that the rain at the antipodes could fall up, but evidently it was all right to believe that a river was flowing up!

are produced because "the revolution and the course of the heavenly bodies have some slight obliquity[1]."

When Kosmas had finished his five books, he was asked how the sun could possibly be hidden behind the northern part of the earth, if it is many times larger than the earth? He therefore devotes his sixth book to proving that the sun is in reality quite small, and nowhere does he prove himself as incapable of reasoning on the simplest facts as here. Because at the summer solstice the shadow of a man at Antioch or Rhodes (the beginning of the sixth climate of Ptolemy[2]) was half a foot shorter than it was at Byzantium (a little beyond the beginning of the seventh), he concluded that the sun " has the size of two climates[3]." For at Meroe the man would be shadowless, at Syene (one climate north) the shadow would be half a foot to the north, and in Ethiopia (one climate south) half a foot to the south, therefore the diameter of the sun is two climates[4]!

Such was the celebrated system of the world of Kosmas Indicopleustes. He was not a leader in the Church (it is even uncertain whether he was Orthodox or a Nestorian) and his book apparently never rose to be considered a great authority. By the fact that he wrote, so to say, a text-book on the subject, he acquired a certain notoriety; but though it cannot be denied that he displayed a good deal of originality in twisting the innumerable passages from Scripture with which his book bristles into proofs of his assertions, his system had in reality been indicated by the Church Fathers of the preceding two hundred years. This is acknowledged by Kosmas, who in his tenth book collects a number of quotations from the Fathers, especially from Severianus. Poor Kosmas has come in for a good deal of ridicule, but in fairness this ought to be addressed to his predecessors, who had abused the authority of their position in the Church and their literary ability to propagate ideas which had been abandoned in Greece eight hundred years

---

[1] p. 156.

[2] The first climate begins where the longest day is $12^h 0^m$; the second where it is $12^h 30^m$, and so on.

[3] p. 265.

[4] About 1060 miles, or if we take the two climates from Alexandria to Byzantium, about 680 miles.

before. But what Kosmas does deserve to be blamed for is his
not finding out on his travels that the earth is a sphere.

The fanatics who desired to clear away as noxious weeds the
whole luxuriant growth of Greek science had, however, not
altogether the field to themselves. Some writers there were,
even then, who studied the works of the Greek philosophers,
and were not afraid to accept at least some of their doctrines.
Among these was Johannes Philoponus, a grammarian of
Alexandria, who seems to have lived about the end of the sixth
century, and who wrote commentaries to several of the writings
of Aristotle as well as a number of treatises showing a remarkable freedom of thought, which naturally obtained for him the
name of a heretic, and in later times would certainly have
caused him to be sent to the stake. In his book on the creation
of the world[1] he argues against the abuse of Scriptural quotations
by Theodore of Mopsuestia to prove that the heaven is not
spherical[2] or that the stars are moved by angels appointed to
this task; and he asks why God should not have endowed the
stars with some motive power. He even goes so far as to
compare this power to the tendency of all bodies, heavy and
light, to fall to the earth[3]. Owing to his want of orthodoxy,
the opinions of Philoponus could, however, not influence his
contemporaries to any appreciable extent, and it was of much
more importance that a man holding high office in the Western
Church, Isidorus Hispalensis, Bishop of Seville, expressed himself very sensibly on the constitution of the world. Isidore was
born about 570, and became Bishop of Seville already in 601,
probably through his high family connections; but he soon
became widely known by his learning and eloquence, and twice
presided over Church Councils. He died in 636. Among his
numerous writings is an encyclopædic work: *Etymologiarum
libri xx*, in the beginning of which he enumerates the seven
free arts, grammar, rhetorics, dialectics (the *trivium*), and arith-

---

[1] *Bibliotheca veterum patrum cura Andreæ Gallandii*, Venice, 1776, t. XII.
p. 471 sq.

[2] III. 10.

[3] I. 12. He probably got the idea from Empedokles, quoted by Aristotle,
De Cœlo, II. 1, p. 284 a, 24, and Simplicius, p. 375, 29–34 (Heib.).

metic, music, geometry, astronomy (the *quadrivium*), which already long before his time had come to be considered as embracing all human knowledge[1]. The work is called *Etymologies* (sometimes *Origins*), because Isidore generally explains the meaning of a word or an expression by means of its supposed derivation. When dealing with dangerous topics, such as the figure of the world and the earth, he does not lay down the law himself, but quotes "the philosophers" as teaching this or that, though without finding fault with them. Thus he repeatedly mentions that according to them the heaven is a sphere, rotating round an axis and having the earth in the centre[2]. In the same manner he refers to the spherical shape of the earth, saying in the chapter about Africa[3]: "But in addition to the three parts of the orbis (Asia, Europe, Africa) there is a fourth to the south beyond the ocean, which, owing to the heat of the sun, is unknown to us, at the outskirts of which the antipodes are fabulously reported to dwell." This fourth continent has frequently been postulated by geographers of antiquity, some of them even assuming the existence of two other *oekumenes* in the western hemisphere, one north and one south of the equator; and it is creditable to Isidore that he does not, like his predecessors, rave about the iniquity of imagining the existence of people on the opposite side of the earth.

Isidore also wrote a smaller treatise, *De rerum natura*, giving more details about some of the subjects touched on in the larger work, and here again we find him occupying a place midway between "the philosophers" and the bigoted patristic writers. The heaven is a sphere revolving once in a day and a night[4], and though Ambrosius in his *Hexaemeron* says that

---

[1] The names, though not the usual order, of the seven arts may be remembered by the distich :

"*Gram.* loquitur, *dia.* verba docet, *rhe.* verba ministrat,

Mus. canit, *ar.* numerat, *ge.* ponderat, *as.* colit astra."

[2] For instance: "Nam philosophi dicunt cœlum in sphæræ figuram undique esse convexam, omnibus partibus æquale, concludens terram in media mundi mole libratam. Hoc moveri dicunt, et cum motu ejus sidera in eo fixa ab oriente usque ad occidentem circuire." xiii. 5, compare iii. 31–32.

[3] Liber xiv. cap. 5, *De Libya*, § 17.

[4] Cap. xii.

philosophers make out that there are seven heavens of the seven planets, yet human temerity does not dare to say how many there may be. God the Creator tempered the nature of the heaven with water, lest the conflagration of the upper fire should kindle the lower elements. Therefore He named the circumference of the lower heaven the firmament, because it supports the upper waters[1]. The moon is much smaller than the sun[2], and is nearest to us; the order of the planets is Moon, Mercury, Venus, Sun, &c., and they complete their circles in 8, 23, 9, 19, 15, 22, and 30 years[3]! The stars (fixed stars) move with the world, it is not they which move while the world stands still[4]. A strange mixture of truth and error.

But though enlightened students like Philoponus and Isidore might accept some of the teaching of antiquity, the school of cosmographers of the Kosmas type continued to flourish. From about the seventh century we have a cosmography which goes under the name of one Æthicus of Istria and professes to be translated and abbreviated from a Greek original by a priest named Hieronymus; but nothing is known either of the alleged author or of the translator, who has very probably compiled the book himself. He has wonderful things to tell about Alexander the Great, Gog and Magog, centaurs and minotaurs, and dog-headed men; in fact the whole book reads like the ravings of a lunatic. But as he enjoyed a considerable reputation in the Middle Ages, he cannot be passed over in an account of the cosmical opinions of that time. The earth of course is flat, the sun likewise (it is spoken of as a table, *mensa solis*), and it passes through the gate of the east every morning to lighten up the world, and passes in the evening through the gate of the west to return during the night to its starting-point through the south (?),

---

[1] Cap. xiii.   [2] Cap. xvi.

[3] "Nam luna octo annis fertur explere circulum suum, Mercurius annis xxiii., Lucifer annis ix., Sol annis xix., Pyrois annis xv., Phaeton annis xxii., Saturnus annis xxx." Cap. xxiii., *Opera omnia*, Romæ, 1803, t. vii. p. 36. But on a figure showing concentric orbits, the figures 19, 20, 9, 19, 15, 12, 30, are marked (explet cursum annis 19, &c.). Of course Isidore has misunderstood the meaning of these periods, which are not periods of revolution but periods after which the planets occupy the same places among the stars.

[4] Cap. xii.

hidden in the meanwhile by a thick mist which screens it from us but allows some of its light to reach the moon and stars. Under the earth is the abyss of the great waters. The heaven is spread out over the earth like a skin and encloses the sun, moon, and stars, all of them moving freely and separated by it from the six upper heavens, the dwellings of the heavenly host[1].

Another geographer from the end of the seventh century, the "anonymous geographer of Ravenna," whose work is chiefly statistical, views the world quite like the patristic writers. The world is bounded on the west by the ocean, on the east by a boundless desert, which even made Alexander the Great turn back. The sun illuminates the whole world at the same time. To the north, beyond the ocean, there are great mountains, placed there by God to make a screen, behind which sun and moon disappear. Some people indeed had denied the existence of these mountains, and asked if anyone had ever seen them, but it is clear that the Creator has made them inaccessible in order that mankind should know nothing about them[2].

This is, however, the last writer of note who refuses obstinately to listen to common sense. No doubt there continued throughout the Middle Ages to be clerics to whom the sphericity of the earth was an abomination, and, even among those who acknowledged it, very few had the courage to confess openly that there was nothing impossible in assuming the existence of human beings on the other side of the sphere. But in the peaceful retreat of the monastery the study of the ancient Latin writers had long before the time of the Ravennese geographer taken root, and the geocentric system slowly but steadily began to resume its place among generally accepted facts. The next figure among medieval writers after Isidore is the Venerable Bede, and he followed his predecessor in his

---

[1] *Cosmographiam Æthici Istrici*...edidit H. Wuttke, Leipzig, 1853, caps. 6, 8, 13, &c. The editor is very enthusiastic about his hero and greedily swallows all his marvels; he also willingly accepts the translator's identity with St Jerome and assumes that the original was written before the time of Constantine!

[2] Originally written in Greek, but only a Latin translation has come down to us. Edited by Pinder and Parthey, *Ravennatis Anonymi Cosmographia*. Berlin, 1860.

opinions about the world. Born about 673 in the north of
England, Bede spent most of his life in two monasteries in
that neighbourhood, to which a considerable number of books
had been brought from Rome by their founder, and he made
good use of them in preparing his numerous writings. So
great was the reputation which he acquired by these that long
after his death (even four or five centuries later) spurious
tracts were produced and palmed off on the reading world as
products of the great English monk, to whom posterity had
given the title of "Venerable" in token of its admiration.
There is, however, no difficulty in separating these spurious
treatises from the genuine ones, particularly as Bede four years
before his death (which took place about 735) drew up a list
of his writings and appended it to his famous *Ecclesiastical
History*.

Among the undoubted writings of Bede is a treatise *De
natura rerum*[1], which in 51 paragraphs deals with the stars,
the earth and its divisions, thunder, earthquakes, &c. The
contents are taken from Pliny, often almost verbatim; and the
spherical form of the earth, the order of the seven planets
circling round it, the sun being much larger than the earth,
and similar facts are plainly stated[2]. But the unlucky water
around the heaven and the usual explanation of its existence
could of course not be kept out of the book[3], even though
Pliny does not mention it, and Bede had stated that the
heaven was a sphere. Another and much longer book by
Bede deals with chronology (*De temporum ratione*) and shows
a fair knowledge of the annual motion of the sun and the
other principal celestial phenomena. When mentioning the
zones of the earth[4] he says that only two of them are capable
of being inhabited, while no assent can be given to the fables
about antipodes, since nobody had ever heard or read of
anyone having crossed the torrid zone and found human beings
dwelling beyond it.

---

[1] *Venerabilis Bedæ Opera*, ed. Giles, Vol. vi. pp. 100–122 (London, 1848).

[2] He even copies from Pliny (ii. 49) that the moon is larger than the earth
(cap. xix.). Possibly Pliny misunderstood the papyrus of Eudoxus, col. xx. 15.

[3] Cap. vii.   [4] Cap. xxxiv.

That it was worth while to be very cautious in speaking of antipodes appears from the ruin which threatened Fergil, an Irish ecclesiastic of the eighth century, better known as Virgilius of Salzburg. He was originally Abbot of Aghaboe (in the present Queen's County) and started for the Holy Land about 745, but he did not get further than Salzburg, where he became Abbot of St Peter's. In 748 he came into collision with Boniface, the head of the missionary Churches of Germany, about the validity of a baptism administered by a priest ignorant of Latin, and when Boniface reported this to the Pope (Zacharias) he took the opportunity to complain that Virgil in his lectures had taught that there was "another world and other people under the earth." Zacharias replied that Boniface should call a council and expel Virgil from the Church, if he really had taught that. Whether any proceedings were taken against Virgil is not known[1], but in any case he cannot have been condemned as guilty of heresy, since he became Bishop of Salzburg in 767 (when both Boniface and Zacharias were long dead) and ruled that see till his death in 784 or 785. No writings of his are extant, and nothing is known of his doctrines except the words quoted above from the Pope's reply, to which in one edition is added that the other world underneath ours had its own sun and moon[2]. But this is probably a marginal improvement made by some transcriber to emphasize the shocking heresy of Virgil, and we cannot doubt that Virgil merely taught the existence of

[1] In the *Thesaurus Monumentorum* of Canisius, III. 2, p. 273 (Antwerp, 1725), it is said that Virgil did not obey the summons to Rome. The accusation is not mentioned in the *Monumenta Germaniæ* (Script. T. XI. p. 84 sq.).

[2] "De perversa autem et iniqua doctrina, quam contra Deum et animam suam locutus est, si clarificatum fuerit ita eum confiteri, quod alius mundus et alii homines sub terras sint; hunc accito concilio, ab Ecclesia pelle sacerdotii honore privatum." So in Usher's works, ed. by Elrington, IV. p. 464, and in S. Bonifacii *Opera*, ed. I. A. Giles, London, 1844, I. p. 173. But in *Sacrosancta Concilia*, studio P. Labbæi et Gabr. Cossartii, Venice, 1729, t. VIII. p. 256, the words "seu sol et luna" occur after "sub terra sint." In the Benedictine *Histoire littéraire de la France*, t. IV. p. 26 (1738), it is said that Virgil discovered the antipodes "ou un autre monde qui a son soleil, sa lune et ses saisons comme le nôtre." Virgil was canonized in 1233 by Pope Gregory IX. on account of the miracles wrought by his bones after they had been found in 1171; see Canisius, l. c. pp. 399 sq. and Riccioli, *Almag. nov.* II. p. 489.

antipodes[1]. And after all there is nothing very remarkable in the fact that an Irish monk knew the earth to be a sphere. Not only were many Irish monasteries centres of culture and learning, where the fine arts and classical literature were studied at a time when thick night covered most of the Continent and to a less extent England; but devoted missionaries had before the time of Virgil spread the light of Christianity as far north as the Orkneys, while Adamnan, the biographer of St Columba, had had personal intercourse with Arculf, who had made a pilgrimage to the Holy Land. That the sphericity of the earth, asserted by the Greek and Roman writers, was an undoubted fact, must have been made clear by comparing notes with these travellers, whose experience extended over 25° of latitude. In the following century we find another Irishman of note, Dicuil, who finished his geographical compilation, the *Liber de mensura orbis terræ*, in 825. Though he says nothing about the figure of the earth, he tells us of Irish missionaries who thirty years earlier visited Thule (which here undeniably means Iceland), where they saw the sun barely hidden at midnight in midsummer, as if it went behind a little hill, so that there was nearly as much light as in the middle of the day, "and I believe that at the winter solstice and during the days thereabouts the sun is visible for a very short time only in Thule, while it is noon at the middle of the earth[2]." Dicuil must therefore have clearly understood the phenomena of the "oblique sphere."

However dangerous it might be to assert the existence of human beings in what was thought to be an inaccessible part of the earth, beings who could not be assumed to be descended from Adam or to have been redeemed by the death of Christ, the idea that religion and secular learning were of necessity opposed to each other was fast disappearing, and it had by this time become quite a customary thing among men of learning to recognize that the earth is a sphere. Still some

---

[1] So Maestlin understood him; Kepleri *Opera*, I. p. 58.

[2] *Dicuilii liber de mensura orbis terræ*, a G. Parthey recognitus, Berlin, 1870, pp. 42–43. Remains of the Irish settlements in Iceland were found by the Northmen on their arrival in 874.

people chose to say nothing about it, e.g. Hrabanus Maurus,
Abbot of Fulda and afterwards Archbishop of Mainz (d. 856),
who, though he did much to encourage classical studies, yet in
his encyclopedic work *De Universo* merely says that the earth
is situated in the middle of the world[1]. The inhabited land,
he says, is called orbis "from the rotundity of the circle,
because it is like a wheel[2]"; but he sees the necessity of
assuming it to be a square, since Scripture speaks of its four
corners, and he finds it awkward to explain why the horizon
is a circle. But he refers to the fourth book of Euclid and
seems to think that a square inscribed in a circle will save the
situation. His statement that the heaven has two doors, east
and west, through which the sun passes[3], looks, however, as if
his point of view was much the same as that of the patristic
writers. But when an eminent mathematician like Gerbert
ascended the papal throne as Sylvester II. (in 999, died 1003),
the game was up for the followers of Lactantius. The example
of Bede, who had openly taught the sphericity of the earth,
had borne fruit, and so did doubtless that of a Pope who was
familiar with the scientific writings of the ancients[4] and in his
younger days had constructed celestial and terrestrial globes to
assist his lectures on astronomy, and had been in the habit of
exchanging them for MSS of Latin classics. And the horizon
of mankind continued to be widened out by the spread of
geographical knowledge through the intercourse with the Arabs
in Spain on the one side, and the travels and adventures of
the Northmen on the other. Adam of Bremen (about 1076),
whose chronicle is of great importance for the study of the
history of his time, has nothing in common with Kosmas or
the geographer of Ravenna; he understands perfectly the
cause of the inequality of the day and night in different
latitudes and shows himself an apt student of Bede's writings.
The maps of this period also mark a considerable advance.

---

[1] *De Universo*, XII. 1.—B. Rabani Mauri *Opera omnia*, ed. J. P. Migne, Paris,
1864, T. v. col. 331.

[2] Ibid. XII. 2, col. 332–333.

[3] Ibid. IX. 5, col. 265.

[4] Among the sources of his Geometry Gerbert mentions Plato's *Timæus*,
Chalcidius, Eratosthenes, &c. Cantor, *Gesch. d. Math.* I. p. 811.

Beside the ordinary "wheel maps" or *T-O* maps, so called
from the form, which resembles a *T* inscribed in a circle (Asia
being above the horizontal stroke of the *T*, the vertical stroke
of which is the Mediterranean), we find more elaborate maps.
They are mostly founded on a design by Beatus, a Spanish
priest who lived at the end of the eighth century; they
represent Africa as not reaching to the equator, and though
they do not show any sign of the antipodes, they contain
nothing against the rotundity of the earth; and by degrees as
the designers of maps became better acquainted with ancient
works on geography, they made bolder attempts at depicting
the earth.

From about the ninth century the rotundity of the earth
and the geocentric system of planetary motions may be considered to have been reinstated in the places they had held
among the philosophers of Greece from the days of Plato.
The works of these philosophers were still unknown in the
West, where Greek had been an unknown tongue after the
fifth century; but the writings of Pliny, Chalcidius, Macrobius
and Martianus Capella supplied a good deal of information to
anyone who read them, and since the days of Charles the
Great (768–814) Roman literature was rapidly becoming better
known. We possess two works of unknown date which have
been founded on these writers. They go under the name of
Bede, but they are undoubtedly much later productions and
have not been included in the modern edition of his works.
One of them is entitled *De mundi cœlestis terrestrisque constitutione liber*[1], and it is hardly possible that anyone can ever
have believed it to have been written by Bede, as he is quoted
in it and there are several references to the chronicles of Charles
the Great, so that it must at any rate have been written after
the year 814. The author has a fair knowledge of the general
celestial phenomena such as could be gathered from the above-
mentioned writers, but no more. He proves that the earth is
a sphere by the different length of the day in different latitudes,
and by the fact that the various phenomena in the heavens do

---

[1] Ven. Bedæ *Opera*, Col. Agripp. 1612, T. I. cols. 323–344.

not occur at the same time for different localities. He says
that Plato followed the Egyptians in placing the solar orbit
immediately outside that of the moon, but his own opinion
seems to be that Venus and Mercury are sometimes above the
sun and sometimes below it, as it is recorded in the history of
Carolus that Mercury for nine days was visible as a spot on
the sun, though clouds prevented both the ingress and the
egress being seen. When they are below the sun they are
visible in the middle of the day, and he refers to the star seen
at the time of Cæsar's funeral, which he supposes to have been
Venus[1]. The limits of the planets in latitude are also given[2].
The writer shows himself somewhat independent of his authorities by adding a fair sprinkling of astrology, and still more
by giving the various theories current about the unavoidable
"supercelestial waters[3]." One idea is that there are hollows in
the outer surface of the heaven in which water may lie (as it
does on the earth's surface), and notwithstanding the rapid
rotation of the heaven it is not spilt, just as water will remain
in a vessel swung rapidly round! Another idea is, that the
water is only vapour like clouds; another that it is frozen
owing to the great distance from the sun, the principal source
of heat, and that Saturn is called the most frigid star because
it is nearest the water. But the waters are simply held there
by the power of God in order to cool the heaven, and above
them are the spiritual heavens in which the angelic powers
dwell.

The other work, which formerly was counted among the
writings of Bede, is entitled "Περὶ διδάξεων *sive elementorum
philosophiæ libri IV*.[4] It has been ascribed to William of
Conches, a Norman of the first half of the 12th century, and
in any case it cannot have been written much earlier, as it
shows a freedom of thought which would have been impossible
at the time of Bede[5]. This is particularly the case with

[1] It was a comet.
[2] Taken from Martianus Capella, but the sun's range is 2°, as in Pliny.
[3] p. 332.   [4] *Opera* (1612), T. ii. pp. 206–230.
[5] There are two other editions differing very little from the Pseudo-Bede:
Philos. et astron. institutionum Guil. Hirsaugiensis libri iii. Basle, 1531, and
De Philosophia Mundi; *Honorii Opera, Max. Bibl. Pat.* T. xx.

regard to the question whether there is water above the ether[1].
Quoting the passage from Genesis about the water above the
firmament, the writer says that it is *contra rationem*, for if
congealed it would be heavy and the earth would then be the
proper place for it, while the water above would be next the
fire and either put it out or be dissolved by it, as we could not
suppose that there is any boundary between them. The air is
called the firmament because it strengthens and regulates the
earthly things, and above it there is water suspended in the
form of clouds, which are indeed different from the water below
the air. "Although we think it was said more allegorically
than literally." Turning to the planets, the writer is aware of
the difference of opinion as to the position of the solar orbit.
He dismisses the idea that the sun has been placed next after
the moon in order that the heat and dryness of the sun might
thus counteract the cold and humidity of the moon, which
otherwise might become excessive owing to the proximity of
the earth. Also the idea that the sun has to be next the
moon because the latter is illuminated by it. But as the sun,
Venus and Mercury move nearly in the same period round the
zodiac, their circles must be nearly equal in size and are not
contained within each other but intersect each other[2]. The
sun is eight times as large as the earth. The air reaches to
the moon; above that is ether or fire, which is so subtle that
it cannot burn unless mixed with something humid and dense;
while the sun and stars are not made of fire alone, but also of
the other elements, though fire is the predominating material.
In all this there is nothing new.

As regards the earth, the writer says that it is in the
middle, as the yolk in the egg, and outside it is the water like
the white round the yolk; around the water is the air like the
skin round the white, and finally fire, corresponding to the egg-
shell. The two temperate zones are inhabitable, but we believe

---

[1] p. 213.

[2] The range in latitude of the planets is illustrated by a diagram, the figures
resulting from this (not given in the text) being practically the same as in the
book *De mundi constitutione*. But the diagram is perhaps a later addition, and
a very absurd one.

only one to be inhabited by men. "But because philosophers talk about the inhabitants of both, not because they are there, but because they may be there, we shall state what we believe there are, from our philosophical reading." The zone in which we live consists of two parts, of which we inhabit one and our antipodes the other, and similarly the other inhabitable zone consists of two parts, of which the upper is that of our *anthei* and the lower that of their antipodes. Thus we and our antipodes have summer or winter together, but when we have day they have night. In other words, the author adheres to the old idea of the four *oekumenes*, but he uses the word antipodes in a sense which is not the usual one, but which signifies people who live in our hemisphere but 180° distant from us in longitude.

From about the same time we have the *Imago Mundi* by Honorius of Autun, a kind of short encyclopædia from the first half of the twelfth century[1]. The cosmographical part is borrowed from Pliny, but with the necessary additions to suit the taste of medieval readers[2]. The two doors of heaven are duly mentioned, though they do not fit well in the geocentric system of the world. The upper heaven is called the firmament; it is spherical, adorned all over with stars which are round and fiery, and outside it are waters in the form of clouds, above which is the spiritual heaven, unknown to man, where the habitations of the angels are, arranged in nine orders[3]. Here is the Paradise

---

[1] *Mundi Synopsis sive De Imagine Mundi libri tres.* Ab Honorio Solitario Augustudunense. Spiræ, 1583. There are several other editions from the 15th and 16th centuries.

[2] Among the things borrowed from Pliny may be mentioned the greatest latitudes of the planets (i. 79) and the musical intervals of the planets (i. 81) ; one tone = 15,625 miliaria, which is the distance of the moon, from that to Mercury is 7812½ miliaria, and so on, so that the distance of heaven (seven tones) is 109,375 miliaria.

[3] Compare the *Libri Sententiarum* of Peter the Lombard, Bishop of Paris (d. 1164), where the nature of angels and their hierarchy are discussed in the second book. As to the waters above, he quotes the opinion of Bede that they form the solid heaven, as crystal is made of water, and that of Augustine that they are in the form of vapour much lighter than that which we see in clouds. "Anyhow we must not doubt that they are there." The hierarchy of angels was fixed by the Pseudo-Dionysius Areopagita and universally accepted during the Middle Ages ; it is arranged as follows in the *Summa Theologiæ* of Thomas

of Paradises, where the souls of saints are received, and this is
the heaven which was created in the beginning together with
the earth[1]. In the centre of the earth is Hell, which is described
in some detail. The writer does not seem to know where
Purgatory is situated.

The work of Honorius found several imitators, both in prose
and verse, among the latter being the *Image du monde*, written
in 1245 by a certain Omons (otherwise unknown), who mentions
Honorius and William of Conches among his authorities[2]. The
ideas set forth are like theirs. Ptolemy, King of Egypt, invented
clocks and various instruments and wrote several books, one of
which is called the *Almagest*. There are two heavens, the
crystalline and the empyrean; angels dwell in the latter, and
from it the demons were expelled. Children, on account of their
innocence, can hear the celestial music. The air of heaven is
called ether, and the bodies of the angels are formed of it. The
writer says nothing about the planetary system.

The taste for encyclopedic writing became strongly developed in the thirteenth century and was much influenced by
the knowledge of Aristotle's works, which at last had begun to
spread in the western countries. About the middle of the
twelfth century Arabian translations of Aristotle began to be
introduced into France from Spain, and with them came the
commentaries of Alexander and Simplicius and works of other
Greek philosophers. These had to be translated into Latin;
and though the translations were not very accurate, having
passed through Syriac and Arabic before putting on the Latin
garb, still they opened up to a wondering world the treasures of
Greek thought. At first the Church was hostile to this movement, a natural consequence of the mass of mystical, pseudo-

Aquinas (I. 108). Seraphim, Cherubim, Thrones form the uppermost, empyrean hierarchy; the Thrones pass on the commands of God to the first order
of the second hierarchy, the Dominations, next to whom come the Virtues,
who guide the motions of the stars and planets, and the Powers who remove
anything which might hinder these motions. The third hierarchy, Princi-
palities, Archangels, Angels rule the earthly affairs. Compare Dante, *Convito*,
II. 6.

[1] Honorius, I. 87–90, 138–140.
[2] *Notices et Extraits des manuscrits*, T. v. pp. 243–266.

neoplatonic and Arabian speculations, which had been imported
under the guise of Aristotelean treatises; and at a provincial
council, held at Paris in 1209, it was decreed that neither
Aristotle's books on Natural Philosophy nor commentaries on
them should be read either in public or privately in Paris.
In 1215 this prohibition was renewed in the statutes of the
University of Paris. But by degrees the fears of the Church
wore off, so that in 1254 official orders were issued, prescribing
how many hours should be used in explaining the physical
treatises of Aristotle; and the Aristotelean natural philosophy
was from henceforth and for nearly four hundred years firmly
established at the Paris University, and indeed at every seat of
learning. Fresh translations had already earlier been made by
order of the Emperor Frederic II.; others made directly from
the Greek were soon provided at the instance of Albertus
Magnus (1193–1280) and his disciple Thomas Aquinas (1227–
1274), and soon Aristotle had become the recognized ally of the
theologians. Both Albert and Thomas contributed by their
writings greatly to the spread of knowledge of ancient science,
a work in which the gigantic encyclopædia of Vincent of
Beauvais (*Speculum Naturale*, completed in 1256) also had a
great share.

The most representative writer among the scholastics is
Thomas Aquinas, and among his works there is one in particular which must be mentioned here. It is a commentary on
Aristotle's book on the heavens[1], and the spirit in which it is
written shows the vast strides from darkness towards light
which had been recently made. Though Aquinas was deeply
convinced that revelation is a more important source of knowledge than human reason, he considers both to be two distinct
and separate ways of finding truth; and in expounding Aristotle
he therefore never lets himself be disturbed by the difference
between his doctrine and that of the Bible, but assumes both to
be ultimately derived from the same source. His commentary is
very interesting to read[2], much clearer than that of Simplicius,

---

[1] S. Thomæ Aquinatis *Opera omnia*, T. III., *Commentaria in libros Aristotelis
de Cœlo et Mundo*...Romæ, 1886, fol., a magnificent edition.

[2] Especially if one reads it after wading through the patristic writers.

with which he is well acquainted and which he frequently quotes together with the works of Plato, Ptolemy, and others. Wherever necessary, he points out that philosophers after Aristotle have come to differ from him, as for instance in substituting epicycles for the homocentric spheres, or as regards the motion of the starry sphere, which Aristotle assumed to be the uppermost one, while later astronomers say that the sphere of the fixed stars has a certain proper motion (i.e. precession), for which reason they place another sphere above it, to which they attribute the first motion[1]. In speaking about the position of the earth at rest in the centre of the world, he quotes Ptolemy's arguments in its favour[2].

Another and very much humbler writer may also be mentioned here, as his little book on the sphere remained the principal elementary text-book on astronomy for nearly four centuries. We know next to nothing of the life of Johannes de Sacro Bosco, or John of Holywood, except that he died at Paris in 1256. He quotes Ptolemy and Alfargani (the latter had been translated in the middle of the twelfth century) and describes the equants, deferents and epicycles, being the first European writer in the Middle Ages to give even a short sketch of the Ptolemaic system of planetary motions[3]. After the long and undisturbed reign of Pliny and Martianus Capella, Ptolemy at last began to come to the front again.

But to the great majority of scholastics there was no going beyond Aristotle, who was held to represent the last possibility of wisdom and learning. One man there was, however, who was not content to be a mere slave of Aristotle, any more than the Alexandrian thinkers had been. Roger Bacon (1214–1294) in his *Opus Majus* shows himself thoroughly acquainted with the literature of the Greeks and Arabians. But in opposition to the general tendency of the previous thousand years he does not

[1] Lib. II. lect. IX. p. 153 a; compare XVII. p. 189 a, where Ptolemy's 1° in 100 years is mentioned.

[2] Lib. II. lect. XXVI. p. 220 b.

[3] Sacrobosco's knowledge of the Ptolemaic system was evidently of a very elementary nature and only acquired second-hand, for he copies the mistake of Alfargani and Albattani, that the two points on the epicycle in which the planet is stationary are the points of contact of the two tangents from the earth.

think it enough to write wordy commentaries on the ancients: he is able to think for himself, and he lays stress on the importance of experiments as offering the only chance of helping science out of the state of infancy in which he is fully aware it still lies. The scholastic doctors also, after the manner of the ancients, talked finely about experience as the only safe guide in the visible world. But it began and ended in talk; they did not find a single new fact in natural philosophy, they did not determine a single value of any astronomical constant. Roger Bacon was a man of a different stamp, and had he lived under more favourable circumstances we cannot doubt that he would have opened a new era in the history of science, instead of being merely a voice crying in the wilderness, whose wonderful work had to lie in manuscript for nearly five hundred years before it was printed. His object was to effect a reform in natural philosophy by brushing aside the blind worship of authority and by setting forth the value of mathematical investigations. As he was only a poor persecuted student, he had not the means to carry out his ideas; but his treatise on perspective shows what he was capable of, and what he would have done, if he had been the petted son of the Church instead of being its prisoner. In his general ideas about the universe he followed Ptolemy, and we shall therefore here only allude to one or two points. He remarks that the earth is only an insignificant dot in the centre of the vast heaven; according to Alfargani, the smallest star is larger than the earth, a sixth magnitude star being 18 times as large, while a first magnitude star is 107 and the sun 170 times as large (i.e. in volume[1]). Ptolemy has shown that a star takes 36,000 years to travel round the heaven (i.e. by precession), while a man can walk round the earth in less than three years. In the chapter on geography[2] it is interesting to see that he discusses at some length the question how large a part of the earth is covered by the sea, and, from the statements of Aristotle, Seneca, and Ptolemy, comes to the

[1] *Opus Majus*, ed. S. Jebb, London, 1733, p. 112. On p. 143 he gives the dimensions of the orbits in Roman miles according to Alfargani, the diameter of the starry sphere being 130,715,000 miliaria.

[2] pp. 181–236.

conclusion that the ocean between the east coast of Asia and Europe is not very broad. This part of Bacon's work was almost literally copied by Cardinal d'Ailly (Petrus de Alyaco) into his *Imago Mundi* (written in 1410, first printed in 1490) without any mention of Bacon; it was quoted by Columbus in his letter from Hispaniola to the Spanish monarchs in 1498, and it had evidently made a very strong impression on him[1]. It is pleasant to think that the persecuted English monk, then two hundred years in his grave, was able to lend a powerful hand in widening the horizon of mankind.

A reader of Roger Bacon cannot fail to be struck with the vast difference between him and the patristic writers. While they struggled hard to accept the most literal interpretation of every iota in Scripture, Roger Bacon fearlessly points out difficulties in various passages of the Old Testament, and urges that the only way to get over them is by making a thorough study of science, which the Fathers of the Church had failed to do. He mentions as examples the first chapter of Genesis, the sun standing still at the bidding of Joshua, and the shadow on the dial going back ten degrees[2]. Similarly the statement of St Jerome (on Isaiah) that there are twenty-two stars in Orion, nine of which are of the third, nine of the fourth, and the rest of the fifth magnitude, which does not agree with the eighth book of the *Almagest*.

But though a few enlightened men like Thomas Aquinas and Roger Bacon knew the works of Ptolemy, they certainly remained quite unknown to the leading men of the thirteenth century. This fact is strongly illustrated by the cosmographical ideas of Dante, whose *Divina Commedia* represents the prevailing views of his time (around the year 1300) as to the structure of the world. In general it is a risky thing to draw conclusions from astronomical allusions in poetical works to the state of scientific knowledge of the time[3], but in the case of Dante it is

[1] Humboldt, *Kritische Untersuchungen*, I. p. 71 sq.

[2] Isaiah xxxviii. 8.

[3] For instance, the novels of the 19th century would lead one to think that nothing was known at that time about the motion of the moon, since it is quite a common thing to read in them of a young moon rising in the evening, the full moon sailing high in the heavens in summer, &c.

quite legitimate to do so, as he in the *Commedia* as well as in his other writings shows himself fully equipped with the learning then attainable. He was a pupil of Brunetto Latini, who during his residence in France, from 1260 to about 1267, became infected with the mania for encyclopedic writing prevailing in that country, and composed his celebrated work, *Li Livres dou Tresor*, in the North-French language[1]. Like all the other books of its kind, this is a mere compilation from classical and medieval sources, the astronomical part being very meagre. Though Dante had doubtless studied the structure of the world deeper than Brunetto had done, none of his writings show any familiarity with the *Syntaxis* of Ptolemy, while Aristotle (with the commentary of Thomas Aquinas), Pliny, and especially Alfargani, seem to have been the authors by the study of whom he had profited most[2]. He began writing an encyclopedic work, the *Convito*, or Banquet, intended to comprise fourteen books, of which, however, only four were written. In this work his cosmological ideas are set forth more systematically, with the addition of a good deal of astrology and other fancies[3].

In Dante's majestic poem, Hell is a conical cavity reaching to the centre of the earth. Around the sloping sides the places of punishment are arranged in circles of gradually decreasing diameter, so that the worst sinners are placed nearest to the apex of the cone, where Lucifer dwells in the very centre of

---

[1] The French original was not printed till 1863 (Paris, edited by P. Chabaille), but an Italian translation has been printed several times, the chapters on astronomy separately by B. Sorio, *Il Trattato della sfera di Ser Brunetto Latini*, Milano, 1858.

[2] At the end of his poem *Il Tesoretto*, Brunetto tells how he on Mount Olympus met Ptolemy, master of astronomy and philosophy, and asked him to explain about the four elements; upon which Ptolemy "rispose in questa guisa"—and there the poem ends abruptly!

[3] The opinions set forth in the *Convito* differ in a few cases somewhat from the ideas of the *D. C.*, the most notable astronomical instance being the spots in the moon. In the *Conv.* II. 14, Dante says the spots are caused by the rarity of parts of the lunar body, which do not reflect the sun's rays well. In *Par.* II. Beatrice delivers a long lecture showing that this theory is erroneous (because those parts would be transparent and would show themselves to be so during solar eclipses); the moon shines by its own light, which differs in various places under the influence of the various angelic guides, just as the stars in the eighth sphere differ in brightness, owing to the different virtue communicated to them by the Cherubim who rule them.

the earth. When Dante and his guide Virgil have passed to
the bottom of the abyss and continue their journey straight on,
Dante looks back and sees Lucifer upside down, whereupon his
guide explains that they have now commenced their ascent to
the other side of the earth[1]. Purgatory is a large, conical hill,
rising out of the vast ocean at a point diametrically opposite to
Jerusalem, the navel of the dry land. Having passed over the
seven terraces of the mount, and reached the earthly paradise
at the top, the poet is finally permitted to rise through the
celestial spheres. These are, of course, ten in number, first that
of the moon (to which the blue air reaches[2]), then the spheres of
Mercury, Venus (to which the shadow of the earth reaches[3]),
the sun, Mars, Jupiter, and Saturn. In each of these spheres
spirits, though they have not their permanent abode there,
appear to Dante, in order to illustrate to him the gradually
increasing glory which they have been found worthy to enjoy,
and to indicate their former earthly characters and tempera-
ments, which had been chiefly influenced by one of the seven
planets[4]. The eighth sphere is that of the fixed stars, the
ninth is the Primum Mobile, the velocity of which is almost
incomprehensible owing to the fervent desire of each part of it
to be conjoined to the restful and most Divine Heaven, the
tenth or Empyrean, the dwelling of the Deity[5]. The nine
spheres are moved by the three triads of angelic intelligences,
the Seraphim guiding the Primum Mobile, the Cherubim the
fixed stars, the Thrones the sphere of Saturn, and so on down
to the moon's sphere, which is in charge of the angels[6]. In the
eleventh canto of *Purgatorio* (v. 108) there is a distinct allusion

[1] *Inferno*, XXXIV. 87 seq.

[2] *Purg.* I. 15.   [3] *Par.* IX. 118.

[4] In the *Convito* (II. 14–15) Dante explains that the first seven spheres correspond to the Trivium and Quadrivium of the seven liberal arts. For instance,
Mercury, the smallest planet and the one most veiled in the sun's rays, corresponds to Dialectics, an art less in body and more veiled than any other, as it
proceeds by more sophistic and uncertain arguments. The eighth or starry
sphere corresponds to Physics and Metaphysics, the ninth to Moral Science, and
the tenth or Empyrean heaven to Theology.

[5] *Convito*, II. 4.

[6] *Convito*, II. 6; ibid. II. 5, the motive powers of the spheres are said to be
known as angels among "la volgare gente."

to the precession of the equinoxes or, as it was still assumed to be, of the sphere of the fixed stars: "che più tardi in cielo è torto." There is only one slight allusion to epicycles[1], otherwise the planets are merely said to move in the ecliptic[2], and it is curious to find the sun's motion stated to be along spirals[3], just as Plato of old had said in the *Timæus*. Another old acquaintance meets us in the statement that the sphere of the moon has the slowest motion[4].

Dante continued throughout his life to be deeply interested in cosmography. In 1320, the year before his death, he delivered a lecture "De Aqua et Terra" in order to refute the opinion occasionally promulgated in the Middle Ages, and even later, that the water- and land-surface of the earth do not form part of one and the same sphere, but that the earth consists of a land-sphere and a water-sphere, the centres of which do not coincide[5].

We may here close our review of medieval cosmology. Dante died in the year 1321, almost exactly a thousand years after the Emperor Constantine had made the Christian faith the state religion of the Roman Empire. It had been a long and perfectly stationary period, at the end of which mankind occupied exactly the same place as regards culture as at the beginning; scarcely even that, as Greek science, philosophy and poetry were still very imperfectly known in the West, so that no serious attempt could be made to build further on the foundation they offered. For centuries men had feebly chewed the cud on the first chapter of Genesis; then compilers like Pliny and Martianus

---

[1] *Par.* viii. 2,   " Che la bella Ciprigna il folle amore
                    Raggiasse, volta nel terzo epiciclo."
Compare *Convito*, ii. 4 (on the back of this circle in the heaven of Venus is a little sphere which turns by itself in this heaven, and the circle of which astronomers call epicycle), also ii. 6 near the end, where the same is said of the planets in general.
[2] *Par.* x. 7,   " Leva dunque, Lettor, all' alte rote
                    Meco la vista dritto a quella parte
                    Dove l' un moto e l' altro si percote."
[3] *Par.* x. 32.   [4] *Par.* iii. 51, compare above, Chapter iii. pp. 70 and 81.
[5] When Columbus in 1498 near the coast of South America noticed the steady current of water opposing his progress (coming from the Orinoco), he thought he was near the highest point of the sea, from which the water rushed down.

Capella had grudgingly obtained a hearing; finally Aristotle had been discovered, and had almost at once been accepted as the infallible guide.

But in the East the light once issuing from Greece had not been so long obscured. The flame had been kept alive by the very people who at first had seemed destined to trample all civilisation under foot, as the Huns had once done in Europe; and from the Arabs came the first impulse which led to the awakening of the West. We must now turn back and examine how the Eastern nations had used the intellectual treasures they found in the countries which came under their sway.

CHAPTER XI.

ORIENTAL ASTRONOMERS.

THE conquests of Alexander the Great made the Greeks acquainted with the Eastern world, which had up to that time been visited by very few Europeans, and it likewise spread Greek culture to all the countries which the victorious Macedonian had been able to reach. The Indian province of his Empire became independent soon after Alexander's death, and though the spread of Buddhism in the third century B.C. checked the progress of Hellenism in Northern India, the rise of the Greek kingdom of Bactria and its gradual extension south and east continued for a long time to keep alive the connection between India and the West. Whether (as has been asserted) the Indian drama and Indian architecture have been strongly influenced by Hellenistic contact, may be doubtful, but it is beyond a doubt that Indian astronomy is the offspring of Alexandrian science.

In earlier times astronomy had only been cultivated in India to a slight extent. Some idea had been acquired of the periods of the sun and moon and the planet Vrihaspatis (Jupiter), which were used for chronological purposes, the lunar motions being specially connected with the proper times for sacrificial acts; but otherwise early Hindu astronomy seems to have been chiefly astrology, and there is no sign of any accurate knowledge of the planetary motions earlier than about the third century of our era. From thenceforth astronomy, which had hitherto only formed a subject of poetical effusions, appears as a science, treated in the course of the next thousand years in a series of text-books, the *Siddhántas*, the contents of which, though supposed to be derived from divine sources, are strongly influenced by or simply borrowed from Greek authors[1]. The week of seven days (previously unknown) and the dedication of each day to the deity of one of the seven planets, now appear for the first time. The names of the planets have also become Greek, e.g. Âsphudit (Aphrodite), Dyugatiḥ or Jiva (Zeus), Heli (Helios), &c., while the zodiacal signs have superseded the earlier but totally different twelve star-groups connected with the sun's motion, and proclaim their origin by their names:

 Kriya, Tâvuri, Jituna, Karkin, Leya, Pâthena, Jûka,

  Kaurpya, Taukshika, Âkokera, Hṛidroga, Ittha,

corresponding to Κριός, Ταῦρος, Δίδυμος, Καρκίνος, Λέων, Παρθένος, Ζυγόν, Σκορπίος, Τοξότης, Αἰγόκερως, Ὑδροχόος, Ἰχθύς[2].

A great many other Greek terms connected with geometry, astronomy and astrology have also been transferred into Sanskrit works[3], so that the Greek origin of Hindu astronomy has been conclusively proved. This was also distinctly acknowledged by some of the early Hindu writers, e.g. by Varaha Mihira, who quotes the Yavanas or peoples of the west as authorities for the scientific statements he makes[4]. The name of the Romaka Siddhânta (which is at least as old as A.D. 400) also points in an unmistakable manner to its origin in one of the provinces of the Roman Empire.

The astronomers of the Siddhântas[5] taught that the earth is

---

[1] The dates of the principal ones are: The *Romaka* or *Paulisa Siddhânta* not later than A.D. 400, the *Panchasiddhantica* of Varaha Mihira about 570 (he died in 587), the *Brahmasphuta S.* of Brahmagupta about 630 (he was born 598), the *S. Siromani* of Bhâskara Âchârya about 1150. The *Sûrya Siddhânta*, in the form in which we have it, is probably from the 13th century, though founded on an original probably 800 or 900 years older. J. Burgess, "Notes on Hindu Astronomy," *Journ. R. Asiat. Soc.*, October, 1893, p. 742.

[2] J. Burgess, l.c. p. 747.

[3] A long list given by Burgess, p. 748.

[4] Colebrooke, *As. Res.* XII. p. 245 (*Essays*, II. p. 410).

[5] A translation of the *Sûrya Siddhânta* by Rev. E. Burgess appeared at New Haven, 1860. Another (with a translation of the *S. Siromani*) appeared at Calcutta in 1861, but an analysis of the contents by S. Davis had already been published in 1789 in the *Asiatic Researches*, vol. II., reviewed by Delambre, *Hist. de l'Astr. anc.*, T. I. p. 450 sqq. The *Panchasiddhantica* was translated by G. Thibaut, London, 1889. Compare Colebrooke's paper "On the notion of the Hindu astronomers concerning the precession of the Equinoxes and

a sphere, unsupported in space, and they reject the ancient
mythological notion that it is supported by some animal which
in its turn rests on another, and so on, until the support of the
last one after all has to be left unexplained. Bhâskara Âchârya,
about A.D. 1150, who comments on the absurdity of this, also
rejects the idea that the earth is perpetually falling, since it
would fall faster than an arrow shot upwards, on account of
being heavier, so that an arrow could never again reach the
earth[1]. Round the earth the planets are moving, all with the
same linear velocity. The diameter of the earth is 1600 yojans,
the distance of the moon is 51,570 yojans (or 64·5 times the
radius of the earth, nearly equal to Ptolemy's greatest distance, $64\frac{1}{6}$), while the distances of the other planets result from
the assumption of equal velocities[2]. The equation of centre of
the planets is found by an epicycle, and to this arrangement the
Hindus add one of their own invention, by assuming that the
epicycle had a variable circumference, greatest when the planet
is at apogee or perigee and least at 90° from these, when the
equation reaches its maximum. This contrivance of an oval
epicycle was by some astronomers applied to all the planets, by
others (Brahmagupta and Bhâskara) only to Mars and Venus,
by others it was altogether rejected[3]. Why they complicated
the calculation in this way is not clear. Aryabhata of Kusu-
mapura or Pâtaliputra, born A.D. 476, made another deviation
from the Alexandrian doctrines, as appears in the Brahmasphuta
Siddhânta of Brahmagupta, wherein he quotes the following
from Aryabhata: "The sphere of the stars is stationary, and the
earth, making a revolution, produces the daily rising and setting
of stars and planets." Brahmagupta rejects this idea, saying:
"If the earth move a minute in a prana, then whence and what
route does it proceed? If it revolve, why do not lofty objects
fall?" But his commentator Chaturveda Prit'hudaca Swami

Motions of the Planets," *Asiatic Researches*, XII. pp. 209–250, and *Misc. Essays*,
vol. II.

[1] *As. Res.* XII. p. 229 (*Essays*, II. p. 394).

[2] The distances are proportional to the orbital periods of revolution, but
for Mercury and Venus to the periods in the epicycles.

[3] For further details see *As. Res.* II. p. 251 (Davis) and XII. p. 236 (Colebrooke, also *Essays*, II. p. 401).

replies: " Aryabhata's opinion appears nevertheless satisfactory, since planets cannot have two motions at once; and the objection, that lofty things would fall, is contradicted; for every way, the under part of the earth is also the upper; since wherever the spectator stands on the earth's surface, even that spot is the uppermost spot[1]."

It is very interesting to see the theory once advocated by Herakleides of Pontus transplanted on Indian soil, especially when we remember that Seleukus the Babylonian had adopted that theory. From Babylon the theory might easily find its way to India, though it is of course equally possible that Aryabhata, quite independently of his Greek precursors, hit on the same idea. He appears to have accounted for the earth's rotation by a wind or current of aërial fluid, the extent of which, according to the orbit assigned to it by him, corresponds to an elevation of little more than a hundred miles (114) from the surface of the earth, or fifteen yojans, while he put the diameter of the earth equal to 1050 yojans (of 7·6 miles each[2]). This was in accordance with the general opinion of the Hindus, that the planets are carried along their orbits by mighty winds with the same velocity and parallel to the ecliptic (while one great vortex carries all stars round the earth in twenty-four hours), but that the planets are deflected from these courses by certain invisible powers, having hands and reins, with which they draw the planets out of their uniform progress. The power at the apogee, for instance, constantly attracts the planet towards itself, alternately with the right and left hand (like Lachesis in Plato's *Republic*), while the deity at the node diverts the planet from the ecliptic first to one side and then to the other. And lastly the deity at the conjunction causes the planet to move with variable velocity and to become occasionally stationary and even retrograde. This is gravely set forth in the Sûrya Siddhânta, and even Bhâskara gives the theory in his notes, though he omits it from his text. Similarly Brahmagupta, although he gives the theory of eclipses, affirms the existence of an eighth

---

[1] *Asiat. Res.* xii. p. 227; Colebrooke's *Essays*, ii. p. 392.

[2] Colebrooke, *Notes and Illustrations to the Algebra of Brahmagupta*, p. xxxviii., *Essays*, ii. p. 467.

planet, Rahu, which is the immediate cause of eclipses; and he blames Varaha Mihira, Aryabhata and others for rejecting this orthodox explanation of the phenomenon[1].

Hindu astronomy is thus a curious mixture of old fantastic ideas and sober geometrical methods of calculation. The latter, being of foreign origin, could not drive the older notions from the field. As remarked by Colebrooke, the absence of the most characteristic parts of Ptolemy's system, the equant and the details of the theories of the moon and Mercury, seems to indicate that Greek planetary theory must have been introduced in India between the times of Hipparchus and Ptolemy; and with the exception of the deviation of the epicycle from the circular form, the Hindus did not modify the theory or perfect it in any way. The precession of the equinoxes they held to consist in a libration within the limits of 27° (Aryabhata says 24°) east and west of its mean position, but they came much nearer to the truth than Ptolemy did as regards the annual amount, as they supposed the space travelled over in a century to be $1\frac{1}{2}°$.

Notwithstanding the complete isolation of India from Europe during the Middle Ages, Hindu astronomy was destined to exercise an indirect influence on the progress of astronomy. Through the conquest of Persia in the seventh century, the Arabs, like the Greeks a thousand years earlier, came in contact with India, from whence physicians and astrologers found their way to the court of the Caliph already before the reign of Harun al Rashid. We possess a detailed account of the manner in which Indian astronomy was introduced at Baghdad, from the pen of the astronomer Ibn al Adami (who died before 920), confirmed by the celebrated memoir on India by Al Birûni, written in 1031[2]. In the year 156 of the Hijra (A.D. 773), there appeared before the Caliph Al Mansur a man who had come from India; he was skilled in the calculus of the stars known as the Sindhind (i.e. Siddhânta), and possessed methods for solving equations founded on the kardagas (i.e. kramajyâ, sines) calculated for every half degree, also methods for computing eclipses

[1] *Asiat. Res.* XII. pp. 233, 241; *Essays*, II. pp. 398, 407.

[2] Hankel, *Zur Geschichte der Mathematik im Alterthum und Mittelalter*, Leipzig, 1874, p. 229; Cantor, *Gesch. d. Math.* I. p. 656.

and other things. Al Mansur ordered the book in which all this was contained to be translated into Arabic, and that a work should be prepared from it which might serve as a foundation for computing the motions of the planets. This was accordingly done by Muhammed ben Ibrahim Al Fazari, whose works the Arabs call the great *Sindhind*, and from it an abstract was afterwards made for Al Mamun by Abu Giafar Muhammed ibn Musa al Kwarizmi, who made use of it to prepare his tables, which obtained great renown in the lands of Islam. But when Al Mamun became Caliph, he promoted these noble studies and called in the most learned men in order to examine the *Almagest* and make instruments for new observations.

The account of which the above is an abstract shows us clearly the origin of the study of astronomy and mathematics under the Abbasid Caliphs. But though the first impulse came from India, the further development of Arabian science was altogether founded on that of Greece and Alexandria. It was through the court physicians from the flourishing medical school kept up by Nestorian Christians of Khusistan that a knowledge of Greek philosophy and science was first spread among the subjects of the Caliphs; and by degrees the works of Aristotle, Archimedes, Euclid, Apollonius, Ptolemy, and other mathematicians were translated into Arabic. Fresh translations of Ptolemy were made from time to time in the various kingdoms into which the vast empire of the Caliph was soon split up[1], and a thorough knowledge of Ptolemaic astronomy was thus spread from the Indus to the Ebro. There were several special inducements for Muhammedans to pay attention to astronomy, such as the necessity of determining the direction in which the faithful had to turn during prayers, also the importance of the lunar motions for the calendar, and the respect in which judicial astrology was held all over the East. The Caliph Al Mamun, son of Harun Al Rashid (813–833) is the first great patron of science, although the Omayyad Caliphs had much earlier had

---

[1] The earliest is probably that of Al Haggag ben Jûsuf ben Matar early in the ninth century. See Suter, *Die Mathematiker und Astronomen der Araber und ihre Werke*, Leipzig, 1900 (p. 9), which valuable bibliographical summary I follow as regards names and dates.

an observatory near Damascus, and the Jew Mashallah (who died about 815) had already before the reign of Al Mamun won a name as an observer and astrologer. But the Damascus observatory became quite eclipsed by that erected at Baghdad in 829, where continuous observations were made and tables of the planetary motions constructed, while an important attempt was made to determine the size of the earth. Among the astronomers of Al Mamun and his successors one of the greatest was Aḥmed ben Muhammed Al Fargani (afterwards known in the West as Alfraganus), whose *Elements of Astronomy* were translated into Latin in the twelfth century and contributed greatly to the revival of science in Europe[1]. Tâbit ben Korra (826–901) was a most prolific writer and translator, but is chiefly known in the history of astronomy as a supporter of the erroneous idea of the oscillatory motion of the equinoxes. A younger contemporary of his, Muhammed Al Battani (died 929), was the most renowned of all the Arabian astronomers and became known in the West in the twelfth century (under the name of Albategnius) by the translation of the introduction to his tables[2]. Already in his time the power of the Caliphs had commenced to decline, and they soon lost all temporal power. The study of astronomy was, however, not influenced by this loss of patronage, as the Persian family of the Buyids, who in 946 obtained possession of the post of Amir-al-Omara (corresponding to the Frankish Major Domus), took over the *rôle* of patrons of science, so long and so honourably carried on by the Abbasid Caliphs. Sharaf al Daula built in 988 a new observatory in the garden of his palace, and among the astronomers who worked there was Muhammed Abu 'l Wefa al Bûzjani (959–998), who wrote an *Almagest* in order to make the contents of Ptolemy's work accessible to the less learned. In the nineteenth century this book gave rise to a long

---

[1] First printed at Ferrara in 1493. I quote the edition of Golius, Amsterdam, 1669.

[2] Translated by Plato of Tivoli. First printed in 1537 after the book of Alfargani. I have used the edition of Bologna, 1645, and a new edition, which is now being published by C. A. Nallino, of which the Arabic and a Latin translation of the text have already appeared (*Pubbl. d. R. Osservatorio di Brera in Milano*, No. 40, 1899–1903).

controversy, which we shall presently consider somewhat in detail.

In the eleventh and twelfth centuries we do not find any names of conspicuous astronomers in Muhammedan Asia. But the western countries under Islam had in the meantime become ready to do their share of the work of keeping the mathematical sciences alive. In the Fatimite kingdom of Egypt Ali ben Abi Said Abderrahman ben Ahmed ben Jûnis, generally called Ibn Jûnis (died 1009), was distinguished both as an astronomer and a poet. At Cairo a liberally equipped observatory enabled him to verify the planetary theories which had once been developed in the neighbouring Alexandria, and in token of his gratitude to the reigning sovereign, Al Hakim, he named his work the Hakemite Tables[1]. We have to pass to the farthest west to find the next astronomer of mark in the person of Ibrahim Abu Ishak, known as Al Zarkali (in Europe afterwards called Arzachel). He was a native of Cordova, lived about 1029–1087, and edited planetary tables called the Toledo Tables[2]. In the following century we find two celebrated astronomers of Seville, Gabir ben Aflah, known as Geber (died 1145, often mistaken for the great alchemist, Gabir ben Haijan, in the eighth century[3]), and Nur ed-din al Betrûgi (Alpetragius), both of whom raised objections to the planetary theories of Ptolemy, though they failed to produce anything better of their own. Spanish astronomy continued to flourish for a while, although the power of the Arabs in the peninsula was rapidly declining, and it produced in the thirteenth century a very remarkable man, who, although a Christian king, must be included in this account of Arabian astronomy, as he owed all he knew about the science to the example and the teaching of Muhammedans and Jews. King Alfonso X., of Castille, named el Sabio (1252–1284), followed

---

[1] Caussin has published an extract in vol. VII. of the *Notices et Extraits des manuscrits* (*Le livre de la grande table Hakémite*). Other chapters, translated by the elder Sédillot but never published, are reviewed by Delambre, *Hist. de l'astr. du Moyen Age*, p. 95 sqq.

[2] Never published. Delambre, l. c. p. 176, and Steinschneider, *Études sur Zarkali, Bullettino Boncompagni*, T. xx. p. 1.

[3] The word algebra has also sometimes erroneously been connected with his name.

the example of the Caliphs and called astronomers to his court to assist in the preparation of the renowned Alfonsine Tables.

With Alfonso the study of astronomy disappeared from Spain, but not before it had been revived in the East. In 1258 the still existing but shadowy Caliphate of Baghdad was swept away by the Mongol conqueror Hulagu Khan, grandson of Genghis Khan; but already in the following year this great warrior listened to the advice of his new vizier, Nasir ed-din al Tûsi (born at Tûs in Khorasan in 1201, died in 1274), and founded a great and magnificent observatory at Merâgha, in the north-west of Persia. In this observatory, which was furnished with a large number of instruments, partly of novel construction, Nasir ed-din and his assistants observed the planets diligently and produced, after twelve years' labour, the " Ilokhanic Tables." Among the astronomers of Merâgha seems to have been Juhanna Abu 'l Faraj, called Bar Hebrâyâ, or the son of a Jew. He was a Christian, born in 1226, and from 1264 till his death in 1286 Maphrian or Primate of the Eastern Jacobites. He left a well-known chronicle and an astronomical work, both written in Syriac, as well as other writings[1]. The observatory at Merâgha had not a long life, and Asiatic astronomy had to wait a century and a half, until the grandson of another terrible conqueror erected another observatory. Ulug Begh, grandson of Tamerlan, drew learned men to Samarkand and built an observatory there about the year 1420, where new planetary tables and a new star catalogue, the first since Ptolemy's, were prepared. Ulug Begh died in 1449, he was the last great Oriental protector of astronomy; but just as the Eastern countries saw the star of Urania setting, it was rising again for Europe.

In this rapid review of Arabian astronomers we have only mentioned those whose work we shall have to allude to in the following pages, omitting several names of distinction, whose owners devoted themselves to other branches of astronomy.

[1] *Le livre de l'ascension de l'esprit sur la forme du ciel et de la terre.* Cours d'Astronomie rédigé en 1279 par Grégoire Aboulfarag, dit Bar Hebræus. Publié par F. Nau, Paris, 1899–1900 (2 parts, Syriac and French). His chronicle is the chief authority for the fable about the burning of the Alexandrian library by order of the Caliph Omar. For a very thorough refutation of this see Butler, *The Arab Conquest of Egypt*, Oxford, 1902, pp. 401–426.

Though Europe owes a debt of gratitude to the Arabs for keeping alive the flame of science for many centuries and for taking observations, some of which are still of value, it cannot be denied that they left astronomy pretty much as they found it. They determined several important constants anew, but they did not make a single improvement in the planetary theories. It will therefore be sufficient to enumerate the improvements attempted and the opinions held by Arabian astronomers without keeping strictly to the chronological order, although we are here dealing with a period of about six hundred years and men belonging to very different nations, who had little in common except their religion and the language in which they wrote.

Turning first to the question of the figure of the earth, we find a remarkable contrast between Europe and Asia. In the world under Islam there was an entire absence of that hostility to science which distinguished Europe during the first half of the Middle Ages. Though we learn from Kazwini's *Cosmography*[1] that some of the earlier Arabs believed the earth to be shaped like a shield or a drum, still there is no record of any Arabian having been persecuted for asserting that the earth is a sphere capable of being inhabited all over. Whether this was in consequence of the warriors of the Caliphs having carried their arms to the centre of France on one side and to the borders of China on the other, while their merchants travelled southward to Mozambique and northward to the centre of Asia, is another question; anyhow, the fact of the earth being a sphere of very small dimensions in comparison to the size of the universe was accepted without opposition by every Arabian scholar, and the very first scientific work undertaken after the rise of astronomy among them was a determination of the size of the earth. It was carried out by order of the Caliph Al Mamun in the plain of Palmyra. According to the account given by Ibn Jûnis, the length of a degree was measured by two observers between Wamia and Tadmor and by two others in another

---

[1] *Zakarija Ben Muhammed Ben Mahmûd El Kazwini's Kosmographie*, deutsch von H. Ethé, Leipzig, 1868, p. 295.

locality, we are not told where. The first measure gave a degree equal to 57, the second one equal to 56¼ Arabian miles of 4000 black cubits, and the approximate mean, 56⅔ miles, was adopted as the final result, the circumference of the earth being 20,400 miles and the diameter 6500 miles. Another report, by Ahmed ben Abdallah, called Habash, an astronomer under Al Mamun (quoted by Ibn Jûnis), states that a party of observers (no names given) proceeded along the plain of Sinjar until they found a difference in meridian altitudes, measured the same day, equal to one degree, while the distance travelled over was found to be 56¼ miles[1]. Probably two different determinations were made. If the "black cubit" is the Egyptian and Babylonian cubit of 525 mm.[2], the mile would be = 2100 m. and 56⅔ miles = 119,000 meters, rather a large result.

The doctrine of the spherical earth remained undisputed in the Muhammedan learned world, though the curious error of assuming that the level of the sea was higher on some parts of the earth than on others appears to have found some adherents among Arabian writers as well as in Europe[3]. We may, therefore, at once pass on to the motions of the heavenly bodies. Al Battani determined the longitude of the sun's apogee and

---

[1] Caussin, *Not. et Extraits*, VII. pp. 94–96; Delambre, *Hist. de l'astr. du Moyen Age*, pp. 78 and 97; Shems ed-din, *Manuel de la cosmographie*, traduit par Mehren, Copenhague, 1874, p. 6. Suter, p. 209, mentions a third report (from Ibn Challikân's *Biographical Dictionary*), according to which the sons of Mûsâ first measured in the plain of Sinjar and afterwards as a test at Kûfa, by order of Al Mamun. The eldest of the sons of Mûsâ died 41 years after Al Mamun, and the names of the observers in the first report are different, so that the third report is not to be relied on. Al Fargani merely gives 56⅔ miles as the result of Al Mamun. According to Shah Cholgii *Astronomica... studio et opera Ioh. Gravii*, London, 1652, p. 95, Ala ed-din Al Kûsgi (one of Ulug Begh's astronomers) gives the circumference of the earth = 8000 parasangs. As a Persian parasang = 30 stadia (Hultsch, *Griech. u. Röm. Metrologie*, p. 476) this would seem to be the value of Posidonius, 240,000 stadia. Kazwini (p. 298) gives the circumference = 6800 parasangs on the authority of Al Birûni.

[2] Hultsch, p. 390.

[3] It deserves to be mentioned that Shems ed-din of Damascus (1256–1327) explains the great preponderance of dry land in the northern hemisphere by the attraction of the sun on the water, which is greatest when the sun is in perigee, at which time it is nearly at its greatest south declination. That this accumulation of water would not be a permanent one does not occur to him (*Cosmographie*, p. 4).

found it = 82° 17′[1], or 16° 47′ more than Ptolemy had given. As he believed that Ptolemy's value had been found by himself, and as he adopted 54″ (or 1° in 66 years) as the annual amount of precession, there remained (assuming that 760 years had passed since the time of Ptolemy) an outstanding error of 79″ − 54″ = 25″ per annum. In reality the annual motion of the solar apsides is $11\frac{1}{2}$″; still we may say that the discovery of this motion is due to Al Battani, though he did not announce it as such; in fact he merely gives his own value as an improvement on that of Ptolemy. Even Ibn Jûnis (who found 86° 10′) did not suspect that the apogee was steadily moving, but merely says that it must be corrected for precession (1° in 70 years), and remarks that the longitude of the apogee is very difficult to determine accurately[2]. On the other hand, Al Zarkali found a smaller value, 77° 50′, and as he also found a smaller value of the excentricity, he thought it necessary to let the centre of the sun's excentric orbit describe a smaller circle, after the example set by Ptolemy in the case of Mercury[3]. The inclination of the ecliptic which the Greeks had found = 23° 51′ 20″ was by the astronomers of Al Mamun found = 23° 33′ (in 830), by Al Battani (in 879) and by Ibn Jûnis = 23° 35′[4]. When Al Zarkali found 23° 33′, he, and afterwards Abu 'l Hassan Ali of Morocco, concluded that the obliquity oscillated between 23° 53′ and 23° 33′, an idea to which the prevailing belief in the "trepidation" of the equinoxes lent countenance[5].

If we now turn to the moon, we do not find that the Arabs made any advance on Ptolemy. Several of them noticed that the inclination of the lunar orbit was not exactly 5°, as stated by Hipparchus. Thus, Abu 'l Hassan Ali ben Amagiur early in the tenth century says that he had often measured the greatest

[1] *Scient. Stell.* cap. xxviii. Bologna, 1645, p. 72; Nallino, p. 44. At the end of cap. xlv. he says the apogees of sun and Venus are both in 82° 14′, and Ibn Jûnis also gives 82° 14′ as the value found by Al Battani (Caussin, p. 154).

[2] Caussin, pp. 232 and 238. Abu 'l Faraj gives 89° 28′ for the year 1279 (p. 22).

[3] Sédillot, *Prolégomènes aux tables astron. d'Olough Beg* (1847), pp. lxxx–lxxxii.; Riccioli, *Almag. Novum*, I. p. 157.

[4] Caussin, p. 56. For A.D. 900 Newcomb gives 23° 34′ 54″, with a diminution of 46″ per century, so that the Arabian astronomers erred less than 1′.

[5] Aboul Hassan Ali, *Traité des instruments astron. des Arabes*, T. I. p. 175; Sédillot, *Mémoire sur les instr. astr. des Arabes*, p. 32.

latitude of the moon and found results greater than that of Hipparchus, but varying considerably and irregularly. Ibn Jûnis, who quotes this, adds that he has himself found 5° 3′ or 5° 8′, while other observers are said to have found from 4° 58′ to 4° 45′[1]. Want of perseverance and of accurate instruments caused them to miss a remarkable discovery, that of the variation of the lunar inclination.

But an even more remarkable discovery has been claimed for an Arabian astronomer. In 1836 the younger Sédillot announced that he had found the third inequality, the variation, distinctly announced in Abu 'l Wefa's *Almagest*. A fierce controversy raged for a number of years as to the reality of this discovery, Sédillot alone defending his hero with desperate energy and refusing to listen to any arguments, while Biot, Libri and others as strenuously maintained that Abu 'l Wefa simply spoke of the second part of the evection, the prosneusis of Ptolemy. The fight had died out when, in 1862, Chasles suddenly took up the cudgels for Sédillot and pointed out what seemed to him to be some contradictions in Ptolemy's statement[2]. Nobody answered this until Bertrand did so in 1871; he called attention to several inaccuracies in the text of Abu 'l Wefa as we possess it now, and also showed that Abu 'l Wefa did not *add* his "mohazat" to the prosneusis, the latter not being included in his "second anomaly[3]." It is unnecessary to enter into a more detailed account of the controversy; but to show that any weapon was considered good enough with which to defend Abu 'l Wefa, it may be mentioned that Sédillot and Chasles tried to prove that Tycho Brahe must have copied his discovery from Abu 'l Wefa, because he calls it *hypothesis redintegrata*. Tycho used this same phrase in speaking of his

---

[1] Sédillot, *Prolégomènes*, p. xxxviii., *Matériaux pour servir à l'hist. des sciences chez les Grecs et les Orientaux*, T. I. p. 283. The sons of Mûsâ ben Sakir (about 850) seem to have been the first to find a value differing from that of the ancients. Abraham ben Chija, a Jewish writer who lived about A.D. 1100, says that Ptolemy found 5°, but that according to the opinion of the Ishmaelites it is 4½° (*Sphæra mundi*, Basle, 1546, p. 102).

[2] *Lettre à M. Sédillot sur la question de la variation lunaire*, Paris, 1862, 15 pp. 4°, and *Comptes Rendus*, vol. 54, p. 1002.

[3] *Comptes Rendus*, vol. 73, pp. 581, 765, 889; *Journal des Savants*, 11 Oct. 1871.

own planetary system, which he most emphatically claimed as
an original discovery, and which he vigorously defended against
other claimants. In future it will be hopeless for anybody to
claim the discovery for Abu 'l Wefa, as the matter has now
been thoroughly sifted, both by mathematicians and orientalists.

The *Almagest* of Abu 'l Wefa has never been published in
full, but there are three translations of the chapters in question[1],
which only differ in some trivial points. In no part of the book
does he make any advance on Ptolemy or claim to have made
any new discovery, and in speaking of *three* inequalities he
merely does what the other Arabian astronomers do[2]. He begins
by describing the first (equation of the centre) and the second
(evection) and states when they reach their maxima. He then
says that we have found[3] a third inequality, which takes place
when the centre of the epicycle is between the apogee and the
perigee of the excentric, and which reaches its maximum when
the moon is about a *tathlith* or a *tasdis* from the sun, while it is
insensible in syzygy and quadrature. The maximum is $\frac{3}{4}°$. He
explains that this is caused by a deviation of the line of apsides
of the epicycle, and he describes quite correctly the construction
adopted by Ptolemy (whose name he does not mention), letting
the line of apsides be directed, not to the earth but to another
point on the line of apsides of the excentric. It is difficult for
an unbiassed reader to understand how anyone could fail to see
that Abu 'l Wefa is simply copying Ptolemy. Sédillot maintained that the words *tathlith* and *tasdis* mean the octants
(where the variation reaches its maximum); but every other
orientalist who has expressed an opinion, states that by their

[1] By Reinaud, Munk, and de Slane (for Biot) in the *Journal des Savants*,
March, 1845 (14 pp., the whole section on the moon); by Sédillot, *Matériaux*,
I. pp. 45–49; and by Carra de Vaux, "L'almageste d'Abû 'l Wéfa Albûzdjani,"
Journal asiatique, 8ᵉ Série, T. XIX. (1892), pp. 408–471 (translation on pp.
443–44). Most of the chapters on the planets are lost.

[2] The unknown author of a short *résumé* of astronomy (in the Bibl.
Nationale) even calls the inequality of prosneusis the *first* equation (Carra de
Vaux, l.c. p. 460). This is not unreasonable, since the equation of the centre
must be taken from the lunar tables, using as argument not the mean anomaly
but the latter corrected for the effect of the prosneusis.

[3] He uses exactly the same expression when speaking of the first and second
inequalities.

roots the words correspond to the numbers 6 and 3, in other words, to elongations 60° and 120° from the sun. This is in accordance with facts, as Biot has shown from Ptolemy's numerical data that the deviation of the line of apsides reaches its maximum value of $\pm$ 13° 8'·9 in elongations 90° $\mp$ 32° 57'·5[1]. But it must be acknowledged that the words in question are also used very vaguely, e.g. by Abu 'l Wefa himself, who says that the velocity of the superior planets after emerging from the sun's rays diminishes gradually till their distance from the sun is about a *tathlith*, when they become stationary. It looks almost as if these words might be used to denote any elongation outside syzygy and quadrature[2].

If Abu 'l Wefa had made a new discovery, we should have expected later Arabian astronomers to have alluded to it. But not one of them gives anything but interpretations of the lunar theory of Ptolemy, and in expressions very similar to those employed by Abu 'l Wefa. Attention was at once called to this fact, and Isaac Israeli of Toledo (about 1310) and Geber of Seville were quoted as examples[3], though it would, of course, have been quite possible for these two writers to have remained ignorant of whatever progress astronomy might have made in the school of Baghdad. But this objection does not apply to Nasir ed-din al Tûsi, in whose review of the *Almagest* and *Memorial of Astronomy* the inequalities known to Ptolemy, and no others, are described and credited to Ptolemy[4]; nor to

---

[1] *Journal des Savants*, 1843, p. 701 ("Sur un traité arabe relatif à l'astronomie," Reprint, p. 47). This deviation does not represent the amount of the correction to the moon's place as seen from the earth, so that there is not any contradiction in Ptolemy's account.

[2] Carra de Vaux, l. c. p. 466. The Arabs had no word for "octants." Nasir ed-din on one occasion wants to mention them, and has to call them "the points midway between syzygy and quadrature." See below, p. 270.

[3] Isaac Israeli repeatedly speaks of these inequalities discovered by Ptolemy, two of which are not found at conjunction and opposition. *Liber Jesod Olam seu Fundamentum Mundi auctore R. Isaac Israeli Hispano*, section III. ch. 8 and sect. v. ch. 16, Part I. p. xxiv., Part II. p. xxxi. (Berlin, 1848 and 1846 ; this publication is not mentioned by Carra de Vaux).

[4] C. de Vaux, "Les sphères célestes selon Nasir Eddin Attûsi," Appendix to P. Tannery's *Recherches sur l'astr. anc.* p. 342, and *Journ. asiat.* 1892, p. 459 : "The third anomaly is that of the prosneusis; it is called the equation of the proper motion" (i.e. of the motion on the epicycle).

Mahmud al Jagmini (about 1300), who wrote a compendium (*mulachchas*) of astronomy[1]. Nor can any objection be raised to Abu 'l Faraj (Bar Hebraeus), and it would be impossible to explain more clearly than he does the effect of the prosneusis. He says: "The third inequality is the angle formed at the centre of the epicycle by two lines which are drawn, one from the centre of the universe and the other from the point called the prosneusis, at the end of which is the apogee of the epicycle, at which commences the proper motion, and which is called the mean apogee. The apogee which is at the end of the line drawn from the centre of the universe is called the apparent one. The point prosneusis is on the side of the perigee of the excentric, 10 parts 17 minutes from the centre of the world[2], which is itself at the same distance from the centre of the excentric. The maximum value of this angle is 13 parts 9 minutes when the moon is a crescent or $\frac{3}{4}$ gibbous, that is, near the hexagon or trigon with the sun. In fact, when the epicycle is four or eight signs distant from the apogee of the excentric, the sun is itself two or four signs distant from [the centre of] the epicycle, because it is half way between this centre and the apogee. In the tables, this inequality of the two apogees is called the first angle and is included in the motion of the centre[3]." While this describes the construction of Ptolemy as clearly as possible, at the same time the agreement of the account with that of Abu 'l Wefa is perfect. Abu 'l Faraj even (like Nasir ed-din) describes as a fourth inequality in longitude that caused by the motion along an orbit inclined to the ecliptic, so that he would not have neglected to describe the variation, if it had been found by an astronomer of Baghdad. We may add that the Jewish writer Abraham ben Chija (A.D. 1100), in his *Sphaera Mundi*, also describes the

[1] Translated by Rudloff and Hochheim, *Zeitschrift der Deutschen Morgenländ. Ges.* XLVII. pp. 213–275. He describes (p. 249) how the line of apsides is directed to a point called "the corresponding point," and gives its position correctly. The inequality he calls the deviation.

[2] Nasir ed-din gives 10° 9'.

[3] *Le livre de l'ascension*, &c. T. II. pp. 29–30. Two codices add after the word prosneusis: "This is the point mohazat."

"aberration" of the apside of the epicycle, chiefly "in sexta et tertia parte mensis[1]."

Therefore, Abu 'l Wefa did not know a single thing about the motion of the moon which he had not borrowed from Ptolemy. But the prosneusis of Ptolemy is not the variation discovered by Tycho Brahe. The latter depends solely on the elongation of the moon from the sun, as it is $= + 39'\cdot5 \sin 2\epsilon$, while it is beyond the power of mortal man to express the effect of the prosneusis without the anomaly. Ptolemy's expression for all the inequalities in longitude assumed by him, when developed analytically, is found to contain, in addition to terms representing the equation of the centre and the evection, the latter being

$$+ 1° 19'\cdot5 \sin (2\epsilon - m),$$

a very considerable term

$$+ 17'\cdot8 \sin 2\epsilon [\cos (2\epsilon + m) + 2 \cos (2\epsilon - m)],$$

where ϵ is the elongation and m the mean anomaly[2]. Obviously this term has nothing in common with the variation, except that it disappears in the syzygies and quadratures. Tycho Brahe did not hang his new term on to the unaltered lunar theory of Ptolemy, and by doing that we should in fact only spoil the latter and make its maximum error rise to more than a degree[3]. Owing to the insufficiency of the observations at his disposal, Ptolemy could only perceive that there was some outstanding inequality after allowing for the evection, only appearing outside the syzygies and quadratures, but he was neither able to find the law which governed the phenomenon, nor was he aware what a large quantity it represented; he could only tinker up his constructions a little, and in this he

---

[1] *Sphæra Mundi* (1546, ed. Schreckenfuchs), p. 75. Münster's commentary to the Hebrew text (p. 116) has "cum centrum est in sextili aut trino aspectu [id est, quando abest a sole duobus signis aut quatuor]"; the words in brackets are not in the Hebrew original. The words "sixth" and "third" are unmistakable (shithith and shelishith). Apparently no one has hitherto thought of consulting Abraham ben Chija.

[2] P. Tannery, *Recherches*, p. 213. Another expansion of Ptolemy's lunar inequalities in a series was given by Biot, *Journal des Savants*, 1843, p. 703 (Reprint, p. 49).

[3] P. Kempf, *Untersuchungen über die Ptolemäische Theorie der Mondbewegung*, Berlin, 1878 (Inaug. Diss.), p. 37.

was most faithfully followed by the Arabs, who added nothing to what he had done, and left it to the reviver of practical astronomy to discover the third lunar inequality.

Passing to the five planets, we find that, generally speaking, very few attempts were made to improve the work of Ptolemy. But the Arabs were not content to consider the Ptolemaic system merely as a geometrical aid to computation; they required a real and physically true system of the world, and had therefore to assume solid crystal spheres after the manner of Aristotle. Above the moon is the Alacir, the fifth essence, which is devoid of lightness and heaviness, and is not perceptible to the human senses; of this substance the spheres and planets are formed[1]. Already in the book of Al Fargani we find the principle adopted which we have seen dates from the fifth century (Proklus) and which became universally accepted in the Middle Ages, that the greatest distance of a planet is equal to the smallest distance of the planet immediately above it, so that there are no empty spaces between the spheres[2]. The semidiameter of the earth is by Al Fargani given as 3250 miles, which corresponds very nearly to Al Mamun's $56\frac{2}{3}$ miles to

Greatest Distance of	Al Fargani	Al Battani	Abu 'l Faraj[3]
Moon	$64\frac{1}{6}$	$64\frac{1}{6}$	$64\frac{1}{6}$
Mercury	167	166	174
Venus	1120	1070	1160
Sun	1220	1146[4]	1260
Mars	8876	8022	8820
Jupiter	14405	12924[5]	14259
Saturn	20110	18094	19963

[1] Al Battani, cap. 50 (p. 195).

[2] Al Fargani, cap. 21 (ed. Golius, p. 80). Much later, Maurolycus in his *Cosmographia* (Venice, 1543, f. 20 a) proves that Mercury and Venus must be below the sun, by pointing out that there would otherwise be a large vacant space between sun and moon.

[3] pp. 189–191.

[4] So in Nallino's ed. (Milan, 1903, p. 121); the ed. of 1645 has 1176.

[5] The ed. of 1645 has 12,420; obviously an error, as the ratio of greatest to smallest distance is given as 37 : 23, for Saturn 7 : 5 (misprinted 7 : 2), or "quantitas unius et duarum quintarum ad unum" (p. 199). Nallino's ed. (Milan, 1903) has 12,924. Abraham ben Chija has 12,400.

a degree, if we put $\pi = \frac{22}{7}$. Starting from Ptolemy's distances of the moon and sun, it was easy to express the other distances in semidiameters of the earth, the ratios between the greatest and smallest distances being in substantial agreement with the theory of ' Ptolemy. Al Battani also gives a similar set of figures, though with some slight differences. He does not mention the peculiar treatment given by Ptolemy to the theory of Mercury. The above table gives the distances expressed in semidiameters of the earth.

Al Kûsgi, one of the astronomers of Ulug Begh, gives a list of the semidiameters of the " concavities " of the planetary spheres (i.e. the smallest distances of the spheres) expressed in parasangs, the diameter of the earth being 2545 parasangs[1]. Expressed in semidiameters of the earth, the figures turn out somewhat different from those given above, e.g. the smallest distance of the sun being 1452 and the greatest of Saturn 26,332, but he does not supply any means of making out how these figures were found.

Before leaving this subject, we shall also give the diameters of the planets according to Al Fargani, as they became known in Europe at an early date and were quoted by Roger Bacon and others[2]. With trifling variations the same values are given by Al Battani, Abu 'l Faraj, and Abraham ben Chija.

		Apparent Diameter	True Diameter (earth's $=1$)
Moon in apogee	....	$31\frac{2}{5}'$	.... $1 : 3\frac{2}{5}$
Mercury, mean dist.	....	$\frac{1}{15}$ of sun's	.... $\frac{1}{28}$
Venus	,, ,,	.... $\frac{1}{10}$,,	.... $1 : 3\frac{1}{3}$
Sun	,, ,,	.... $31\frac{2}{5}'$	.... $5\frac{1}{2}$
Mars	,, ,,	.... $\frac{1}{20}$ of sun's	.... $1\frac{1}{8}$
Jupiter	,, ,,	.... $\frac{1}{12}$,,	.... $4\frac{1}{2} + \frac{1}{16}$
Saturn	,, ,,	.... $\frac{1}{18}$,,	.... $4\frac{1}{2}$

The system of the spheres is set forth in greatest detail in three treatises of later date, the cosmography of Zakarija ben

---

[1] *Astronomica Shah Cholgii*, pp. 95–97.

[2] There are some slight differences between the figures given in the various editions (I have compared those of 1493, 1546, and 1669), but those give above agree with the cubic contents according to Al Fargani. The figures of Kazwini seem to have been greatly corrupted.

Muhammed ben Mahmud al Kazwini (about 1275), the astronomy of Abu 'l Faraj, written in 1279, and that of Mahmud ibn
Muhammed ibn Omar al Jagmini, whose date and nationality are
equally uncertain, but who probably wrote in the thirteenth or
fourteenth century. We find in these text-books an elaborate
system of spheres designed to account for every particular of
planetary motion, in perfect agreement with each other as to
the general arrangement of the spheres, and offering nothing
new as to lunar or planetary theory. The accompanying figures
(taken from Jagmini) will illustrate the ideas better than a
lengthy description[1]. The sun is a solid spherical body, fitting
between two excentric spherical surfaces, which touch two

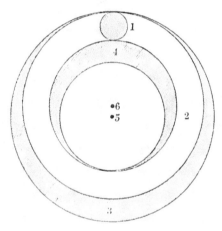

1 The sun.   2 Excentric sphere.   3 The surrounding sphere.   4 The
complement of the surrounding sphere.   5 Centre of the world.
6 Centre of the excentric sphere.

other surfaces, in the common centre of which the earth is
situated, and which between them enclose a space (or intersphere, as Abu 'l Faraj calls it), named by Jagmini al-mumattal,
or the equably turning sphere, which has the same motion from
west to east as the fixed stars, i.e. precession. The spheres of
the three outer planets and Venus are arranged on the same

[1] Al Kûsgi gives very similar diagrams of the spheres of Saturn, Mercury,
and the moon.

plan, except that the place of the body of the sun is taken by
the epicycle-sphere of each planet, to the inner surface of which
the planet (a solid spherical body) is attached, or (as Abu 'l
Faraj says[1]) " fixed like a pearl on a ring, touching the surface
in one point." The axis of the excentric sphere is inclined to
that of the mumaṭṭal sphere, which causes the motion in latitude.
The lunar system comprises an additional sphere outside the
others, the centre of which coincides with the centre of the

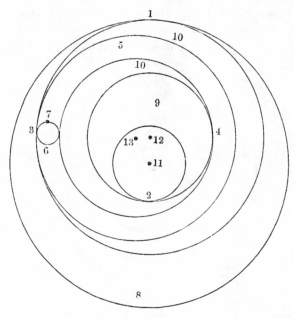

Spheres of Mercury.

1 Upper Apsis.   2 Lower Apsis.   3 Upper Apsis of deferent sphere.
5 Deferent sphere.   4 Lower Apsis of deferent sphere.   6 Epicycle.
7 Mercury.   8 Surrounding complement.   9 Surrounded part of
Mumaṭṭal sphere.   10 Mûdir sphere.   11 Centre of the world.
12 Centre of Mûdir.   13 Centre of deferent sphere.

world, and which is called *al-ǵauzahar*, signifying the constellation Draco, as this sphere provides for the revolution of the
lunar nodes ("the head and tail of the dragon") round the

[1] p. 37.

zodiac[1]. The inner one of the two concentric spherical surfaces, between which the excentric sphere lies, surrounds immediately the fire sphere of the earth. The system of Mercury is more complicated, as a space had to be provided for the revolution of the centre of the excentric sphere. The figure shows the excentric sphere enclosed in a sphere, *al-mûdir*, or the turning one, which allows the upper apsis or apogee of the excentric or deferent sphere (*J* in the figure) to move right round the outer surface of the *mûdir*. The inner surface of the *mumaṭṭal* sphere immediately surrounds the *ǧauzahar* sphere of the moon.

It was a necessary consequence of the large solar parallax of 3′ accepted by Ptolemy, that Mercury and Venus must be very near the earth, since they are assumed to be nearer than the sun. Thus Abraham ben Chija says that the shadow of the earth extends beyond the orbit of Mercury but does not reach that of Venus[2]. Ptolemy never mentions the parallaxes of Mercury and Venus, as to which nothing was known, though they ought, of course, to be greater than 3′. But on the assumption that the smallest distance of Mercury is equal to the distance of the moon at apogee, the parallax of Mercury ought to rise to 54′, which must have been felt to be too large a quantity, though it does not seem to have struck Al Battani as anything surprising, perhaps because Mercury cannot be seen when in inferior conjunction. It may have been this necessarily large parallax of Mercury, which induced Ibn Jûnis (without any explanation) to reduce the solar parallax from 3′ to 2′, or rather to 1′ 57″[3]. Geber[4] blames Ptolemy for having said that the parallaxes of the planets are insensible, and remarks that

---

[1] Precession is supposed to be included in this, "the first motion." The second one is that of the concentric oblique intersphere (called the mâil sphere or the sphæra deflectens) round the centre of the world, 11° 9′ per day, by which amount the lunar apogee moves towards the west. The third motion is that of the excentric, carrying the centre of the epicycle 24° 22′ towards the east. The fourth is the motion on the epicycle. Abu 'l Faraj, p. 27.

[2] *Sphæra mundi*, ed. Osw. Schreckenfuchs, Basle, 1546, pp. 84–86.

[3] Unpublished chapters of Ibn Jûnis, reviewed by Delambre, *Hist. de l'astr. du Moyen Age*, p. 101.

[4] *Instrumentum primi mobilis a P. Apiano...Accedunt ijs Gebri filii Affla Hispalensis...libri IX. de astronomia*, Norimbergæ, 1534, fol. (Introd. p. 3 and lib. vii. p. 104).

he ought therefore logically to have placed Venus and Mercury above the sun. He takes great pains to show that Venus may be exactly on the line joining the sun and the earth. Indeed, Geber neglects no opportunity of criticising Ptolemy's methods of finding the elements of the orbits[1], and he is generally very unjust to him, but he does not venture to substitute any other system and does not object to the general principles of the Ptolemaic system[2].

Geber's attempts to pick holes in the work of Ptolemy were, perhaps, not unconnected with the rapid rise of Aristotelean philosophy in Spain in the twelfth century, which, though not destined to last long, nevertheless exercised a considerable influence on the spread of knowledge of Aristotle in the Christian world, while it cast a halo round the Caliphate of Cordova, which at that time, under the enlightened rule of the Almohades, seemed to have reestablished the glory of the best days of the Moslem world. Three names are specially associated with this movement: Abu Bekr Muhammed Ibn Jahya al Sayeg, called Ibn Badja (of Saragossa, died 1139), known as Avempace among the Scholastics; his pupil Muhammed ben Abdelmelik Ibn Tofeil (of Granada, died 1185–1186), called Abubacer by the Scholastics; and finally the greatest philosopher of Islam, Ibn Rošd Abu Welid, known as Averroes (1126–1198). In studying Aristotle they laid special stress on his scientific works, and did not, like their Christian successors, think of little but dialectics. The acceptance of the system of homocentric spheres or some modification of it must, therefore, have seemed a necessity to the Arabian philosophers, and this, of course, led them to reject the theory of epicycles. The little we know of the opinions of Ibn Badja on this subject is found in the famous work *The*

---

[1] See the long indictment on pp. 2–3 of his introduction. He blames Ptolemy among other things for assuming that the centre of the deferent is half-way between the centres of the zodiac and of the equant, while he himself deduces this from the movements.

[2] Copernicus possessed a copy of Geber's book, which is now in the University library at Upsala. On the title-page, after the author's name, he has written: "Egregii calumniatoris Ptolemæi," while a number of marginal notes show that he has read the book carefully. Curtze, *Mittheilungen des Coppernicus Vereins*, I. p. 37.

Guide of the Perplexed of the great Jewish scholar Moses ben
Maimun of Cordova, better known as Maimonides, who tells us
that he had his information from a pupil of Ibn Badja. Like
Geber (with whose son he had been familiar), Maimonides
doubted that Mercury and Venus were nearer than the sun,
though he would not venture to say how they actually moved[1].
But what is more important, he declared the motion of a planet
on an epicycle to be contrary to physical principles, because
there are only three motions possible in this world: around its
centre, or towards it, or away from it; while he also maintained
that according to Aristotle circular motion can only take place
round a real, central body[2]. Though Aristotle in reality did not
object to epicyclic motion with a mathematical point as centre,
for the simple reason that it had not been proposed when he
wrote, while, as we have seen, his moving principle had nothing
to do with the centre of motion, it is easy to see that Ibn Badja's
real difficulty was the same which afterwards produced so many
obstacles to the advance of science in Europe: whatever could
not be found in Aristotle's books must be unworthy of notice.
According to Maimonides (who, however, makes the reservation
that he had not heard it from disciples), Ibn Badja constructed
a system of his own, in which he only admitted excentric circles
but no epicycles. We are not given any particulars as to this
system, but there can hardly be any doubt that its author confined himself to generalities and did not attempt to represent
phenomena like the lunar inequalities by it. Maimonides
remarks that there is nothing gained by Ibn Badja's reform,
since the excentric hypothesis is as objectionable as the epicyclic one, as it also supposes motion round an imaginary point
outside the centre of the earth. The centre of the excentric, on
which the sun is supposed to move, is outside the convexity of
the lunar sphere and inside the concavity of that of Mercury;
the centre of Mars' motion and that of Jupiter's are between the
spheres of Mercury and Venus, and the centre of Saturn's excentric is between the spheres of Mars and Jupiter. He adds

---

[1] *Rabbi Mosis Majemonidis Liber...Doctor Perplexorum*, Basileæ, 1629,
Pars II. cap. IX.

[2] Ibid., Pars II. cap. XXIV.

that the revolution of a number of concentric spheres around a common axis is conceivable, but not the revolution round different axes inclined to each other, as the spheres would disturb each other unless there are other spherical bodies between them. This attempt to revive and modify the system of (movable ?) ·excentrics did, therefore, not mend matters[1].

Ibn Tofeil, the second of the three Moslem philosophers of Spain, vizier and physician at the court of Jusuf ben Abd el Mumin of Morocco, seems to have walked in the footsteps of his master; but the only extant work of his, a kind of religious mystic romance about the emancipation of a soul from the trammels of this material world, does not give any clue to his ideas as to the planetary system. But Averroes, who also objected to the excentrics and epicycles, says in his commentary to Aristotle's *Metaphysics* that Ibn Tofeil possessed on this subject excellent theories[2]; and Ibn Tofeil's pupil, the astronomer Al Betrugi, in the introduction to his theory of the planets, says of him : " You know that the illustrious judge Abu Bekr Ibn Tofeil told us that he had found an astronomical system and principles of the various movements different from those laid down by Ptolemy and without admitting either excentrics or epicycles, and with this system all the motions are represented without error." Ibn Tofeil was therefore probably the real author of the fairly elaborate system, which his pupil worked out and handed down to us in a work on the planets, which was translated into Hebrew in the following century and from that again into Latin, and published in 1531[3].

---

[1] Maimonides also remarks (in the same chapter) that the supposed inclinations of Mercury and Venus in the Ptolemaic system are difficult or impossible to comprehend or imagine as really existing. Therefore, if what Aristotle says is true, there is neither epicycle nor excentric, and everything turns round the centre of the earth.

[2] Munk, *Mélanges de philosophie juive et arabe*, Paris, 1859, p. 412.

[3] *Alpetragii Arabi Planetarum theorica phisicis rationibus probata, nuperrime latinis litteris mandata a Calo Calonymos, Hebreo Neapolitano*, Venice, 1531, 28 ff. folio (published with Sacrobosco's *Sphæra*). A translation by the famous Michael Scot has never been printed, but is still extant in Paris (Munk, *Mélanges*, p. 519). The principle of the system is described by Isaac Israeli, who, however, does not mention the author's name (*Liber Jesod Olam*, II. 9, Part I. p. xi.).

The object of this system was to explain the constitution of
the universe as it really is, and not merely to represent the
motions of the planets geometrically, so as to be able to foretell
their places in the heavens at any time; and the author (be he
Ibn Tofeil or Al Betrugi *alias* Alpetragius) specially disclaims
any intention of testing the theory by comparing it with observations or of accounting for minor details of the motions[1]. The
leading idea is that of the homocentric spheres, each star being
attached to a sphere, and the motive power is the ninth sphere,
the sphere outside that of the fixed stars. The Spanish philosopher ought therefore to have been content with the system
of Eudoxus or its modification by Aristotle (whom he never
mentions by name, but only as "the sage"), but unfortunately
he became possessed with the notion that the prime mover
must everywhere produce only a motion from east to west, and
he had therefore to reject the independent motion of the planets
from west to east, and revert to the old Ionian idea that the
seven planets merely perform the daily revolution with a speed
slightly slower than that of the fixed stars. The true speed of
the *primum mobile* is a little faster than this; the eighth sphere
performs a revolution in a slightly longer period (24 hours), and
the effect of the prime mover is gradually weakened more and
more, with increasing distance, until we find the sphere of
the moon, being furthest from the prime mover, taking nearly
twenty-five hours to complete a revolution. This was the old
primitive Ionian idea, but Al Betrugi (or his teacher) saw that
this was not sufficient, as not only is the pole of the ecliptic
different from that of the equator, which prevents the planets
from moving in closed orbits, but the planets do not even keep
at the same distance from the pole of the ecliptic but have each
their motion in latitude, as well as a variable velocity in longitude; and all this had yet to be accounted for. The ninth
sphere has but one motion, but the eighth has two, that in
longitude (precession) and another which is caused by the pole
of the ecliptic describing a small circle round a mean position,
thereby producing the supposed oscillation or trepidation of the

---

[1] Fol. 8 b.

equinoxes[1]. Similarly, the pole of each planet describes a small circle round a mean position (i.e. the pole of the ecliptic), thereby producing inequalities in longitude and motion in latitude[2]. Whenever the actual orbit-pole of a planet is on the parallel of the mean pole, it is obvious that the planet will perform its daily revolution with its mean velocity, while the velocity is increased or lessened when the actual pole is respectively at its minimum or maximum distance from the pole of the heavens (the motion of the pole of the orbit being added to or subtracted from the motion of the planet), so that the epicycle is hereby rendered superfluous. The lengths of the radii of these small circles are not given, except in the case of Saturn, where the radius is $3° 3'$[3], while the mean pole of the moon is $5°$ (the inclination of the lunar orbit) distant from the pole of the ecliptic[4], and the small circle is so exceedingly small as to produce no retrograde motion, which is also the case with the sun. The periods of the poles of the outer planets are given by the following figures. Saturn makes 57 revolutions in 59 years and $1\frac{1}{2} + \frac{1}{4}$ days, in which period the mean pole lags behind 2 revolutions and $1\frac{2}{3}° + \frac{2}{9}°$. Jupiter makes 65 revolutions in 71 years, the mean pole lagging behind 6 revolutions. Mars makes 37 revolutions in 79 years and $3\frac{1}{4} + \frac{1}{15}$ days, the pole lagging behind 42 revolutions and $3\frac{1}{6}°$[5].

In other words, the motions on these small circles are completed in the synodic periods of the planets. Similarly, the pole of Venus makes 5 revolutions in 8 years less $2\frac{1}{4}^d + \frac{1}{20}$, lagging $1\frac{5}{8}$ revolutions in one year; and Mercury 145 revolutions in 46 years and $1\frac{1}{30}^d$[6]. It is curious that Alpetragius alters the order of the planets, placing Venus between Mars and the sun, because the *defectus* (lagging) of Venus is smaller than that of the sun[7]. He also says that nobody has given any valid reason for accepting the usually assumed order of the planets, and that Ptolemy is wrong in stating that Mercury and Venus

[1] Fol. 9 b. [2] Fol. 14 b, sq.
[3] Fol. 16 a. [4] Fol. 25 a.
[5] Fol. 16 a, 18 a, 19 b. [6] Fol. 21 b, 24 b.
[7] "Nam reperimus defectum eius primum minorem defectu orbis solis et maiorem defectu orbis martis, et sequitur juxta radices nostras ut sit inter eos ambos." Fol. 21 a.

are never exactly in a line with the sun (a remark already made by Geber); and as they shine by their own light they would not appear as dark spots, if passing between us and the sun. That they do not receive their light from the sun is proved, he thinks, by the fact that they never appear crescent-shaped[1].

There is no need to dwell any longer on this quaint theory of spiral motion, as it has been rather improperly called[2]. It represented a retrograde step of exceedingly great magnitude, totally unjustified, as the theory could not seriously pretend to be superior to the Ptolemaic system, which had only become so very complicated because it took into account every single known detail of irregular motion, but which could also be made very simple if one was content with representing only the principal phenomena. We are told by the Jewish astronomer Isaac Israeli of Toledo, that the new system made a great sensation, but that it was not sufficiently worked out to be taken seriously, and that the system of Ptolemy, founded on the most rigorous calculations, could not be superseded by it[3]. Another Jewish author, Levi ben Gerson, in a work written in 1328, entered into a lengthy refutation of the hypotheses of Al Betrugi[4]. But the latter certainly represented a general desire on the part of the Spanish Aristoteleans to overcome the physical difficulties in accepting the Ptolemaic system; thus Averroes says that the astronomy of Ptolemy is nothing *in esse*, but is a convenient means of computing, and that he himself in his youth had hoped to prepare a work on the subject[5].

While ineffectual attempts were being made in the far west to devise a new astronomical theory, the astronomers of the east did not remain blind to the desirability of finding a system, in which the planets were not supposed to move unsupported in

---

[1] Fol. 21 a.

[2] e.g. by Riccioli, *Almag. Nov* T. i. p. 504, where Kepler's figure of the real motion of Mars in space from 1580 to 1596 (supposing the earth to be at rest) is copied, as if that had anything to do with the " spirals " of Alpetragius.

[3] He adds that he was not qualified himself to sit in judgment on the proposed system (*Liber Jesod Olam*, ii. 9, p. xi.).

[4] Munk, *Mélanges*, pp. 500 and 521.

[5] *Commentary to Aristotle's Metaphysics*, Munk, p. 130. Already quoted by Rheticus in the *Encomium Borussiae* at the end of his *Narratio prima*.

space in such a wonderfully complicated manner; and in the
thirteenth century we find one of the greatest Arabian astronomers, Nasir ed-din Al Tûsi, advocating a system of spheres
which he supposed to be more acceptable than excentrics and
epicycles[1]. In addition to a review or digest of the *Syntaxis*
of Ptolemy he wrote a shorter work entitled *Memorial of
Astronomy*, in various passages of which he shows his dis-
satisfaction with the Ptolemaic system. In the chapter on the
moon (to which we have already alluded) he counts up the
various anomalies, among which he mentions the anomaly of
illumination, that is, the spots on the moon, which he believes
to be caused by other bodies moving in the lunar epicycle and
unequally exposed to the moon's light. He then says that we
should expect in a simple theory to find the centre of the
epicycle in equal times describing equal arcs on the deferent,
and the diameter of the epicycle joining the pericentre and the
apocentre pointing to the centre of the deferent. But neither
of these conditions is fulfilled. In the theories of the planets
he makes the same objections, which it must be said are very
just, since the introduction of the equant was a very unnatural
arrangement. But this is nothing to the artificial machinery
designed by Ptolemy to account for the motion in latitude
of the five planets, especially of Mercury and Venus. Nasir ed-
din describes the marvellously complicated movements of the
deferents and epicycles of these planets, and remarks that
" these motions require the introduction of a system of guiding
spheres, about which the ancients have not said anything."
He next proceeds in the following chapter to explain a system
of his own which allows us to discard these combinations.

First he proves that if there are two circles in one plane,
one touching the other internally and of a diameter equal
to half that of the other, and if the greater one rotates, and a
point moves along the circumference of the smaller one in the
opposite direction with twice the velocity and starting from the
point of contact, then that point will move along a diameter of

---

[1] " Les sphères célestes selon Nasir-Eddin Attûsi. Par M. Carra de Vaux."
Appendix VI. to Tannery's *Recherches sur l'astr. anc.* pp. 337–360. Includes a
translation of the chapter in which the new theory is set forth.

the greater circle[1]. These two circles may now be assumed to
be the equators of two spheres, and for the point we may
substitute a sphere representing the moon's epicycle (*1* in the
figure). Nasir ed-din assumes another sphere (*2*) surrounding
the epicycle and destined to keep the diameter from apogee to
perigee in its place, always coinciding with the diameter of the
sphere (*4*); "let us give it a suitable thickness, but not too

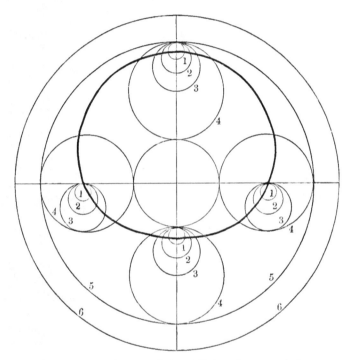

The thick line is *not* a circle. All the others are circles.

great, so as not to take up too much space." He next assumes
two more spheres, one (*3*) which corresponds to the smaller
sphere in the above proposition, and the diameter of which is
equal to the distance of the centre of the deferent in the
Ptolemaic system from the centre of the earth; and another
sphere (*4*) with a diameter twice as great. Finally (*4*) is

[1] Compare Copernicus, *De revolutionibus*, III. 4 (Secular ed. 1873, p. 166).

placed in the interior of a carrying sphere (*5*) concentric with
the world and occupying the concavity of the sphere (*6*), the
equator of which is in the plane of the lunar orbit. (*2*) and (*4*)
and (*5*) revolve in the same period, that in which the centre of
the epicycle performs a revolution ; (*3*) revolves in half that
time, while (*6*) revolves in the opposite direction with the same
speed as the apogee of the excentric. The figure now shows
how the epicycle moves to and fro along the diameter of (*4*),
and during the revolution of the circle (*5*) describes a closed
curve, about which Nasir ed-din justly says that it is somewhat
like a circle but is not really one, for which reason it is
not a perfect substitute for the excentric circle of Ptolemy.
He estimates the greatest difference between the lunar places
given by the two theories as one-sixth of a degree, half-way
between syzygy and quadrature. Except for the action of
the guiding sphere (*2*), it would not be the centre of the
epicycle but the point of contact of circles (*3*) and (*4*), which
describes the curve resembling a circle. The same method may
be adopted for Venus and the three outer planets, and Nasir
ed-din promises to explain the new theory of Mercury in an
appendix, but this appears to have been lost.

Nasir ed-din also endeavours to improve on the machinery
proposed by Ptolemy to illustrate the manner in which the
epicycle remains parallel to the plane of the ecliptic. He
mentions that the celebrated Ibn al Haitham (afterwards known
in the west as Alhazen, author of a well-known book on optics)
had written a chapter on this subject, adding to each epicycle
two spheres to account for the inclination of the diameter
perigee-apogee, and two additional ones for the inferior planets
for the diameter at right angles thereto[1]. Nasir ed-din makes
use of the same principle which guided him in his demonstration about the motion in longitude, and he shows how in this
way we may by means of two spheres make the extremities
of the diameter of the epicycle move backwards and forwards

---

[1] Ibn al Haitham said that by using discs instead of spheres one might
complete the demonstration ; but Nasir ed-din objects to this arrangement
(about which he gives no details) that a non-spherical system is not in accordance with the principles of astronomy.

along an arc of a sphere[1]. He claims that this arrangement is
superior to that of Ptolemy by not introducing any error in
longitude[2], but he acknowledges that he has not been able
to get rid of the strong objection to Ptolemy's auxiliary circle,
viz. that the irregular motion in longitude with regard to the
centre of the deferent necessitates the introduction of a corresponding irregularity in the motion on the auxiliary circle by
letting the motion be uniform with regard to an equant. It
baffled Nasir ed-din's ingenuity to find an arrangement of
spheres which could obviate the necessity of having recourse to
this expedient.

All the attempts at rebellion against the Ptolemaic system
had thus turned out failures. And they deserved nothing else,
since it was impossible to find anything better than what
Ptolemy had produced, until it was perceived that where
Ptolemy was wrong was not in his mathematical methods,
which were perfect, but in the fundamental idea of the earth
being at rest. The time was apparently not ripe for a radical
change with regard to this idea. Though the doctrine of the
earth's motion does not seem to have been mentioned by
Arabian writers, we have evidence that the hypothesis of the
daily rotation of the earth was not unknown among them,
a natural consequence of their familiarity with the writers
of antiquity. One of Nasir ed-din's fellow-workers at the
Merágha observatory, Ali Negm ed-din al Katibi, who died in
1277, wrote a book, the *Hikmat al-ain*, on philosophy, in
which he combats this opinion, which he attributes to "some
philosophers." "I do not," he says, "advance as an argument
against it that, if this were the case, a bird flying in the
direction of the motion of the earth would not be able to keep
up with it, because the motion of the earth would be much
faster than that of a bird, inasmuch as it returns to its place
in a day and a night. Such an argument is not conclusive,
because it may be urged that the atmosphere which is close

---

[1] It is not quite clear whether this plan is his own or is the same as Ibn
al Haitham's.

[2] Due to disturbance of the position of the diameter from perigee to apogee,
from which the anomaly is counted.

to the earth partakes of its motion as the ether partakes
of the motion of the heavenly sphere. But I reject this theory,
because all terrestrial motions take place in a straight line,
and therefore we cannot admit that the earth should move
in a circle[1]."

What reformation of astronomy could be hoped for, as long
as this kind of argument could be used? We cannot see from
this remark of Katibi's whether there really were any Arabian
philosophers who believed in the rotation of the earth. It
is however stated in the *Zohar*, the great Kabbalistic work
attributed to Mosheh ben Shemtob of Leon (died 1305), that a
certain Rabbi Hamnuna the Elder (otherwise unknown) taught
that "the earth turns like a sphere in a circle round itself, and
that some people are above and others below[2]." Though this
passage as well as others in the *Zohar* may have been interpolated much later, it would after all not be very surprising if
some learned Jews had been influenced by the opinion of
Herakleides, since it is an established fact that the doctrines of
the Kabbalists were intimately connected with the later Greek
philosophy. But anyhow nothing came of this isolated case,
and the daily rotation of the heavens continued to be universally
accepted as a self-evident fact.

Arabian astronomers who really wished to follow in detail
the celestial motions were therefore obliged to adopt the
Ptolemaic system altogether. New planetary tables had long
been found to be a necessity, and this important work was at
last undertaken by King Alfonso X. of Castille and several
Jewish and Christian astronomers working under him at
Toledo, who prepared the celebrated Alfonsine Tables. Appar-

[1] A. Sprenger, "The Copernican System of Astronomy among the Arabs,"
Journ. Asiat. Society of Bengal, vol. xxv. (1857), p. 189. Katibi's contemporary,
Abu 'l Faraj (ii. p. 10) deems it necessary to prove that the earth cannot be in
motion, neither rectilinear nor circular, but his arguments (about birds and
stones flung upwards) seem merely taken from Ptolemy, lib. i. cap. 6. Kazwini
(*Kosmographie*, p. 296) says that among the ancients there were some adherents
of Pythagoras who maintained that the earth continually moves round in a
circle; but whether these adherents were Greeks or Arabians cannot be seen
from the context.

[2] *Sohar*, Amsterdam, 1728, T. iii. f. 10 a; Günther, *Studien z. Gesch. d.
math. Geogr.*, p. 113.

ently the King must have had his doubts about the physical
truth of the system, judging from his well-known saying that if
God had consulted him when creating the world, he would
have given Him good advice. The tables were prepared under
the direction of the Jew Ishak ben Said, called Hasan, and
a physician, Jehuda ben Mose Cohen, and were finished in
1252, the year in which Alfonso ascended the throne of Castille.
They continued in great repute for three hundred years as the
best planetary tables ; they were first printed in 1483, but had
been spread all over Europe long before that time in numerous
MS copies, many of which are still in existence. Twenty-six
codices are counted up in the *Libros del Saber de Astronomia
del Rey D. Alfonso X. de Castella*, Madrid, 1863–67 (5 vols. fol.).
This compilation, a series of chapters on spherical and theoretical
astronomy followed by tables, must have been made up from
several codices, as there are numerous repetitions even of very
elementary matters. In the third volume the theories of the
planets are dealt with, but one looks in vain for any improvement on Ptolemy; on the contrary, the low state of astronomy
in the Middle Ages is nowhere better illustrated. In general
the elements of the orbits are those of Ptolemy, though sometimes only approximations are given, while different values are
given in different chapters. Though Ptolemy places the centre
of the deferent midway between the centre of the equant and
the earth, the *Libros del Saber* places the centre of the equant
(*cérco del alaux*[1]) midway between the earth and the centre of
the deferent (*cérco del levador*[2]), as in Ptolemy's theory of
Mercury, which the authors would seem to have extended to
the other planets, omitting the motion of the centre of the
deferent on a small circle; this they have, however, correctly
given in the case of Mercury[3]. There is a very curious figure[4]
of the deferent of Mercury in the form of an ellipse (the axes
being as 6 to 5 nearly), with what looks like the sun in the

---

[1] *al* is the Arabic article, *aux* (apside) is a corruption of the Arabic Oudj
(Abu 'l Faraj, II. p. 25). The equant is also called the *cérco del yguador*.

[2] Vol. III. pp. 246–253.

[3] Vol. III. pp. 253 and 278. In the latter place the radius of the small
circle is $\frac{1}{2}r$, as in the "Hypotheses" of Ptolemy.

[4] Vol. III. p. 282.

centre. This curve has been constructed from a number of small circular arcs[1], and it is obviously nothing but the curve described by the centre of the epicycle of Mercury in Ptolemy's theory. For according to the latter the centre of the deferent describes a small circle with radius $= \frac{1}{21}$ of that of the deferent, in direction from east to west, in the same time which the centre of the epicycle takes to pass round the circumference of the deferent from west to east. This makes the centre of the epicycle describe a closed curve resembling an ellipse, the axes of which are in the ratio 11 : 10, almost exactly the same as in the Spanish diagram, and there is therefore in the latter no anticipation whatever of Kepler's great discovery, since in the case of the inferior planets it is the epicycle which is the real orbit[2]. The small sun-like object in the centre of the ellipse represents the centre of Ptolemy's small circle, and it has either been inserted in the manuscript centuries after the essay had been written, or, more likely, it has been caused by a small blot on the place in the parchment where the stationary leg of the draughtsman's compasses had made a small hole. An oval deferent of Mercury occurs in several books published in the sixteenth and seventeenth centuries[3].

[1] See the lengthy description on pp. 278–280.

[2] The editor, Don Manuel Rico y Sinobas, on p. xxxiii. of his preface, even goes so far as to suggest that Kepler may have known of this great discovery of Alfonso's, or rather of Arzachel's, as the text attributes the construction to him. This and other similar diagrams were intended to be used instead of planetary tables in the manner afterwards adopted by Apianus.

[3] First (about 1460) in Purbach's *Theoricæ novæ Planetarum* (ed. of Basle, 1573, p. 82) : " Ex dictis apparet manifeste, centrum epicycli Mercurij, propter motus supradictos non (ut in alijs planetis fit) circumferentiam deferentis circularem, sed potius figuræ, habentis similitudinem cum plana ovali, peripheriam describere." Next by Albert of Brudzew in 1482 in his *Commentariolum super theoricas novas*, printed at Milan in 1495 (ed. Cracow, 1900, p. 124), where it is remarked that the centre of the lunar epicycle describes a similar figure. This is also stated by E. Reinhold in his commentary to Purbach, 1542, fol. P 7 verso (ed. of Paris, 1558, fol. 78); by Vurstisius in his *Questiones novæ in theoricas*, &c., Basle, 1573, p. 233; and in Riccioli's *Almagestum novum*, T. I. p. 564. The last three writers (who give a figure) also take the equable angular motion round the centre of the equant into account, which centre lies on the point of the circumference of the small circle nearest the earth. The curve described by the centre of the epicycle thus becomes egg-shaped, and not like an ellipse.

Though the somewhat confused collection of essays entitled the *Libros del Saber* would not, if published in the thirteenth century, have advanced astronomical science, it cannot be denied that the Alfonsine Tables were very useful in their day. The actual elements are not given, nor is anything said about any observations by which somewhat more correct values of the mean motions must have been found[1].

Having finished our review of the planetary theories of the Arabs, we have to say a few words about their ideas as to the nature and motion of the fixed stars. The exaggerated notion which prevailed before the invention of the telescope with regard to the apparent angular diameters of the stars naturally led to erroneous estimates of their actual size, founded on the assumption that the sphere of the fixed stars (the eighth sphere) was immediately outside that of Saturn[2]. The stars of the first magnitude were supposed to have an apparent diameter equal to $\frac{1}{20}$ of that of the sun, from which it followed that their actual diameters were about $4\frac{3}{4}$ times that of the earth, or about equal to Jupiter and Saturn; while those of the sixth magnitude stars are about $2\frac{1}{2}$ times that of the earth, or about twice that of Mars[3]. As to the nature of the stars, they seem generally to have been assumed self-luminous, being condensed parts of the sphere, though Abraham ben Chija says that the eighth sphere does not shine with a uniform light, but has denser spots, which are illuminated by the sun and appear to us as the fixed stars[4].

[1] The tables in vol. IV. of the *Libros del Saber* are quite different from the Alfonsine Tables, and are apparently only intended for astrological purposes.

[2] Al Battani (cap. 50) gives the greatest distance of Saturn = 18,094, and the distance of the fixed stars = 19,000 semidiameters of the earth. Al Fargani (p. 82) puts them exactly equal. Al Kûsgi gives the semidiameters in parasangs, of the concavity of the stellar sphere = 33,509,180, of the ninth sphere 33,524,369, of its convexity "no one but God knows" (Shah Cholgi, p. 97).

[3] Al Fargani (p. 85, Golius) gives the cubic contents of the six classes as 107, 90, 72, 54, 36, 18 times that of the earth. Abu 'l Faraj, p. 199, gives a similar series from 93 to 15¼ for the average star of each class. Shems ed-din of Damascus in his *Cosmography* (p. 3) merely says that the smallest fixed star is much larger than the earth.

[4] According to Suter, p. 77, a writer called Ibn Zura wrote a treatise "On the cause of the light of the stars, though they and the spheres consist of one single substance."

To account for the apparent slow motion of the stars parallel to the ecliptic, from west to east, whereby their longitudes increase while their latitudes remain unaltered, it became necessary to introduce a ninth sphere (*primum mobile*), turning in twenty-four hours and communicating this motion to the eighth sphere, while the latter moved extremely slowly round its own axis, forming an angle of 23° 35′ with that of the ninth[1]. But the simple phenomenon of precession was by many Arabian astronomers complicated by being assumed variable. We mentioned (in Chapter IX.) that according to Theon and Proklus it had been assumed by some astronomers, apparently before the time of Ptolemy, that the precessional motion of the stars was not progressive, but was confined to an oscillation along an arc of 8°, along which the equinoctial points moved backwards and forwards on the ecliptic, always at the same rate of 1° in 80 years. The absurdity of the sudden change of direction must have become obvious as soon as astronomy began to be cultivated among the Arabs, for we find that one of the earliest astronomers, Tâbit ben Korra, substituted a physically less objectionable theory[2]. He imagines a fixed ecliptic (in the ninth sphere) which intersects the equator in two points (the mean equinoxes) under an angle of 23° 33′ 30″, and a movable ecliptic (in the eighth sphere), attached at two diametrically opposite points to two small circles, the centres of which are in the mean equinoxes and the radii of which are = 4° 18′ 43″. The movable tropical points of Cancer and Capricorn never leave the fixed ecliptic, but move to and fro to the extent of 8° 37′ 26″, while two points on the movable ecliptic 90° from the tropical points move on the circumferences of the small circles, so that the movable ecliptic rises and falls on the fixed one, while the

---

[1] The outermost sphere is by the philosopher Ibn Sina (Avicenna) defined as a spherical, single (not composite) body, emanating directly from God and not subject to dissolution, endowed innately with circular motion as an expression of its praise of the Creator (Mehren in *Oversigt, K. Danske Vid. Selskab*, 1883, p. 70).

[2] The treatise "On the motion of the 8th sphere" has never been printed; an abstract is given in Delambre's *Hist. de l'astr. du Moyen Age*, p. 73. Compare a quotation by Ibn Jûnis, Caussin, *Notices et Extraits*, VII. p. 116.

points of intersection of the equator and the movable ecliptic advance and recede to the extent of 10° 45′ either way. This is a motion of the eighth sphere, common to all stars, and the sun will therefore sometimes reach its greatest declination in Cancer, sometimes in Gemini. Tâbit does not say that the obliquity of the ecliptic is variable, and perhaps it did not occur to him that this would be a necessary consequence of his theory; he only notices the change in direction and amount of the motion of the equinoxes, which, he says, has increased since the days of Ptolemy, when it was only 1° in 100 years, while later observers have found 1° in 66 years. The erroneous value given by Ptolemy was therefore mainly responsible for the continuance of the imaginary theory. It is to be observed that Tâbit expresses himself with a certain reservation, and seems to think that further observations are necessary to decide if the theory is true or not. His younger and greater contemporary Al Battani was even more cautious, for though he repeats the account of the trepidation given by Theon (which he says that Ptolemy *manifeste in suo libro declarat*[1]), he does not make use of it, but simply adopts 1° in 66 years (or 54‴·5 a year), which he finds by a comparison between his own observations and some made by Menelaus. In rejecting the erroneous value of Ptolemy, which Al Fargani alone had accepted[2], Al Battani was followed by Ibn Jûnis, who came still nearer to the truth by adopting 1° in 70 years or 51‴·2 a year, and who does not allude to trepidation. It is greatly to the credit of several other Arabian writers that they were not led astray by this imaginary phenomenon; among them are Al Sûfi, the author of the only uranometry of the Middle Ages[3], who followed Al Battani, also Abu 'l Faraj and Jagmini[4],

[1] Cap. 52 (p. 205). Plato's translation gives the period as 84 years, but Nallino's ed. has 80 (p. 127).

[2] *Elem.* c. 13, p. 49.

[3] Schjellerup, *Descr. des étoiles fixes*, p. 43.

[4] Abu 'l Faraj, p. 12, simply says that the motion is 1° in 100 years according to Ptolemy, or 1° in 66 years according to others. But on p. 88 he says that if the ancient Chaldeans gave the tropical points a motion backwards and forwards, and if ancient astrologers adopted this, then the motion of the fixed stars must have been unknown to them. Jagmini (p. 229) says that most people adopt 1° in 66 solar years.

while Nasir ed-din mentions it but seems to doubt its reality[1]. By others it was willingly accepted, for instance by Al Zarkali, who made the period of oscillation of 10° either way equal to 2000 Muhammedan years (or 1940 Gregorian years, i.e. 1° in 97 years or 37″ a year). The motion is in a circle of 10° radius; at the Hijra the movable equinox was at 40′ in increasing precession, and in A.D. 1080 at 7° 25′[2]. The diminution of the inclination of the ecliptic, which the astronomers of Al Mamun had found = 23° 33′, no doubt lent countenance to the idea of trepidation, and the next step in the development of this curious theory was the combination of progressive and oscillatory motion. Al Betrugi, who gives a sort of history of the theory, beginning with a mythical Hermes, makes out that Theon (or Taun Alexandrinus as he calls him) combined the motion of 1° in 100 years with the oscillation[3]. A century later this was actually done, and the theory received its last development by King Alfonso or his astronomers, who perceived that the equinoxes had receded much further than Tâbit's theory allowed. The equinoxes were now supposed to pass right round the heavens in 49,000 years (annual motion = 26‴·45), while the period of the inequality of trepidation was 7000 years, so that in a sort of Great Jubilee year everything was again as it had been in the beginning[4]. The progressive motion belongs to the ninth sphere; the annual precession varies between 26‴·45 ± 28‴·96, or from + 55‴·41 to − 2‴·51[5].

[1] *Sphères célestes*, p. 347.

[2] Sédillot, *Mémoire sur les instr. astr. des Arabes*, pp. 31–32. Abraham ben Chija (p. 196 of Münster's *Sphæra mundi*, Basle, 1546) gives the period as 1600 years without quoting any authority. He adds that the ancient Indians, Egyptians, Chaldeans, Greeks, and Latins first proposed the theory ; Ptolemy neither approved nor disapproved of it, but Al Battani confuted it.

[3] Alpetragius, f. 12 a. He says that Al Zarkali did the same.

[4] A later writer, Augustinus Ricius, *De motu octavæ sphæræ*, Paris, 1521, who traces the theory back to Hermes, 1985 years before Ptolemy (!), credits this development to a Jew of Toledo, Isaac Hassan (see above, p. 273), adding that Alfonso four years after the completion of the tables became convinced of the futility of the theory by reading the book on the fixed stars by Al Sûfi. Riccioli, *Almag. novum*, I. p. 166.

[5] In the Alfonsine Tables the maximum took place at the birth of Christ. In Essler's *Speculum astrologicum*, p. 224 (appended to Purbach's *Theoricæ novæ*, Basle, 1573) the epoch is A.D. 15, diebus 137 completis. Reinhold in his

It was now necessary to assume the existence of a tenth sphere, which as *primum mobile* communicated the daily rotation to all the others, while the ninth produced the progressive and the eighth the periodical motion on the small circles, which are situated "in the concavity of the ninth sphere." This was a nice and comfortable theory on account of the long periods involved and the slow changes it produced in the amount of annual precession; and oblivious of the fact that the theory had no foundation except the circumstance that the obliquity of the ecliptic was now about 20′ less than it had been stated to be by Ptolemy, and that he had given the amount of precession as 36″ a year instead of about 50″, and often shutting their eyes to several of the necessary consequences of it, such as the changes in the latitudes of stars which it ought to produce[1], astronomers continued to accept the theory until at last a real observer of the stars arose and wiped it out by showing that the obliquity of the ecliptic had steadily diminished, and that the amount of annual precession had never varied. We have in this place only alluded to it because it involved some rearrangement of the spheres and because it is eminently characteristic of the period during which no persistent observations were taken, and hardly an attempt was made to improve the theories of Ptolemy. The theory of *trepidatio* or *titubatio*, as it was sometimes called, was one attempt, and it would have been better left alone. But it forms a not uninteresting chapter in the history of astronomy.

Here we finish our review of Oriental astronomy. We have omitted as not coming within our province several valuable contributions to science, which did not deal with cosmology or planetary theory. But even with this limitation enough has been said to show that when Europeans again began to occupy

Commentary to Purbach (Paris, 1558, f. 163 b) explains that 26″·45 is the space passed over by the sun in 10 mins. 44 secs., by which amount the Alfonsine Tables made the tropical year smaller than 365¼ days.

[1] Abraham ben Chija (p. 103, Schreckenfuchs) says that trepidation does not change the latitudes. Perhaps he refers to the earliest form of the notion, that described by Theon of Alexandria.

themselves with science they found astronomy practically in the same state in which Ptolemy had left it in the second century. But the Arabs had put a powerful tool into their hands by altering the calculus of chords of Ptolemy into the calculus of sines or trigonometry, and hereby they influenced the advancement of astronomy in a most important manner.

CHAPTER XII.

THE REVIVAL OF ASTRONOMY IN EUROPE.

SCHOLASTICISM was at its height about the end of the thirteenth century. It had done much to enlighten mankind by introducing the works of Aristotle into the western countries. But no amount of study of Aristotle or of the scholastic writers could by itself advance science. New work was wanted, but workers in science would have to start from the foundation laid by the mathematicians of old, and, if astronomy was to progress, the first thing to be done was to obtain a thorough knowledge of the astronomy of the Alexandrian school as exhibited in the *Syntaxis* of Ptolemy. The desire of obtaining this work at first hand, without depending on Arabian paraphrases translated into Latin, was only one phase of the general desire, through a wider knowledge of Greek literature, to loosen the bonds in which man's ideas lay bound, and learn to look at the world as it is, and not as the theologians thought it ought to have been constructed. Though the Greek language was generally un-known in Europe in the Middle Ages, some study of Greek had been kept up in Irish monasteries and in a few other places, while now and then a learned man might be met with who was conversant with Greek, e.g. Roger Bacon, Richard Grosseteste, Bishop of Lincoln, and the Flemish Dominican, William of Moerbecke, who translated the works of Archimedes, Simplicius, and others. But it was not till the fourteenth century that the desire to know Greek began to spread. Petrarch attempted to learn it, Boccaccio studied it eagerly, and soon natives of Greece began to come to Italy as teachers. Manuel Chrysoloras lectured at Florence in 1397–1400, and he was succeeded by others, who brought Greek manuscripts with them and translated them, so that even before the capture of Constantinople by the Turks in 1453 the Greek language and literature were well known in Italy[1]. Manuscripts were anxiously sought for and collected in great libraries, such as the Vatican at Rome, the Medicean Library at Florence, and Cardinal Bessarion's at Venice.

The breeze from the shores of Hellas cleared the heavy scholastic atmosphere. Scholasticism was succeeded by Humanism, by the acceptance of this world as a fair and goodly place given to man to enjoy and to make the best of. In Italy the reaction became so great that it seemed destined to put paganism once more in the place of Christianity; and though it produced lasting monuments in art and poetry, the earnestness was wanting which in Germany brought about the revival of science, and later on the rebellion against spiritual tyranny. Germany had already, during the second half of the fourteenth century, commenced to prepare for this work by founding one university after another, while Paris was losing the privilege it had so long enjoyed as the greatest seat of learning, and was suffering under the calamities brought over France by the hundred years' war with England. Astronomy profited more than any other science by this revival of learning, and about the middle of the fifteenth century the first of the long series of German astronomers arose, who paved the way for Copernicus and Kepler, though not one of them deserves to be called a precursor of these heroes.

Nicolaus de Cusa was born in 1401 at Cues, a village on the Moselle, where his father, Johannes Chrypffs (or Krebs) was a shipowner and winegrower. Roughly treated by his father, who does not seem to have been a poor man, he fled from home and entered the service of a nobleman, who sent him to be educated in the school of the "Brethren of the Common Life" at Deventer, where he became imbued with the mystic theology

---

[1] In the Byzantine Empire astronomy had been very little cultivated. A few codices are extant which contain notes on spherical astronomy, astrology and chronology, and the principal works of the Arabs appear to have been known at Constantinople, but the outcome of a study of Byzantine astronomy is a poor one. There is not even a commentary on Ptolemy. See Usener's papers: *Ad Historiam Astronomiæ symbola*, Bonn, 1876, 37 pp. 4°; and *De Stephano Alexandrino Commentatio*, Bonn, 1880, 58 pp. 4°.

of this community. He afterwards studied at Heidelberg, Bologna, and Padua, and was introduced to the study of mathematics and astronomy by the celebrated geographer, Paolo Toscanelli, who in the last years of his long life is supposed to have encouraged Columbus to seek a westerly route to the Indies. Cusa played a considerable part in the Council of Basle, where he at first was an advocate of the power of the Council, but later on changed sides and became a firm adherer of the Pope, whose power he steadfastly laboured to restore and increase. His friend Pius II. created him a Cardinal and conferred on him the bishopric of Brixen in the Tyrol, where he had rather a stormy life owing to various quarrels caused by his desire to reform the religious houses there. He died in 1464, having bequeathed to a hospital he had founded in his native town the books he had collected on his extensive travels in Germany and Italy, and a considerable portion of his library is still preserved there. We must here pass over his fruitless attempt to get the Council of Basle to undertake the reform of the Calendar[1] as well as his mathematical writings[2], as we have only to do with his speculations about the position and motion of the earth. These are intimately connected with his philosophical system, a mixture of neo-Platonic and Christian mysticism, set forth in his book *De docta ignorantia*, or on the acknowledged ignorance, i.e. the inability of the human mind to conceive the absolute, which to him is the same as mathematical infinity. This lands him in contradictions, when he considers the properties of mathematical figures and lets them become infinitely great; he proves that when a line is infinite it is at the same time a straight line, a triangle, a circle, and a sphere. These contradictions become theologically important, as the infinitely great triangle is a symbol of the Divine Trinity; but it is of greater importance for his views on the *rôle* played by the earth, that he is led to see that the universe must be infinite in extent and therefore devoid of a centre and of a

---

[1] About Cusa's "reparatio calendarii" see Kaltenbrunner, "Die Vorgeschichte der Gregorianischen Kalenderreform," *Sitzungsberichte d. k. Akad. d. Wiss. zu Wien*, 1876, p. 336 sq.

[2] Cantor, *Gesch. d. Math.* II. p. 192 (2nd ed.).

circumference. Therefore the earth cannot be in the centre of the world, and as he supposes motion to be natural to all bodies, the earth cannot be devoid of all motion. It is simply an illusion, when we think we are in the centre of the world, for if a person stood at the north pole of the earth and another stood at the north pole of the celestial sphere, then the celestial pole would to the former appear to be in the zenith, while the centre would appear in that place to the latter person, and thus both would believe themselves to be at the centre. Therefore we perceive by our intellect (for which the *docta ignorantia* alone is of importance) that we cannot conceive the world, its motion and figure, for it appears as a wheel in a wheel, a sphere in a sphere, nowhere having a centre nor a circumference[1].

To these notions (says Cusa at the beginning of his twelfth chapter) the ancients did not rise, because they were deficient in the learned ignorance. But to us it is manifest that the earth is really moving, only this is not apparent, since we only perceive motion by comparison with fixed objects; for how would one in the middle of the sea know that his ship was moving? And therefore, whether we stood on the earth or on the sun or on any other star, we would think we were in the immovable centre, and that everything else was moving. One motion is more circular and perfect than another, and likewise the figures (of bodies) are different, the figure of the earth is noble and spherical, but it might be a more perfect one[2].

All this is pure speculation, not in any way founded on observation, nor is there any distinct reference to observations or their results, except vague ones, as when he says that the sun is larger than the earth and the earth larger than the moon[3]. But he reasons very sensibly about the nature of the heavenly

---

[1] *De docta ignorantia*, liber II., end of cap. XI. ("Correlaria de motu"). I have used the first edition, Paris, 1514, 3 vols. fol. (f. xxi b).

[2] Cap. XII. (ed. 1514, f. xxi b).

[3] "Et quamvis terra minor est quam sol, ut ex umbra et eclypsibus hoc notum nobis est : tamen non est nobis notum quanto regio solis sit maior aut minor regione terræ, æqualis autem præcise esse nequit, nulla enim stella alteri æqualis esse potest. Neque terra est minima stella: quia est maior luna, ut experientia eclypsium nos docuit. Et mercurio etiam, ut quidam dicunt. Et forte aliis stellis." f. xxii a.

bodies. The earth, the sun, and the other stars have the same elements, they only differ as to the way these are mixed and by the preponderance of one or the other element; each heavenly body has its own light and heat and its own particular influence, different from that of others. He even goes so far in generalising as to suppose that if a person stood on the sun, he would not find it as bright as we see it, as the sun has, as it were, a more central earth (*quasi terram centraliorem*) and a fiery circumference, while between the two there is a sort of aqueous cloud and purer air, so that it is only from the outside that the sun appears very bright and hot. A surprising anticipation of Wilson's theory of the constitution of the sun, proposed more than three hundred years after Cusa's time. But what are we to think of his assertion that the earth is in motion; did he then anticipate the discovery of Copernicus? That Cusa was not thinking of any progressive motion appears from another passage[1], in which he (apparently forgetting that the universe has no centre) says that "God gave to every body its nature, orbit and place; He put the earth in the middle, and decided that it should be heavy and move at the centre of the world (*ad centrum mundi moveri*), so that it would always remain in the middle and neither deviate upwards nor sideways." He can therefore only have thought of a rotatory motion, but as he thought everything to be in motion, he cannot merely have supposed the apparent revolution of the heavens to be caused by a rotation of the earth in twenty-four hours. But that is all we can make out from Cusa's published writings, and it is therefore fortunate that a note in his own handwriting has been found in his library at Cues, in which he clearly sets forth his ideas[2]. It is written on the last leaf of an astronomical treatise issued at Nürnberg in 1444, and is therefore later than the book

---

[1] *De venatione sapientiæ*, cap. xxviii. (vol. ii. ff. ccxii b–ccxiii a).

[2] First printed by the discoverer, Clemens, in his book, *Giordano Bruno und Nicolaus von Cusa*, Bonn, 1847; reprinted several times, by Apelt in his charming book, *Die Reformation der Sternkunde*, Jena, 1852, p. 23; by Schanz in his paper, *Die astron. Anschauungen des Nic. von Cusa*, Rottweil, 1873, and by Deichmüller, "Die astron. Bewegungslehre und Weltanschauung des Kardinals N. von Cusa," *Sitzungsber. d. Niederrhein. Gesellsch. zu Bonn*, 1901 (on the occasion of the 500th anniversary of Cusa's birth).

De docta ignorantia, which was finished in 1440, and it merely carries out in detail the ideas sketched vaguely in that book.

In this note Cusa begins by remarking that it is not possible for any motion to be exactly circular, therefore no star will describe an exact circle from one rising to another, and no fixed point in the eighth sphere will be a permanent pole. The earth cannot be fixed, but it moves like the other stars, wherefore it revolves round the poles of the world, "as Pythagoras says," once in a day and a night, but the eighth sphere twice, and the sun a little less than twice in a day and a night, that is, apparently by $\frac{1}{364}$th part of a circle.

In other words, the starry sphere revolves from east to west in twelve hours, and the earth revolves in the same direction in twenty-four hours, which to an observer on the earth produces the same effect as if the earth was immovable while the starry sphere revolved once in twenty-four hours. To explain the annual motion of the sun, Cusa (like the Ionians) lets the sun lag behind in the daily revolution; but in fixing the amount of this retardation he makes a slight mistake: he overlooks the difference between sidereal and solar time, as the starry sphere turns 366 times round the earth in a year, while the sun turns 365 times round it, so that he ought to have made the retardation $\frac{1}{365}$.

Furthermore: We must imagine other poles situated in the equator, round which the earth revolves in a day and a night, and the eighth sphere in a slightly shorter time, while the body of the sun is about 23° distant from one of these poles; and by the revolution of the world the sphere of the sun is also carried round once in a day and a night, less $\frac{1}{364}$ of its circle, "and from that retardation arises the zodiac." The motion of the eighth sphere round the second pair of poles is so much slower than that of the earth, that in a hundred years a point stays behind as much as the sun does in a day[1].

---

[1] " Punctus autem in octava sphera, qui in loco poli mundi motus ab oriente in occasum visus est, continue parum remanet retro polum, ita quod quum polus videtur circulum complevisse, punctus ille nondum circulum complevit, sed remanet a retro, tantum in proportione ad circulum suum in centum annis, vel quasi, quantum sol remanet retro in die uno."

This second revolution round an axis situated in the equator is by Cusa intended to explain two things. Without this revolution of the solar sphere, the sun would perform its annual motion in or parallel to the equator, but the second revolution of the solar sphere being a little slower than the corresponding one of the earth, the sun will in a year not only seem to move round the heavens, but also seem to move $23\frac{1}{2}°$ to the north and the same distance to the south of the equator. This is quite certainly Cusa's meaning, though he does not express himself very distinctly[1]. Secondly, the starry sphere in performing this second revolution also lags a little behind, but only to the extent of $1°$ in a hundred years. Obviously Cusa supposes that this will explain the changes in the positions of the stars due to precession, but it is hardly necessary to say that no rotation round an axis situated in the celestial equator can possibly represent the phenomena of precession, viz. the steady increase in the longitude of a star, while its latitude remains unaltered. He was probably influenced by reminiscences of Eudoxus when he wrote down this part of the theory, and the rotation of the eighth sphere round the axis lying in the plane of the equator was perhaps intended to represent, not precession itself, but its supposed inequality or trepidation, though in that case the axis should have been placed in the zodiac, and not in the equator.

It has been well said, that the good people who rummage among a dead man's papers and publish any of them they choose, have added another terror to death. As this note of Cusa's may not represent his final opinion on every detail, but is probably merely a very rough and incomplete sketch of what he intended afterwards to work out more carefully, we ought not to blame him for the shortcomings of his theory with regard to precession. But as he in his published work expressed him-

---

[1] First he says that "solare corpus distat ab uno polorum illorum quasi per quartam partem quadrantis scilicet per 23 gradus vel prope." And immediately after the sentence ending in "die uno" he says: "Et sicut punctus unus sphæræ solis semper remanet sub uno et eodem puncto octavæ, qui sub polo motus revolutionis ab occidente fixe persistit, ita punctus unus sphæræ terræ et solis remanet cum polo mundi fixe." The centre of the circle of 23° radius lies in the equator, and by its slower motion round the celestial pole it seems to pass round the heavens in a year.

self so very vaguely, we are certainly justified in adding the testimony of this note to prove that his opinions do not represent any advance, and especially that when speaking of the motion of the earth he did not dream of attributing to it a progressive motion in space, either round the sun or round any other body. He was solely guided by his preconceived notion, that motion is natural to all bodies, and by thus settling the affairs of the universe out of his inner consciousness he reminds us of the early Greek philosophers, who had done the same over and over again without being overburdened with too great a store of observed facts. All the same, he was not afraid to speculate freely on the constitution of the world without being a slave either to theology or to Aristotle, but he probably did not think his ideas ripe for publication and therefore in his books confined himself to generalities.

The general revival of learning in the fifteenth century soon made it clear to anyone interested in astronomy that in order to build further on the foundation laid by the Alexandrian astronomers, it was first of all necessary to obtain a thorough knowledge of that foundation, by a study of the great work of Ptolemy. While Cusa was writing on learned ignorance, a youth was growing up who had a strong desire to acquire learned knowledge. Georg Peurbach, or Purbach, was born in 1423, and took his name from his birthplace, a small town on the Austro-Bavarian border. Already before reaching the age of twenty he had studied at the University of Vienna and had spent some time in Italy, where he, among others, associated with the aged Giovanni Bianchini, author of a modified edition of the Alfonsine Tables. Appointed to a professorship at Vienna soon after his return home, he threw himself with energy into the study of Ptolemy, and perceiving the advantage of using sines instead of chords (as the Arabs had already done), he computed a table of sines for every 10'. To facilitate the study of the planetary theory of Ptolemy he wrote an excellent text-book, *Theoricæ novæ planetarum*, which in the course of the next hundred years was frequently printed and commented on by various editors. There are no new developments of theory in it, it merely describes clearly and concisely the constructions of

Ptolemy; but Peurbach adopted from the Arabs the solid crystalline spheres with sufficient room between them to allow free play to the excentric orbit and epicycle of each planet. His great desire was, however, to become more accurately acquainted with the text of Ptolemy's *Syntaxis* than it was possible to be as long as only second-hand translations through the Arabic were available, as only in that way could any hope be entertained of improving the Alfonsine Tables, the glaring errors of which even the crudest observations revealed. To obtain Greek manuscripts of Ptolemy and other mathematical writers of antiquity, it was necessary to go to Italy, and it was therefore specially fortunate that Peurbach became acquainted with Cardinal Bessarion, a Greek by birth, who was equally anxious to make the Greek literature better known in the West. Before Peurbach could start for Italy he died (in 1461), but his place in the Cardinal's friendship was at once taken by his distinguished pupil, Regiomontanus, who had for some years shared his master's labours and had commenced the study of the Greek language.

Born in 1436 at Königsberg, a village in Franconia, as the son of a miller, Johann Müller, better known as Johannes de Monte Regio, or (after his death) as Regiomontanus, was twenty-six years of age when he started for Italy with Bessarion in 1462. He spent six years there, visiting the principal cities and losing no chance of collecting Greek manuscripts. Some years after his return home he settled at Nürnberg, where he erected an observatory and commenced publishing on a large scale. Among the books printed at Nürnberg none created greater sensation than the astronomical ephemerides of Regiomontanus, which a few years later rendered invaluable services to the intrepid Portuguese and Spanish navigators. Even more important than these were his treatise on trigonometry, the first systematic work on this subject, and his *Tabulæ directionum*, which included a table of sines for every minute and a table of tangents for every degree. Though these works, which have given him a high rank among mathematicians, were not printed during his lifetime[1], the renown of Regiomontanus had spread

---

[1] It has been stated that the tables were already printed at Nürnberg in

far and wide, probably through his ephemerides, and he was therefore, in 1475, by the Pope summoned to Rome in order to carry out the long-contemplated reform of the Calendar. But he died already the following year at Rome, and the chance was thus lost of getting the proposed reform accomplished while the whole of Christendom still acknowledged the supremacy of the Pope.

Regiomontanus did a great deal of valuable work, but he made no advance as regards planetary theory. He completed a text-book commenced by Peurbach, *Epitome in Ptolemæi Almagestum* (first printed at Venice in 1496), in which he accepts the Ptolemaic system in every detail. All the same, several writers of distinction have credited him with a most important discovery, that of the daily rotation of the earth, and have proclaimed him as a precursor of Copernicus[1]. In 1533 Johann Schoner published at Nürnberg a memoir entitled *Opusculum geographicum*[2], the second chapter of which is headed, *An terra moveatur an quiescat, Joannis de Monte regio disputatio*; and it looks as if the writers who on this chapter have founded a claim for Regiomontanus as a precursor of Copernicus have either not read it at all, or have been content to read the heading and the first few lines only. For in this chapter there is not one word in favour of any kind of motion of the earth. First there is a sneer at "certain of the ancients," who taught the rotation of the earth and imagined that the earth was like meat on a spit and the sun like the fire, and who said that it was not the fire which was in need of the meat but the reverse, and likewise the sun did not require the earth but rather the earth required the sun. After this attempt at wit the usual old arguments against the rota-

1475, but it seems very doubtful (Cantor, II. p. 274). They were printed at Augsburg in 1490, the book *De Triangulis* not till 1533 (Nürnberg).

[1] Doppelmayr, *Historische Nachricht von den Nürnbergischen Mathematicis und Künstlern*, Nürnberg, 1730, p. 22 ; Weidler, *Hist. Astr.* p. 310 ; Montucla, *Hist. des Math.* I. p. 543 ; Bailly, *Hist. de l'Astr. moderne*, T. I. p. 318; G. H. Schubert, *Peurbach und Regiomontan*, Erlangen, 1828, p. 38. Several of these writers even say that R. taught the motion of the earth round the sun!

[2] Ioannis Schoneri, Carolostadii, *Opusculum geographicum ex diversorum libris et cartis collectum*, 20 ff. 4°. Reprinted in Schoner's *Opera Mathematica*, Nürnberg, 1551 (and 1561), fol.

tion are dished up: birds and clouds would be left behind, buildings would tumble down, &c. Truly this is not the language of a precursor of Copernicus. And if anyone should say that these were perhaps the arguments of Schoner, and not of Regiomontanus, let him read the *Epitome in Almagestum*, wherein the old arguments of Ptolemy are found, so that it is impossible to doubt that Regiomontanus rejected the rotation of the earth altogether. It is also distinctly affirmed in the *Epitome* that the earth occupies the centre of the world[1]. Doppelmayr, who was the first to circulate the myth, adds that Johannes Prætorius, in a manuscript found after his death, states that Georg Hartmann, a mathematician of Nürnberg (1489–1564), possessed a note written by Regiomontanus, in which he draws the conclusion: "Therefore it is necessary that the motion of the stars must be altered a little (*paululum variari*) on account of the motion of the earth." But how is it possible to found any serious claim for Regiomontanus on evidence so very vague as this, when it is distinctly contradicted by the published writings of the great astronomer? And what kind of a motion of the earth could he have thought of, which only affected the motion of the stars "a little"?

That Regiomontanus thought it necessary in the *Epitome* to put together the arguments against any motion of the earth, does not by any means prove that a doctrine of that kind had been current in his day, since he only followed the example of Ptolemy in doing so. Still, he must have known of the mystical speculations of Cusa, and may have thought it useful to emphasize the arguments of Ptolemy; and he would no doubt have been very much surprised if he had been told that he should some centuries after his death be held up as an advocate of the diametrically opposite opinion. Yet he is not the only great man who has been proclaimed a precursor of Copernicus. Another is Lionardo da Vinci, the versatility of whose genius was indeed so wonderful that the mistake is perhaps pardonable in his case. Libri says of him: *En astronomie, il a soutenu avant Copernic la théorie du mouvement de*

---

[1] *Epitome*, Venice, 1496, fol. a 5 recto.

la terre[1]. In a manuscript written about 1510 Lionardo showed
how a body, describing a kind of spiral, might move towards a
revolving globe such as the earth, so that its apparent motion
with regard to a point in the surface might be a straight line
through the centre. But to propose a problem of that kind is
a very different thing from maintaining that the earth really is a
revolving globe[2]. We might as well accuse him of believing
that falling bodies describe spirals. All we can learn from this
note of Lionardo's (one of some thousands of mathematical
problems and notes recorded in his note books) is, that he had
a very clear idea of the parallelogram of motions.

There was only one man, living at that time, of whom we
know for certain that he taught the daily rotation of the earth
before the book of Copernicus was published. Celio Calcagnini
(1479–1541) was a native of Ferrara, and served in his youth
in the armies of the Emperor and Pope Julius II.; he then
entered the church and became a Professor in the University
of Ferrara, but travelled extensively in Germany, Poland and
Hungary on various diplomatic missions. In 1518 he made
a prolonged stay at Cracow on the occasion of the marriage
of the King of Poland with a Princess of Milan. Nothing
is more likely than that the learned Italian during his visit to
the capital of Poland heard that a Canon of the cathedral of
Ermland (a dependency of Poland) and Doctor of the University
of Ferrara (whom he perhaps remembered as an old college
friend) was working out a new system of the world which was
founded on the idea that the earth is not at rest but in motion.
This is only a supposition, but anyhow Calcagnini (apparently
before 1525) wrote an essay, *Quod caelum stet, terra moveatur,
uel de perenni motu terrae.* None of his writings were printed
in his lifetime, but in 1544 they were collected and printed at
Basle in a folio volume, in which the said essay occupies eight

---

[1] *Hist. des sciences math. en Italie*, T. III. p. 47. He refers to Venturi's
Essai sur les ouvrages de L. da Vinci (Paris, 1797), p. 7, but does not give any
particulars.

[2] Whewell, who quoted the note in the first edition of his *History of the
Inductive Sciences* (vol. II. p. 122), has apparently been of this opinion, since he
omitted it from his third edition.

pages[1]. The writer begins by announcing that the whole
heavens with sun and stars are not revolving in a day and a
night with incredible velocity, but that it is the earth which
is revolving; and he refers to flowers and plants which are
continually turning to the sun, so that it is quite natural that
the different parts of the earth should in their turn face the
sun. The earth is placed in the centre[2] and cannot descend
further; but its mass and weight impelled it, and it began to
move its parts so that without leaving its place it is carried
round, its navel, which we call the centre, being at rest, and
its orb reverting without cessation in itself; for having once
received an impulse from nature it can never stop without
going to pieces. On the other hand the lightness and purity
of the fifth element, of which the heaven is composed, renders
the latter immovable.

This is really all Calcagnini has to say on the subject, but
he manages to clothe it in a great many words, quoting Plato
and Aristotle without caring whether his quotations are
à propos or not, and trotting out Greek words now and then
to embellish his sentences. But towards the end of his essay
he seems to have felt that the rotation of the earth cannot
quite explain everything, and he pulls himself together for a
further effort[3]. That the earth does not only revolve with one
perpetual motion, but inclines now to one side, now to the
other, is shown by the solstices and equinoxes, the increase and
decrease of the moon, the varying lengths of the shadows.
Those people who live near the pole and have a day of six
months' duration and an equally long night, must understand
all this better than anybody else. And if anybody insists on

---

[1] *Caelii Calcagnini Ferrariensis, Protonotarii Apostolici, opera aliquot*,
Basileae, 1544, pp. 387–395, reprinted in Hipler's paper on Calcagnini in the
Mittheilungen des Coppernicus-Vereins zu Thorn, IV. Heft (1882), pp. 69–78.
Dr Hipler does not discuss the contents of the essay, and seems to consider its
author to be on a par with Copernicus. The essay is not dated, but in the
Opera it is placed before one dated January, 1525, from which fact Hipler
concludes that it was written earlier. Tiraboschi (*Storia della letteratura
Italiana*, Milan, 1824, T. VII. p. 706) does not give any date, and suggests that
Calcagnini may have heard of the theory of Copernicus through the lecture
given by Widmannstad before Pope Clement VII. in 1533.

[2] Calcagnini, p. 390.   [3] p. 393.

all this being explained to him, let him explain the reason of
the obliquity of the ecliptic, or why the moon can recede five
degrees from the zodiac, not to speak of the trepidation of
the eighth sphere, or the various motions of epicycles and
deferents. All which things are modern inventions, and people
have sought for the causes of phenomena in the heavens
instead of in the earth. It would be absurd and unworthy
of the generosity of Providence to let the earth turn in one
uninterrupted course only, for then part of the earth would
always be in darkness. Finally Calcagnini says that as
Archimedes promised to move the earth if he had a place
to stand on, he must have thought it mobile, and after quoting
Cicero's remark about Nicetas (Hiketas) and Plato's *Timæus*,
he winds up with an allusion to Cusa, whose writings he should
have liked to see.

These last references show that Calcagnini was aware that
other people before his time had taught the rotation of the
earth. But his feeble attempts at showing it to be quite
possible that some sort of unknown motion of the earth
(without its leaving the centre of the world) may account for
all celestial phenomena without having recourse to any motion
of the heavenly bodies, make it evident that his knowledge of
astronomy must have been extremely limited. It almost looks
as if he had vaguely heard that the Canon of Frauenburg had
been able to explain everything by assuming the earth to be
in motion, but that he had not heard any particulars as to
how this was done, so that he had to confine himself to a
few unmeaning phrases. If this was not the way in which
Calcagnini's essay originated, we can only assume that he knew
nothing of astronomy except the one fact of the apparent
revolution of the heavens in twenty-four hours. Had he been
content to explain this alone by the earth's rotation, he would
have deserved to be called a precursor of Copernicus (assuming
him to have been unaware of the labours of the latter); but
by attempting to account for everything in that way he almost
destroyed any claim he might have had to that honour.

Though not printed till 1544, Calcagnini's essay was probably
known in Italy in his lifetime, so it is quite possible that it is

to him that Francesco Maurolico of Messina, the well-known
astronomer and mathematician (1494–1575), alludes in his
Cosmographia (Venice, 1543).   In this book, which is in the
form of a dialogue, the teacher says that he has now finished
what he has to say about the earth, unless human perversity
should go so far that someone believed the earth to revolve on
its axis.   On the pupil replying that so strange an opinion
could scarcely enter anyone's head, the teacher remarks that
many people teach even greater absurdities, and it may therefore be well to prove that the earth cannot possibly move[1].
The preface to this book is dated February, 1540, but the year
of publication is 1543, so that it must remain an open question
whether Maurolico alludes to Calcagnini or to Copernicus[2].
We may add that in every respect this book is perfectly
medieval in its ideas.   The sun's orbit is in the midst of the
planetary orbits, because the inferior and the superior planets
are quite different as regards the periods of their epicycles
and deferents, the solar period of one year being for the former
the period in the deferent, for the latter the period in the
epicycle; Venus inclines more to the north, therefore it has
more dignity and must be above Mercury, while the latter in
the variety of its motions resembles the moon most and must
therefore be next it.   Saturn and the moon have the smallest
epicycles, and the vertices of their deferents are far from that
of the sun, while those of the planets next the sun, Venus and
Mars, are very near it[3].   We shall see in a following chapter
that Maurolico throughout his long life remained a violent
enemy of the Copernican doctrine.

It is unnecessary here to review the not inconsiderable
number of books "on the sphere" or other text-books on
astronomy which appeared during the first half of the sixteenth
century.   They show that the work of the Alexandrian astron-

---

[1] Folio 12 a.

[2] De Morgan gives this book as an example in his paper " On the difficulty
of correct description of books " (*Comp. to the Alm.* 1853), and comes to the
conclusion that the distinct statement of the author, that he finished the book
in 1535, is not to be relied on (l. c. p. 13).

[3] ff. 20 b to 21 b.   Regiomontanus is his authority for the remark about the
sun's orbit; the others are his own.

omers was now well known and appreciated in Europe, but they show at the same time that no attempt had yet been made to continue and extend that work. The first Latin edition of Ptolemy's *Syntaxis* was printed at Venice in 1515; but it was only the old translation from the Arabic by Gherardo of Cremona, dating from the twelfth century; next came a translation from the Greek by Georgios of Trebizond (Paris, 1528, and Basle, 1551), and at last the Greek original was printed at Basle in 1538 from a codex once in the possession of Regiomontanus, together with the commentary of Theon, so that anyone capable of reading Greek could now test the Latin translations for himself. Only five years after the Greek Ptolemy appeared the work which was to be the corner-stone of modern astronomy, but in the meantime one last despairing effort had been made to revive the theory of solid spheres and thus to try once more to meet the old objection to the Ptolemaic system, that though a convenient means of computation, it was difficult to accept it as the physically true system.

This attempt was made almost simultaneously by two Italian writers, Fracastoro and Amici, of whom the former has obtained a certain amount of celebrity by his works, while the latter is almost perfectly unknown. We shall first examine the ideas of the former.

Girolamo Fracastoro was born in 1483 at Verona. After having studied at the University of Padua he held a professor-ship of logic there from 1501 to 1508, and as Copernicus studied at Padua from the autumn of 1501 for some years, there can hardly be any doubt that the two young men, both interested in astronomy and medicine, must have known each other at Padua, and possibly may have discussed with each other the difficulties of the Ptolemaic system[1]. In 1508 Fracastoro went back to Verona, where he spent the rest of his life till his death in 1553, devoting himself to medicine, astronomy and poetry. His principal work, *Homocentrica*, appeared at Venice in 1538, though it is possible that an

---

[1] Favaro, "Die Hochschule Padua zur Zeit des Coppernicus," *Mittheilungen des Coppernicus-Vereins zu Thorn*, III. p. 44 (1881).

earlier edition had appeared already in 1535[1]. At Padua
Fracastoro had been on friendly terms with three brothers,
Della Torre[2], one of whom is known as the collaborator of
Lionardo da Vinci in his studies of anatomy, while another,
Giovanni Battista, devoted himself specially to astronomy and
designed a plan of representing the motions of the planets
without excentrics and epicycles, using solely homocentric
spheres. He died at an early age, but on his deathbed he
begged Fracastoro to work out his ideas into a new astronomical system; and in fulfilment of the promise given on
that occasion Fracastoro prepared his work *Homocentrica*,
without, however, following strictly the methods of Della
Torre. So much he tells us in his dedication to Pope Paul III.
(the same Pope to whom the great work of Copernicus was
dedicated a few years later), but how much of the system
belongs to Della Torre he does not specify. It is to be hoped
that Fracastoro understood his own system in every particular,
but he certainly had not the gift of making his readers get a
clear idea of every detail of the cumbersome machinery which
he offered as a substitute for the elegant geometrical system
of Ptolemy. The obscurity of the description may have had
something to do with the total want of success of the book,
but in any case the time was long past for making an attempt
to resuscitate the ideas of Eudoxus and Kalippus. A hundred
years earlier, when only the broad outlines of the Ptolemaic
system were known in Europe, there might have been some
sense in proposing to take up the system of Eudoxus. But in
the din of battle, which soon after the publication of Fracastoro's
book began to rage about the motion or non-motion of the
earth, the voice of this ancient spectre was utterly drowned,

---

[1] Favaro (l. c.) says that the edition of 1535 is mentioned in Riccardi,
Biblioteca matematica italiana, but that he had never seen it himself. Delambre,
Moyen Âge, p. 390, gives the year of publication as 1535. It is not given in
Lalande's or Scheibel's Bibliographies. I have used Fracastorii *Opera omnia*,
Venice, 1584, 4°, in which the Homocentrica fill the first 48 ff.

[2] *Fracastorii Vita*, fol. 1 b (*Opera omnia*, Venice, 1584). Moréri's *Dict. hist.*
(Suppl. T. III. p. 867) gives the names of four brothers, sons of a physician who
died in 1506. "Jean-Batiste étoit Philosophe, Astronome et Médecin. Il
mourut jeune."

and nobody has ever thought it worth while to make a
thorough study of the *Homocentrica*. We shall, however,
shortly indicate the leading features of the system.

Kalippus, it will be remembered, had succeeded fairly well
in representing the planetary motions, as far as he knew them,
by means of homocentric spheres[1]. But now there were more
phenomena to be accounted for, viz. the zodiacal inequality of
the planets and precession with its (imaginary) adjunct, trepidation, and these required an increase of the number of spheres.
Fracastoro, however, does not merely proceed on the same lines
as Eudoxus and Kalippus, but following the lead of Della Torre
he wants all his spheres to have their axes at right angles to
each other. He shows how every motion in space can be
resolved into three components at right angles to each other,
while conversely three motions at right angles will produce
"motions in longitude as well as in latitude[2]." He assumes
that an outer sphere may communicate its motion to an inner
one, while an inner one does not influence an outer one, and he
is therefore able to let the Primum Mobile communicate its
daily rotation to all the planets without having with Eudoxus
to assume one sphere for each planet to produce the daily
rotation. A set of spheres generally consists of five spheres
which he calls (beginning with the outermost) circumducens,
circitor, contravectus, anticircitor, and ultimus contravectus, of
which the fourth and fifth revolve in opposite directions to
respectively the second and third, and generally with different
velocities. He shows how the second and third sphere can
produce an oscillation or "trepidation," and points out that the
equinoctial points really describe small "ovals" and not circles[3].
For the fixed stars he assumes five spheres under the primum
mobile, the period of inequality of precession (4° to each side)
being 3600 years, in which time the circitor makes one revolution. The fifth sphere is above the "Aplane," to which the
stars and the Milky Way are attached, and which moves 1° in
100 years. Below that comes the system of Saturn, consisting

---

[1] Fracastoro repeatedly refers to Eudoxus and Kalippus, and in the dedication
also to "Albateticus," by which name he probably means Alpetragius, whose
book had then recently been published (1531).

[2] Sect. I. cap. 12, sect. II. cap. 4.   [3] Sect. I. cap. 14.

of two sets of five spheres each, the special business of the outer group being to account for the zodiacal inequality in longitude by an oscillation of the node, while the inner group provides for the inequality depending on the elongation from the sun, of which the planet's synodic revolution is the period. In both groups the two inner spheres are intended to counteract the exaggerated latitudes which the other spheres would otherwise produce. Below the sphere carrying the pla... t Saturn come those of Jupiter; first one to prevent the complicated motions of Saturn from being communicated to Jupiter, and then two groups of five each; then the spheres of Mars, nine in number, the two groups consisting of five and three spheres; then the sun with four spheres, one for excluding the motions of Mars and three for the annual motion of the sun and its inequality[1]. The two inferior planets have eleven spheres each, the difference between them and the outer planets being the substitution of a year for the sidereal period of revolution as the period of rotation of the outer circumducens. The moon has seven spheres; the first is "that which others call the Deferens Draconis," which produces the retrograde motion of the nodes, while it also prevents disturbance from the spheres of Mercury; then a circumducens which turns in 27 days and 8 hours; under it the circitor turning in 27d. 13h. and making the moon move alternately faster and slower; then come the contravectus and the anticircitor to counteract the motion in latitude of the circitor; then the second contravectus and the sphere carrying the moon. Finally below the moon is a sphere which is not homogeneous but denser in some places than in others. Fracastoro is of course obliged to admit that every planet is subject to changes of brightness, which looks as if they were not always at the same distance from us. This seemingly fatal objection to the homocentric idea he meets by assuming that the media through which we see the planets are denser in some places, and that objects seen through a dense medium look larger than when seen through a thinner medium[2]. The

---

[1] Kalippus used five spheres for the sun, one for the daily rotation and one for the imaginary motion in an orbit slightly inclined to the ecliptic. The latter was of course not required by Fracastoro.

[2] Sect. II. cap. 8. In this chapter he remarks that if you put one lens over

variations in the duration of eclipses he explains by means
of the last sphere below the moon, which makes the moon
appear larger and throw a larger shadow when shining through
the denser parts. He also contents himself with an explanation
of this kind to account for the occasional outstanding error of
the moon's longitude at quadrature (caused by the evection).
This is certainly an easy way of getting over a serious difficulty,
and it is the more remarkable that Fracastoro contented himself
with this poor expedient, since his remark, that the deferents of
Mercury and the moon are ovals in the Ptolemaic theory[1], show
that he must have been well acquainted with the *Almagest*.

With regard to the last sublunary sphere Fracastoro
remarks that it is not an innovation to propose it, since already
Seneca and other philosophers had assumed its existence in
order to account for the motions of comets. Indeed, he did
well to assume comets to move below the moon, since it became
afterwards (when it was conclusively proved by Tycho Brahe
that comets are more distant than the moon) one of the
strongest arguments against the solid spheres, that comets
would have to pass through them in all directions. Fracastoro
describes several comets observed by himself and makes the
important remark that comets' tails are always turned away
from the sun[2]. In this, at any rate, he was right.

The number of spheres assumed by Fracastoro was therefore :

8 carrying stars and planets,
6 for the daily rotation and precession,
10 for Saturn,
11 for Jupiter,
9 for Mars,
4 for the sun,
11 for Venus,
11 for Mercury,
6 for the moon,
1 sublunary sphere.

another you see more distinctly than through one only. There is a long way
from this discovery (made by many an old woman using two pair of spectacles)
to the invention of the telescope.

[1] Sect. III. cap. 17 and 21.   [2] Sect. III. cap. 23.

Seventy-seven in all. But he adds that it would do the sun a great deal of good to get two more spheres, which would make a total of 79. And this system was supposed to be more reasonable than that of Ptolemy!

Simultaneously with Fracastoro, and apparently quite independently of him, the homocentric system was advocated by a young man, Giovanni Battista Amici, in a little book published at Venice in 1536[1]. At the end of the book the author describes himself as a native of Cosenza, a posthumous son of a father of the same names, being in his twenty-fourth year. He was murdered at Padua in 1538[2], and has published nothing but this little book, which has been totally ignored by every historian of astronomy, perhaps because the author has not like Fracastoro acquired notoriety by other writings. And yet his book deserved a better fate, for it is very clearly written, and he does not confine himself to the use of spheres with axes at right angles to each other, but treats the problem in a more general manner. Otherwise he has some points of resemblance with Fracastoro; he does not assume one sphere for each planet to provide for the daily rotation, and his explanation of the change of apparent size of sun and moon and of the brightness of the planets is the same. In winter the sun looks larger because its light has to pass along a longer path to reach an observer on the surface of the earth, and at quadrature the moon looks larger because it cannot then dissolve vapours as well as at full moon, and this renders the air more full of mist[3]. Having in his first six chapters reviewed the theories of Eudoxus, Kalippus and Aristotle, he remarks that nature does not know such things as epicycles and excentrics, and he then proceeds to explain his own ideas. He first demonstrates that if we have two contiguous homocentric spheres with their axes at

[1] Ioannis Baptistae Amici Cosentini, *de Motibus corporum cælestiū iuxta principia peripatetica sine excentricis & epicyclis*, Venetiis, 1536, small 4°, 27 ff. Weidler, p. 357, merely gives the title, with year 1537; Lalande's *Bibliographie* correctly gives the year 1536, but says it was reprinted in 1537. Riccioli, *Almag. Nov.* II. p. 286, only says a few words about it, so does Tiraboschi (*Storia della lett. Ital.* VII. p. 715), who gives the year 1537.

[2] Tiraboschi, l. c. Amici is not mentioned in Moréri's Dictionary.

[3] Fol. F 2 verso.

right angles to each other, and the poles of the outer one move a certain distance to either side of a mean position, then the motion on the inner sphere will be alternately accelerated and retarded. But he next shows that if the poles of the two spheres are n degrees apart and one rotates twice as fast as the other in the opposite direction, they will produce an oscillation on an arc of $4n°$, and herein he shows himself an apt pupil of the ancients[1]. Four spheres suffice for the sun, but for the moon and the five planets (which he clubs together) more are required. First he says that if the moon moved in an epicycle it would not always show us the same face (because, according to medieval ideas, a body ought always to turn the same side to the centre of motion), and therefore the other planets cannot either have epicycles, since the celestial bodies are analogous in every way. He gives them first four spheres to do the work of the epicycle. The highest has its poles in the plane of the orbit (in the oblique circle, as he calls it) and moves from north to south with the speed with which the epicycle would move. Under this is another, whose poles are distant from those of the former one-fourth of the angular diameter of the epicycle at apogee in the Ptolemaic system, and which moves in the opposite direction to the first one with twice the velocity. Then a third sphere, whose poles are under those points of the second one which are carried backwards and forwards, and which moves from south to north, and lastly a fourth sphere with its axes at right angles to the oblique circle (in which the poles of the third sphere are situated), on the greatest circle of which sphere the planet is fixed and will from time to time for a while appear to have a retrograde motion. Only the moon will, owing to the very fast motion of its fourth sphere, not become retrograde but will merely be retarded[2].

Amici has next to account for "the changes in the moon's motion and the varying amount of the retrograde motion of the planets." For this purpose he places between the spheres of access and recess three others to carry the poles of the lower one to and fro and thereby produce changes in the arc of

---

[1] Fol. C verso *et seq.*   [2] Fol. E verso *et seq.* (cap. xi.).

retrograde motion. To prevent excessive latitudes he has to add three more spheres, making ten in all for each of the planets; and the moon has an eleventh sphere outside the others to provide for the motion of the nodes[1]. But he evidently feels that all this may not in the long run prove sufficient, for he remarks that more observations of the five planets are wanted in order to fix the values of what corresponds in his system to the inclinations of the epicycles. The inclination of the diameter in anomaly $0°-180°$ he accounts for by three more spheres, and for the obliquatio of the diameter $90°-270°$ (reflexio, as he calls it), he adds another set of three[2]!

As to the fixed stars, Amici believes in a very slow motion of the ninth sphere (annual amount not stated), and in a motion of the eighth sphere called titubatio, its equinoxes turning in 7000 years in small circles of $9°$ radius round those of the ninth.

It is sad to think that this evidently exceedingly talented young man, to whom a cruel fate only left this one chance of distinguishing himself, should have wasted his powers on a fruitless attempt to adapt a theory, proposed when science was in its infancy, to modern requirements. His age prevents us from thinking that he may have been a pupil of Della Torre, but all the same it is not unlikely that Amici may have heard of the ideas of the Paduan professor and may have worked them into a complete system in his own way. It is useless to speculate on what he might have done if his life had been spared; whether he would have tried to work out the numerical details of his system, or whether he would have been struck by the new light emanating from the Canon's cell at Frauenburg. For while in Italy, in the centre of civilisation, Fracastoro and Amici were vainly endeavouring to put life into a mummy, while Calcagnini in a most self-satisfied manner was pretending that some motion of the earth,

---

[1] Fol. E 4 verso (cap. XII.).

[2] Fol. F 3 verso (cap. XIV.). He quotes the values 10′ for Venus and 45′ for Mercury from Ptolemy. These are the only numerical values he gives anywhere in connection with the planets, except 59′8″ for the daily motion of the sun.

without its leaving the centre of the world, could solve every riddle presented by the stars, and while Maurolico was proving to the meanest intelligence that the earth could not possibly have any motion, a quiet student at the shore of the Baltic, on the very outskirts of civilisation, was preparing to kindle the light which was to illuminate the universe and show to astonished humanity the earth moving through space.

CHAPTER XIII.

COPERNICUS.

NIKLAS KOPPERNIGK was born on February 19, 1473, in the city of Thorn on the Vistula, where his father (about whose descent nothing is known with certainty, but who had emigrated from Cracow to Thorn previous to 1458) was a merchant of some social standing[1]. Together with West Prussia and Erm-land, Thorn had in 1466 come under the suzerainty of the King of Poland, but was not yet incorporated in Poland. Coppernicus or Copernicus, as he afterwards wrote his name[2], went in 1491 to study at the University of Cracow, thereby following the example of many other students from Central Europe, but no doubt also attracted thither by the fact that mathematics and astronomy were specially cultivated at Cracow. He received instruction in astronomy from Albert of Brudzew (Brudzewski), probably in the form of private lessons, since his teacher appears only to have lectured on Aristotle after 1490. In 1482 Albert (his family name is not known) had written a commentary

---

[1] The little that is known about the private life of Copernicus has with great skill been woven together with much interesting information about Prussia and Ermland into a lengthy biography by L. Prowe (*Nicolaus Coppernicus*, 2 vols. in 3 parts, Berlin, 1883–84).

[2] His name was certainly Koppernigk. There are 29 signatures of his ... Of these, the first sixteen (from the years 1512–28) are spelt with *pp* ...thout the termination *us*), five are undated, written in books belong-de of which four have *pp* (three with *us*, one without), while one is h a single *π*. In 1537 Nicolaus Coppernic signed the election of a But seven letters from his last years (1537–41) are all signed ...ernicus, which is also the spelling adopted by his only disciple would therefore seem that Copernicus six years before his death ... write his name in Latin with a single *p*, and there appears to be ...o change the spelling universally followed for 300 years.

20

to Peurbach's *Theoricœ novœ Planetarum* for the use of his pupils, which was printed at Milan in 1495, and is the first of the series of commentaries to that favourite text-book which were published during the next hundred years[1]. Considering the state of mathematical teaching at that time it was no doubt a very useful undertaking to smooth in every way the path of a student anxious to penetrate the tangles of planetary theory, and the writer shows himself capable of drawing all the conclusions logically arising from the constructions of Ptolemy ; as for instance when he notices that the centre of the epicycle, not only of Mercury but also of the moon, must describe an oval figure[2]. But though Albert of Brudzew may claim the credit of having been the first instructor in astronomy of the future reformer of that science, there is no reason whatever to think that he may have suggested to his pupil the possibility or probability of the earth's motion. In his book he accepts the Ptolemaic system altogether, as indeed he was bound to do as a public lecturer ; but it is quite possible that he may have pointed out to the pupil the extraordinary *rôle* played by the sun in the planetary theories, which may have set the great mind of the young student thinking whether this would not eventually furnish the key to the riddles of the planets. But the great discovery had been within the grasp of any daring thinker for so long, that it seems unlikely that Copernicus owed his inspiration to any teacher or friend.

Albert of Brudzew left Cracow in 1494[3], and Copernicus probably returned home in the course of the same year. But his maternal uncle, Lucas Watzelrode, Bishop of Ermland since 1489[4], who intended to provide for his nephew by

---

[1] Reprinted at Cracow in 1900: "Commentariolum super theoricas novas planetarum Georgii Purbachii per Mag. Albertum de Brudzewo...denuo edendum curavit L. A. Birkenmajer."

[2] Ed. 1900, p. 124. Compare above, Chapter xi. p. 274, note 3.

[3] He died in the following year.

[4] The diocese of Ermland, though under the protectorate of the Teutonic Knights and from 1466 of the King of Poland, was practically an independent principality, and the Bishop was considered as a Prince of the Empire. Ermland was situated between East- and West-Prussia, the cathedral was at Frauenburg, on the shore of the Frische Haff (north-east of Elbing, latitude 54° 21′ 34″), about half-way between Danzig and Königsberg.

conferring on him at the first opportunity a canonry in the
Cathedral of Frauenburg, wished him first to extend his
studies by a prolonged stay at Italian universities; and
accordingly Copernicus started for Italy in 1496 and was
enrolled among the students of the *Natio Germanorum* in
the University of Bologna on January 6, 1497. During the
period of about three and a half years which he spent there,
Copernicus not only studied Greek and became acquainted
with the writings of Plato, but it was doubtless of special
importance for his study of astronomy that he became closely
associated with an astronomer of some standing, Domenico
Maria da Novara (1454–1504), " rather as a friend and assistant
than as a pupil," as his disciple Rheticus tells us[1]. Novara was
a practical astronomer; he had for instance in 1491 determined
the obliquity of the ecliptic equal to a trifle over $23° 29'$[2],
and his example probably encouraged Copernicus to watch
the heavens, as his first recorded observation (of an occultation
of Aldebaran) was made on March 9, 1497. In the history
of astronomy Novara's name is only known by an imaginary
discovery announced by him. Having determined the latitudes
of several cities and found values differing more or less from
those given by Ptolemy, especially in the case of Cadiz, where
the difference amounted to nearly a degree, he concluded that
the pole had moved $1° 10'$ so as to approach the zenith of these
cities (*versus punctum verticalem delatum*)[3]. The idea continued for more than a hundred years to attract a good
deal of attention, though most writers decided against the
reality of the alleged change[4]. However useful the acquaintance with Novara may have been to Copernicus, we may take
it for granted that neither he nor any other Italian savant

---

[1] "Non tam discipulus quam adiutor et testis observationum." Rheticus,
Narratio prima, p. 448 of the Thorn edition of Copern. *De revol.* 1873.

[2] According to a note in the original MS of the work *De revolutionibus* of
Copernicus; see the ed. of 1873, pp. 171–172. Compare Gassendi, *Vita
Copernici*, p. 293.

[3] Novara's statement was reprinted by Magini; see Favaro's *Carteggio
inedito*, p. 81.

[4] Tycho (*Mechanica*, fol. H) and Gilbert (*De magnete*, lib. vi. cap. 2) both
mention it with strong disapproval. Kepler, on the other hand, believed in it
during the earlier part of his career. See below, Chapter xv.

sowed the seed which eventually produced the fruit known as the Copernican System[1].

From Bologna Copernicus proceeded to Rome in the spring of 1500, the great year of jubilee, and remained there for about a year. The only particular known about his stay at Rome is, that according to his disciple Rheticus, he gave a course of lectures there on "mathematics," by which probably astronomy is meant[2]. In 1501 he returned home to take possession of the canonry at Frauenburg, to which he had been admitted (almost certainly by proxy only) three years before. He took his seat in the Cathedral Chapter on the 27th July and was granted further leave of absence for the purpose of continuing his studies, among which he undertook to include that of medicine. In the same summer he went back to Italy, this time going to Padua, where he continued to study both law and medicine for about four years with a short interruption in 1503, when he went to Ferrara and obtained the degree of a Doctor of Canon Law on the 31st May. At latest in the beginning of 1506 Copernicus left Italy, where he had spent about nine years, and although we know next to nothing about the people he may have associate with or the manner in which he prosecuted his studies, we cannot doubt that his long residence at the two most renowned Italian universities had put him in full possession of all the knowledge accessible at that time, in classics, mathematics and astronomy, as well as in theology.

From 1506 until his death in 1543 Copernicus lived in Ermland, generally at Frauenburg, where his light duties at the Cathedral gave him plenty of leisure for his scientific work, though he had to lay it aside occasionally to take his share in the administration of the little principality. A good many details are known about this side of his life, but it is much

---

[1] Tiraboschi, *Storia della lett.* T. vi. p. 590, after speaking about Novara and Copernicus, adds that some writers attribute the first idea of the system to Girolamo Tagliavia of Calabria, who lived about that time, and he gives as his authority Tommaso Cornelio, a writer of the seventeenth century. He is, however, just enough to remark that there seems no reason to accept this tradition.

[2] *Narratio prima*, ed. of 1873, p. 448.

to be regretted that we know nothing as to whether he was in correspondence with any of his learned contemporaries, though he cannot in any case have had anything to learn from them. Somehow or other the fame of the earnest student of astronomy must have spread from his very distant home to more central parts of Europe. In 1514, when the question of the reform of the Calendar was brought before the Lateran Œcumenic Council, Copernicus was invited to give his opinion on the subject by Paul of Middelburg, Bishop of Fossombrone ; but though supported by a personal friend and colleague in the Chapter of Ermland (Bernhard Sculteti) the invitation was declined, as Copernicus did not think that the motions of the sun and moon had yet been sufficiently investigated to allow of a final settlement of the question. Some years later, however, another acquaintance succeeded in getting him to give an opinion on another scientific matter. In 1522 Johann Werner of Nürnberg published a small treatise *De motu octavæ sphæræ*, dealing with the problem of precession and trepidation. Wapowski, a Canon of Cracow, called the attention of Copernicus to this paper and asked for his opinion about it. The lengthy reply, though not printed, appears to have been intended for circulation among friends, but it naturally afterwards fell into oblivion until it was finally printed in 1854[1]. It contains a very sharp criticism of Werner's treatise, making use now and then of rather strong language. To the alleged variability of the amount of annual precession Copernicus does not expressly object (in his great work he indeed accepts it), but he points out a chronological error of eleven years made by Werner in fixing the date of an observation by Ptolemy, and shows how unfounded is his conclusion that the motion of the eighth sphere had been more rapid during the period from Ptolemy to Alfonso than since the time of the latter, while it was uniform during the four hundred years before

---

[1] Tycho Brahe had a copy of it and mentions it in his book, *De mundi ætherei rec. phæn.* p. 362. It was first printed in the Warsaw edition of the work of Copernicus (1854), teeming with errors. A critical edition by M. Curtze from two MSS preserved at Berlin and Vienna was published in the *Mittheilungen des Copp.-Vereins zu Thorn*, Heft I., and in vol. II. of Prowe's work, pp. 172–183.

Ptolemy. It is particularly interesting to see Copernicus showing that when the equinoctial point in its passage round the small circle of trepidation crosses the ecliptic, the annual amount of precession changes most rapidly, while the change is nil at the two points 90° from the points of intersection, the actual amount reaching a maximum and minimum there. Werner had imagined that the very opposite would be the case, i.e. that a function would vary most rapidly at a maximum or minimum.

With this one exception we do not find that Copernicus allowed himself to be drawn aside from the steady pursuit of the astronomical work which he appears to have planned soon after his return from Italy. Observations of the heavens on an extensive scale did not form a part of his plans, nor would they have been of much use, as he did not improve on the instruments then existing. He merely from time to time took a few observations, chiefly of eclipses or oppositions of planets, which enabled him to redetermine some of the elements of the orbits. Of observations of that kind twenty-seven, from the years 1497–1529, are quoted in his work, *De revolutionibus*, in addition to which Copernicus mentions that he during thirty years frequently had determined the obliquity of the ecliptic. A few other observations have been found entered in some of the books from his private library which are still in existence[1]. But the work to which Copernicus devoted his life was done, not in the observatory, but in the study; its object was to prepare a new system of astronomy, as complete as that of Ptolemy, on the basis of the idea that the earth is not the centre of the world, but that the earth as well as the planets moves round the sun.

How was Copernicus first led to the idea that the sun

[1] A list of the observations quoted in the book *De revol.* is given in the ed. of 1873, p. 444. Other observations, all from the year 1537, are printed by Curtze in *Mitth. des Coppernicus-Vereins*, I. 35, while two observations from January and March, 1500, of conjunctions of Saturn with the moon will be found in Curtze's *Reliquiæ Copernicanæ*, p. 31 (Reprint from *Zeitschrift für Mathematik und Physik*, XIX–XX.), where determinations of the apogees of Mars, Saturn, Jupiter, and Venus, dating from the years 1523–1532, are also given (p. 29).

is the centre of motion ?   Was he first influenced by the
accounts of those among the ancients who attributed some
motion to the earth, or did he first himself out of the epicyclic
theory deduce the fact that the earth has an annual motion, and
then find comfort and encouragement in recollecting that some
of the ancients had entertained similar ideas ?   He tells us very
little about the path he had trod.   In the noble dedication to
Pope Paul III. with which his book opens, Copernicus says that
he was first induced to seek for a new theory of the heavenly
bodies by finding that mathematicians differed greatly among
themselves on this subject.   After counting up the various
systems of epicyclic, excentric and homocentric motion, and the
difficulties in accepting them, he concludes that something
essential must have been passed over in them, or something
foreign to the subject been introduced, which would not have
happened if sure principles had been followed.   He therefore
took the trouble to read the writings of all philosophers which
he could get hold of, to see if some one of them should not have
expressed the opinion that the motions of the spheres of the
world were different from what is assumed by those who teach
mathematics in the schools.   And he found it stated by Cicero
that Nicetus [Hiketas] had believed the earth to be in motion,
and by Plutarch [i.e. Pseudo-Plutarch] that others were of the
same opinion.   He gives the Greek text of the *Placita Philosophorum* (III. 13) about Philolaus, Herakleides and Ekphantus[1],
and continues : "Occasioned by this I also began to think of
a motion of the earth, and although the idea seemed absurd,
still, as others before me had been permitted to assume certain
circles in order to explain the motions of the stars, I believed it
would readily be permitted me to try whether on the assumption of some motion of the earth better explanations of the
revolutions of the heavenly spheres might not be found.   And
thus I have, assuming the motions which I in the following
work attribute to the earth, after long and careful investigation,
finally found that when the motions of the other planets are
referred to the circulation of the earth and are computed for
the revolution of each star, not only do the phenomena

---

[1] See above, Chapters II. and VI.

necessarily follow therefrom, but the order and magnitude of the stars and all their orbs and the heaven itself are so connected that in no part can anything be transposed without confusion to the rest and to the whole universe."

According to this statement, Copernicus first noticed how great was the difference of opinion among learned men as to the planetary motions; next he noticed that some had even attributed some motion to the earth, and finally he considered whether any assumption of that kind would help matters. We might have guessed as much, even if he had not told us. It must then have struck him as a strange coincidence that the revolution of the sun round the zodiac and the revolution of the epicycle-centres of Mercury and Venus round the zodiac should take place in the same period, a year, while the period of the three outer planets in their epicycles was the synodic period, i.e. the time between two successive oppositions to the sun. This curious relationship between the sun and the planets must have struck scores of philosophers, but at last the problem was taken up by a man of a thoroughly un- prejudiced mind and with a clear mathematical head. Probably it suddenly flashed on him that perhaps each of the deferents of the two inner planets and the epicycles of the three outer ones simply represented an orbit passed over by the earth in a year, and not by the sun! His emotion on finding that this assumption would really "save the phenomena," as the ancients had called it, that it would explain why Mercury and Venus always kept near the sun and why all the planets annually showed such strange irregu- larities in their motions, his emotion on finding this clear and beautifully simple solution of the ancient mystery must have been as great as that which long after overcame Newton when he discovered the law of universal gravitation. But Copernicus is silent on this point.

This *may* have been the way followed by Copernicus, but we cannot be sure that he actually found his system in this manner. In the beginning of his first book he shows how much more reasonable it is to suppose the earth to rotate on its axis in twenty-four hours than to believe that all the

heavenly bodies travel with an incredible velocity in the same
period; and he easily refutes the objections raised by the
ancients to the rotation of the earth by pointing out that if
the air partakes of the daily motion, the latter will not disturb
anything either in the air or at the surface of the earth[1]. In
the ninth chapter he then enquires whether the earth might
have more than one motion, and concludes that in that case
these motions must be such "as appear outside in a corresponding manner in many ways, from which we recognize the
annual revolution[2]." This annual motion of the earth round
the sun, he says (without here going into particulars), will
explain the stations and retrograde motion of the planets.
In the tenth chapter Copernicus discusses the order of the
planets. Having mentioned that some of the ancients had
placed Mercury and Venus below the sun, others above it, he
shows how the theory described by Martianus Capella, according to which these two planets really move round the sun, will
account for their close dependence on the sun. "And if one
takes occasion of this to refer Saturn, Jupiter and Mars to
the same centre, remembering the great extent of their orbits
which surround not only those two but also the earth, he will
not miss the explanation of the regular order of their motions[3]."

This might seem to indicate that Copernicus from the
investigation of the orbits of Mercury and Venus was led to
the conclusion that the outer planets likewise moved round the
sun, though he does not expressly say so. His disciple Rheticus
says that it was the remarkable difference in the brightness of
Mars when rising in the evening and when rising in the
morning which showed Copernicus that this planet did not
move round the earth, since the epicycle could not account for
the great change of distance indicated by the great change of
brightness[4]. But if he had reasoned in this manner, we should

[1] *De revol.* liber I. caps. 5 and 8 (pp. 15, 22 of the ed. of 1873, always quoted
in the following). In chapter 5 he again alludes to Herakleides, Ekphantus,
and Hiketas as having taught the rotation of the earth.

[2] "Qui similiter extrinsecus in multis apparent, e quibus invenimus
annuum circuitum." *De revol.* liber I. cap. 9, page 25.

[3] Lib. I. cap. 10, p. 27.

[4] Rheticus, *Narratio prima*, ed. of 1873, p. 461.

have expected that the annual motion of the earth would have
been the last thing to occur to him; that is to say, it would
have been natural first to conclude that the five planets move
round the sun, and then to add as a finishing touch that the
earth does the same. But he begins by assuming the motion of
the earth as one which "appears outside similarly in many
ways." This seems to point to Copernicus having been first
struck with the idea that each epicycle of the outer planets, its
plane always parallel to the plane of the ecliptic, and carrying
its planet round its circumference in a period closely connected
with the sun, was nothing but an image, so to say, of an orbit
described by the earth, and that the deferents of Mercury and
Venus were identical with that orbit.

However this may be, it is certain enough that Copernicus
owed very little, if anything, to the ancients. He did not
make the mistake (so persistently made after his time down to
the present day) of believing that Philolaus had taught the
heliocentric theory. In the dedication he correctly quotes the
statement of Aëtius that Philolaus let the earth, like the sun
and moon, move round the fire, and when speaking of the
rotation of the earth he merely says that Philolaus taught that
the earth turns, moves along with several motions, and is one
of the planets[1]. We cannot doubt that he clearly understood
the peculiar character of the Philolaic system from the full
accounts of it given by Aristotle and Aëtius; and though
he was probably not acquainted with the commentary of
Simplicius, it is likely enough that he knew that of Thomas
Aquinas[2]. Nowhere else does he mention Philolaus, and in the
printed book *De revolutionibus* there is not a single allusion
to Aristarchus. But in the original manuscript of the book
there is after chapter XI. of the first book a long passage which
has been struck out by lines in very black ink and was therefore
not printed. This passage begins : " Although we acknowledge
that the course of the sun and moon might also be demon-

---

[1] Lib. I. cap. 5, p. 17.

[2] The first edition of Simplicius was printed at Venice in 1526, but it was
only a Greek paraphrase of the Latin translation by William of Moerbecke.
The latter was printed at Venice in 1540.

strated on the supposition of the earth being immovable, this agrees less with the other planets. It is likely that for these and other reasons Philolaus perceived the mobility of the earth, which also some say was the opinion of Aristarchus of Samos, though not moved by that reasoning which Aristotle mentions and refutes. But as these things are such as cannot be understood except by a sharp mind and prolonged diligence, it remained at that time hidden to most philosophers, and there were but few who grasped the reason of the motions of the stars, as Plato tells us. But if they were known to Philolaus or some Pythagorean, it is probable that they were not handed down to posterity, for it was the custom of the Pythagoreans not to commit things to writing, &c.[1]" In proof of the last remark Copernicus next translates one of the many spurious letters concocted by various writers of late Alexandrian times, the one chosen by him being "From Lysis to Hipparchus"; it only deals with the love of secrecy of the Pythagoreans but does not allude to either Philolaus or Aristarchus. As we have already said, nothing of all this occurs in the printed work of Copernicus. Nor does he make any other allusions to pre-Ptolemaic astronomy in the rest of his work, except that he in the fifth book gives the Greek names of the planets and mentions the name of Apollonius in connection with the epicyclic theory, as already done by Ptolemy[2]. But there was nothing more for him to say about the ancients, for only one of them, Ptolemy, had formed a complete system of astronomy, and for him he felt the admiration which is due to him, though his own life-work was to supersede that of the Alexandrian astronomer. The totally erroneous system of Philolaus and the vague statement, that Aristarchus let the earth move round the sun, may have helped at the outset

[1] Printed in the ed. of 1873, pp. 34–35.

[2] Lib. v. introd. and cap. 3, pp. 307 and 324. The name of Eudoxus is not mentioned anywhere, and that of Kalippus only in connection with his era (in the length of the year; but in the beginning of the *Commentariolus* (presently to be mentioned) Copernicus says: "Kalippus and Eudoxus endeavoured in vain on the assumption of concentric circles to account for the cause of the sidereal motions, not only of the phenomena of the revolutions of the planets but also when the latter sometimes are seen to move away from us, sometimes to approach to us, which does not agree with the assumption of concentricity

to turn his mind in the right direction, but they had failed
to do that to many great minds before him.

The book *De revolutionibus* was the result of many years'
labour. In the dedication Copernicus says that he had kept
back the book not nine years but four times nine years; and if
this is to be taken literally, he must have had a clear conception of the new system and commenced to write down his ideas
in or soon after the year 1506, while residing with his uncle at
Heilsberg. The working out of the planetary theories was
no doubt done very gradually, and the manuscript of the whole
work, now preserved in the Nostitz Library at Prague, is
not older than the year 1529, since observations made in that
year are entered in the body of it, though subsequent alterations and re-alterations were made in it[1]. On the other hand
the manuscript cannot have been written later than 1531, as
Copernicus has not made use of a determination of the apogee
of Venus which he made in 1532 and entered on a leaf inserted
in his copy of the *Tabulæ Directionum* of Regiomontanus. On
this leaf are noted among other things the apogees of Saturn,
Jupiter, Mars and Venus, determined respectively in the years
1527, 1529, 1523 and 1532 from the writer's own observations[2].
The three first determinations are duly quoted in their proper
places in the book *De revolutionibus*[3], the fourth (48° 30′)
is neither found in the MS nor in the printed book. The MS
must therefore have been written before 1532, and the author
forgot afterwards to enter this determination, though he seems
to have gone over the whole of the contents twice, touching up
and modifying sentences here and there, altering the division
into chapters, and even correcting figures occasionally.

It must gradually have become rumoured in the learned
world that Copernicus had worked out an entirely new theory
of planetary motion, and probably at the request of some friend
or friends he drew up a short sketch of his system, which was
circulated in manuscript. Even after the publication of the
detailed treatise this short summary (*commentariolus*) continued
only
The

[1] Ed. of 1873, p. xvii.
[2] Curtze, *Reliquiæ Copernicanæ*, p. 29.
[3] Lib. v. caps. 6, 11, and 16 ; p. 337, l. 29, p. 349, l. 32, p. 361, l. 2.

to be valued by admirers of Copernicus; thus Tycho Brahe was in 1575 at Ratisbon presented with a copy of it by the Emperor's physician, Thaddæus Hagecius (Hayck)[1], and a copy probably made from that one is now preserved in the Hof-Bibliothek in Vienna. This interesting relic, of which Tycho in the course of years presented copies to various German astronomers, but which afterwards (like the letter on Werner's tract) had been utterly forgotten, has now been printed more than once[2]. It contains first a short introduction in which the failure of the theory of Eudoxus to account for the varying distances of the planets, and the objectionable character of the equants of Ptolemy are briefly alluded to, which had induced the writer to try to find a new arrangement of circles; and the leading features of the new system are then stated in six "petitiones" or axioms. Then follow seven short chapters dealing with the order of the orbits, the triple motion of the earth, the desirability of referring all motions not to the equinox but to the fixed stars, the circles proposed for the motion of the moon, those for the outer planets, for Venus and for Mercury. The relative sizes of all the proposed circles and epicycles are given, but no proofs or reasons for anything. The tract can therefore only have been intended to give readers acquainted with the details of the Ptolemaic system an idea of what the new one was like; but there is not the smallest attempt to convince the reader of the truth of the startling idea that the earth is in motion.

It was probably this *Commentariolus* which enabled a certain Widmanstad in 1533 to give Pope Clement VII. a verbal account of the new system[3]. Three years later Cardinal Nicolaus von Schonberg, Archbishop of Capua, a very liberal-minded man and a trusted councillor both of Clement and of his successor Paul III., wrote to Copernicus urging him to make his discovery known to the learned world and begging

---

[1] *Astron. inst. Progymn.* p. 479.

[2] First by Curtze (from the Vienna MS) in the *Mitth. des Coppern.-Vereins*, Heft I., then from a copy found immediately afterwards at the Stockholm Observatory, in the *Bihang till K. Svenska Vetensk. Akad. Handlingar*, 1881. A critical edition based on both of these is given by Prowe, vol. II. pp. 184-202.

[3] Tiraboschi, *Storia della lett. Ital.* VII. p. 706.

for a copy of whatever he had written, together with the tables
belonging thereto, all to be copied at the Cardinal's expense[1].
It is too well known to need repetition in detail here, how
Copernicus, dreading the storm his daring theory would
necessarily cause, for years shrank from publishing his great
work, notwithstanding the urgent solicitations of several
friends, among whom Tiedemann Giese, Bishop of Kulm,
deserves special recognition, and how finally a young Professor
in Wittenberg, Georg Joachim Rheticus, longing to get
authentic information about it, braved the risk which must
have attended a prolonged visit from a Professor in that
terrible nest of heresy, the University of Wittenberg, to a
diocese in which a threatening *Mandatum wieder die Ketzerei*
had just been issued. Rheticus went to Frauenburg in 1539
and spent about two years there. He was cordially welcomed
by Copernicus, who gave him leave to study his great work,
and the young enthusiast at once set about composing a
lengthy review of it, addressed to his teacher Johann Schoner,
which was printed at Danzig in 1540[2]. This *Narratio prima*
must have created a great sensation among competent judges,
as we find that Erasmus Reinhold, afterwards the computer of
the first set of tables according to the new system, in his
edition of Peurbach's *Theoricæ* in 1542[3], hails Copernicus as
a most distinguished artist who may be expected to restore
astronomy, and speaks of him in another place as a new
Ptolemy. Perhaps it was the reception given to the *Narratio
prima*[4] which finally induced Copernicus to yield to the prayers
of his friends; he entrusted the precious manuscript to Giese,
who sent it to Rheticus in order to get it printed. It was

[1] The letter is dated 1 Nov. 1536, and is printed at the beginning of the
book *De revolutionibus*. Schonberg died in 1537.

[2] Reprinted at the end of the edition of 1873 and in vol. II. of Prowe's
biography.

[3] *Theoricæ novæ planetarum Georg. Purb. ab R. Reinholdo...auctæ*, in the
preface and in the chapter *De motu octavæ sphæræ* (edition Paris, 1558, fol. 4 a
and 161 b).

[4] A second edition was published at Basle in 1541 by Achilles Pirminius
Gassarus (1505–1577, physician), who added a dedicatory epistle addressed to a
friend, which is also printed in the edition of *De revolutionibus*, Basle, 1566,
and by Prowe, vol. II. p. 288.

published at Nürnberg in 1543, and a copy reached Copernicus on the day he died, May 24, 1543.

The manuscript, which the author had revised and pruned in the course of about twelve years and which is still in existence, was not used in the printing-office, as may be seen from the fact that it does not very closely correspond to the text of the printed book, sentences having been added or re-inserted although the author had struck them out. For instance, in the discussion of the irregularities in the apparent motion of the sun, Copernicus had added in the margin but afterwards struck out the following sentence: "The proof would be quite the same if the earth stood still and the sun moved in the circle round it, as according to Ptolemy and others." This sentence occurs in the printed book[1]. On the other hand the editor has omitted the fine introduction to the first book, on the importance and difficulty of the study of astronomy[2]. The printing was at first superintended by Rheticus, but when he had to leave Nürnberg in 1542 to take up a new professorship at Leipzig, he confided his duties to Andreas Osiander, a well-known Lutheran theologian at Nürnberg, under whose supervision the printing was completed. Osiander was evidently uneasy at the daring character of the new theory of the earth's motion, which was sure to be considered very objectionable by many people on theological and other grounds; and to avoid trouble for the author and perhaps for himself, he added an anonymous preface "To the reader about the hypotheses of this work." In this it is stated, that though many will take offence at the doctrine of the earth's motion, it will be found on further consideration that the author does not deserve blame. For the object of an astronomer is to put together the history of the celestial motions from careful observations, and then to set forth their causes or hypotheses about them, if he cannot find the real causes, so that those motions can be computed on geometrical principles. But it is not necessary that his hypotheses should be true, they need not even be probable; it is sufficient if the calculations founded on them agree with the observations. Nobody would consider the epicycle of Venus

[1] Lib. III. cap. 15, p. 204.   [2] Ed. 1873, pp. 9-11.

probable, as the diameter of the planet in its perigee ought to be four times as great as in the apogee, which is contradicted by the experience of all times. Science simply does not know the cause of the apparently irregular motions, and an astronomer will prefer the hypothesis which is most easily understood. Let us therefore add the following new hypotheses to the old ones, as they are admirable and simple, but nobody must expect certainty about astronomy, for it cannot give it; and whoever takes for truth what has been designed for a different purpose, will leave this science as a greater fool than he was when he approached it.

These opinions had already been set forth by Osiander in two letters to Copernicus and Rheticus, written in 1541 in reply to one written by Copernicus the year before. Kepler, who had these letters of Osiander's before him and quotes their contents, does not say what Copernicus had written in his letter, except that the writer with the firmness of a Stoic believed that he ought to proclaim his conviction before the world even though science should be damaged[1]. But there is otherwise abundant testimony to prove, that to Copernicus the motion of the earth was a physical reality and not a mere working hypothesis. Not to speak of the fact that he nowhere in his work calls it a hypothesis but deals with it as a real motion, the physical objections to which he tries to meet, it is sufficient to refer to the end of his dedication to the Pope. In this place he says that some ignorant people might distort a passage of Scripture for the purpose of attacking his work, which he would treat with contempt, since even Lactantius, a distinguished writer but no mathematician, had spoken very childishly about the figure of the earth, sneering at those who teach that it is a sphere. If Copernicus had merely wanted to add another computing-hypothesis to the many existing ones, he would not have run the risk of offending the Pope by speaking slightingly about a Father of the Church. His personal friends were quite aware that the preface did not express

---

[1] *Kepleri Apologia Tychonis contra Ursum*, Opera, ed. Frisch, i. p. 246. About this essay, which Kepler laid aside after Tycho's death and never published, see my *Tycho Brahe*, p. 304.

the opinion of Copernicus. On receipt of the book Giese wrote indignantly to Rheticus to complain of " the abuse of confidence and the impiety " of the printer or of some envious person, who, regretting to have to give up his old notions, wanted to deprive the book of credibility. In order that this should not go un- punished, Giese suggested that a letter, which he enclosed, should be sent to the senate of Nürnberg to claim its protection for the author. Whether this was done is not known, anyhow the book had already been published long before Giese made his suggestion[1].

That Osiander and not Copernicus was the author of the strange preface, does not seem to have become generally known for a long time, although a careful reader might have noticed that the wording of it was hardly compatible with its having been written by the author of the book. Kepler found out the author's name from a learned colleague at Nürnberg and announced it in a very conspicuous place, on the back of the title-page of his book on Mars issued in 1609 ; but it certainly is to be regretted that Copernicus had until then in the eyes of many people lain under the imputation of having proposed a startling hypothesis while believing it to be false[2].

Having endeavoured to trace the development of the new system and the influences under which it took shape in the mind of its author, we now proceed to consider it more in detail.

The work is divided into six books. The first gives a general sketch of the new system and finishes with two chapters

---

[1] Giese's letter was printed in a small collection of letters published at Cracow in 1615, of the existence of which very few astronomers can have known. It has been reprinted in the Warsaw edition of the work of Copernicus (1854), p. 640, and in Hipler's *Spicilegium Copernicanum* (Braunsberg, 1873), p. 354; in German in Menzzer's valuable translation, *Nic. Coppernicus über die Kreisbewegungen der Weltkörper*, Thorn, 1879, p. 4 of the notes.

[2] It was a statement of the French mathematician Ramus which induced Kepler to defend Copernicus against this accusation, which had also been made by Ursus (*Kepleri Opera*, 1 p. 245). Osiander has also been credited with having added the words "orbium caelestium" to the title *De revolutionibus*, adopted by Copernicus. It must, however, be said that the two words cannot be objected to, since Copernicus uses them repeatedly in the dedication and in first book.

21

on plane and spherical triangles[1]. The second book deals with spherical astronomy. The third discusses the precession of the equinoxes and the motion of the sun (or rather the earth), the fourth the theory of the moon's motion; the fifth the motions of the planets in longitude; and the sixth their motions in latitude.

At the beginning of the first book it is stated that the world has the form of a sphere, the most perfect as well as the most roomy figure, which everything tends to assume, as we may see in drops of water and other fluids. It is next proved that the earth is a sphere, and that land and water form one sphere. Next it is argued that the motion of the heavenly bodies is uniform and circular, or a composition of circular motions, since only a circle can bring a body back to its original position, while a real inequality of motion could only be caused by a change in the motive power or by a variation in the body moved, both of which assumptions are absurd. The questions as to the place of the earth and whether it has a circular motion are next discussed. Nearly all writers, says Copernicus, agree that the earth is at rest in the centre of the world and would think it absurd to maintain the contrary opinion. Yet a little consideration will show that this question has not been settled, since any change observed may either be caused by a motion of the object observed, or by that of the observer, or by different motions of both; so that if the earth had a motion, it would produce an apparent motion of everything outside it in the opposite direction; a turning of the earth from west to east would thus account for the rising and setting of the sun, moon, and stars, as some of the ancients already taught. And if one were to say that the earth is not at the centre of the world, though the distance between them is not great enough to be measured on the starry sphere, but yet sufficient to be comparable to the orbits of the planets, he might perhaps find the true cause of the apparently irregular motions to be, that the motions are referred to a centre outside the earth.

Having thus shaken the reader's faith in the time-honoured

[1] Printed separately by Rheticus at Wittenberg, 1542.

opinions, Copernicus shows that though ..

regarded as a point in comparison to the immense ₁ the tenth
starry sphere, it does not by any means follow there There had
the earth rests in the centre of the sphere; and it ;h performs
reasonable to suppose that the immense sphere shoun, body to
in twenty-four hours. For as the earth is an actual body 10ᵗ
ought to turn with all the rest of the world and in the same
period, but in that case there would always be noon at one
place on the earth and midnight at another, and no rising or
setting. The difficulty is solved when we consider that bodies
describing smaller circles always move more rapidly than those
which describe larger ones; Saturn, the outermost planet, completing its course in thirty years, and the moon, which is certainly
nearest to the earth, in a month, so that it must be conceded
that the earth rotates in a day and a night. He then recounts
the arguments of the ancients against this rotation; that of
Aristotle being that the four elements can only have a motion
in a straight line up or down and the heavenly bodies a circular
motion; the argument of Ptolemy being that a rotation in
twenty-four hours would be so violent a motion that the loose
earth would long ago have been scattered over the heavens,
while falling bodies would never reach the place intended, as
the latter would have been torn away from under them, and
clouds and other bodies in the air would always be moving
towards the west. To this Copernicus remarks that Ptolemy
ought to be more afraid that the immense heavenly sphere
would fly asunder; and as to the clouds, we have only to
assume that not only the earth and water but also a considerable portion of the air rotate, whether the reason be that
the lower layers, mixed with earthy and watery matter, are of
the same nature as the earth, or that the friction with the earth
makes the air partake of the earth's rotation. It has been
pointed out that the highest regions follow the heavenly motion,
which is proved by the fact that those suddenly appearing stars,
which the Greeks call comets or bearded stars, and which are
supposed to originate in those regions, rise and set like the
stars. To this we can only answer that that part of the air
owing to its great distance remains free from the motion of the

rising bodies have a double motion with
on plane ande world, a rectilinear and a circular one; and as long
spherical as remains in its natural place it will only have the
equinoxes a appear to be at rest, while only bodies, which some-
fourth the been taken out of their natural place, have a rectilinear motion.

In the ninth chapter Copernicus considers whether the earth
is in the centre of the world, or whether it is a planet. That it
is not the centre of all the circular motions is proved by the
apparently irregular movements of the planets and their varying
distances from the earth. As there must therefore be several
centres, nobody can be in doubt as to whether the centre of the
world or another is the centre of terrestrial gravity. "I at
least am of the opinion that gravity is nothing but a natural
tendency, implanted by Providence in all particles, to join
themselves into a whole in the form of a sphere. And it is
credible that this tendency is also innate in the sun, moon, and
other planets, by the effect of which they retain their round
shape, while they complete their circuits in various ways."
Copernicus evidently means that the same conditions obtain on
the heavenly bodies as on the earth: whatever is out of "its
natural place" must move in a straight line, the heavy elements
(earth and water) downwards, the light ones (air and fire) upwards, i.e. away from the centre[1]. This idea is apparently set
forth in this place to show that as there is an analogy between
the earth and the planets, it is not unreasonable to assume that
the earth like the planets is endowed with orbital motion. At
any rate it is remarked immediately afterwards (as we have
already mentioned) that if the earth in addition to the daily
rotation has other motions, we must find these revealed in the
motions of the planets, first of all in the annual circuit of the
sun, and, when this is transferred to the earth, in the stations
and retrograde motions of the five planets, which are not real
but only apparent phenomena caused by the earth being in
motion, and lastly we shall believe that the sun itself is the
centre of the world.

[1] There is a long way from this ancient notion, even though here extended
to the planets, to the idea of universal gravitation.

After these preliminaries Copernicus proceeds in the tenth
chapter to fix the order of the planetary orbits. There had
hitherto been perfect unanimity as to the moon, which performs
its revolution in the shortest time, being the nearest body to
the earth ; and Saturn, having the longest period, being the most
distant one ; Jupiter and Mars having their orbits inside that of
Saturn. But the case was different with Mercury and Venus,
Plato having placed these two above the sun, Ptolemy and most
of the later astronomers below the sun, while Alpetragius placed
Venus above and Mercury below the sun. Those who followed
Plato are of opinion that, as the planets are dark bodies illuminated by the sun, those two planets if nearer than the sun
ought to appear half round or at least not perfectly round, while
the sun ought from time to time to be partially eclipsed by
them when they pass between us and the sun. On the other
hand those who place Venus and Mercury below the sun defend
their opinion by pointing to the extent of the space between
sun and moon. The greatest distance of the moon has been
assumed to be $64\frac{1}{6}$ times the semidiameter of the earth, and the
smallest distance of the sun to be 1160; the large interval
between their orbits has been filled up by letting the smallest
distance of Mercury follow the greatest distance of the moon,
and the smallest of Venus follow the greatest of Mercury, while
the smallest distance of the sun as it were touches the greatest
of Venus. For these people suppose that there are 177 semidiameters of the earth between the apsides of Mercury, and that
the remaining space is nearly filled by the orbit of Venus, 910
semidiameters in extent. They also maintain that there is not
any opaqueness in the planets as in the moon, but that the
former either shine by their own light or are saturated with
sunlight and do not eclipse the sun, even when on very rare
occasions they have so small a latitude that they pass across the
sun's disc, because they are very small bodies in comparison
with the sun, Venus (which is larger than Mercury) hardly
covering the hundredth part of the sun. It is therefore concluded that these two planets move below the solar orbit. But
how uncertain this conclusion is, may be seen from the fact that
though according to Ptolemy the smallest distance of the moon

is 38 semidiameters of the earth, or more correctly fully 52[1], still we do not know that there is anything inside this large space except air and, if you like, the so-called fiery element. Also from the fact that the greatest elongation of Venus from the sun is 45°, so that the diameter of its orbit must be six times as great as the smallest distance of Venus from the earth ; and what should fill that great space and the enormous epicycle of Venus ?   The argument of Ptolemy, that the solar orbit is in the middle, between the orbits of those planets which digress to any extent and the orbits of those which only recede slightly from the sun, is disproved by the moon, which can be at any elongation.   And what reason can be given by those who place Venus and Mercury below the sun, why these two planets do not move in orbits as separate from and independent of the sun as the other planets, if the ratio of their quickness and slowness does not falsely represent the order of their orbits ?   Therefore either the earth is not the centre, or there is no reason for the accepted order, nor why one should give Saturn the highest place rather than another.

"Therefore," continues Copernicus[2], "I think the opinion set forth by Martianus Capella and some other Latin writers is not to be despised.   For he supposes that Venus and Mercury travel round the sun and therefore cannot get further away from it than the convexity of their orbits allow, since the latter do not surround the earth.   The sun therefore is the centre of their orbits, and the orbit of Mercury is enclosed within that of Venus, which is more than twice as great.   If we take occasion of this to refer Saturn, Jupiter, and Mars to the same centre, bearing in mind the great extent of their orbits which enclose those two planets as well as the earth, we shall not fail to find the true order of their motions.   For it is certain that these are nearest to the earth when in opposition to the sun, the earth being between them and the sun, but that they are farthest from us when the sun is between them and the earth, which sufficiently proves that their centre rather belongs to the sun

---

[1] 49 in the MS of Copernicus; the printed book has 52 (p. 26) which is correct, since the lunar theory of Copernicus gives $52\frac{17}{60}$ (lib. IV. cap. 17, p. 278).
[2] I. 10, p. 27.

and is the same as that round which Venus and Mercury move.
It is then necessary that the space left between the orbits of
Venus and Mars should be occupied by the earth and its companion the moon and all that is below the moon. For we cannot
in any way separate the moon from the earth, to which it undoubtedly is nearest, particularly as there is plenty of room for
it in that space. Therefore we are not ashamed to maintain
that all that is beneath the moon, with the centre of the earth,
describe among the other planets a great orbit round the sun
which is the centre of the world; and that what appears to be
a motion of the sun is in truth a motion of the earth; but that
the size of the world is so great, that the distance of the earth
from the sun, though appreciable in comparison to the orbits of
the other planets, is as nothing when compared to the sphere of
the fixed stars.   And I hold it to be easier to concede this than
to let the mind be distracted by an almost endless multitude of
circles, which those are obliged to do who detain the earth in
the centre of the world. The wisdom of nature is such that it
produces nothing superfluous or useless but often produces
many effects from one cause.   If all this is difficult and almost
incomprehensible or against the opinion of many people, we
shall, please God, make it clearer than the sun, at least to those
who know something of mathematics.   The first principle
therefore remains undisputed, that the size of the orbits is
measured by the period of revolution, and the order of the
spheres is then as follows, commencing with the uppermost.
The first and highest sphere is that of the fixed stars, containing
itself and everything and therefore immovable, being the place
of the universe to which the motion and places of all other stars
are referred.   For while some think that it also changes somewhat[1], we shall, when deducing the motion of the earth,
assign another cause for this phenomenon.   Next follows the
first planet Saturn, which completes its circuit in thirty years,
then Jupiter, with a twelve years' period, then Mars, which moves
round in two years.   The fourth place in the order is that of
the annual revolution, in which we have said that the earth is
contained with the lunar orbit as an epicycle.   In the fifth

―――――――――
[1] This of course refers to precession.

place Venus goes round in nine months, in the sixth Mercury with a period of eighty days. But in the midst of all stands the sun. For who could in this most beautiful temple place this lamp in another or better place than that from which it can at the same time illuminate the whole ? Which some not unsuitably call the light of the world, others the soul or the ruler. Trismegistus calls it the visible God, the Electra of Sophokles the all-seeing. So indeed the sun, sitting on the royal throne, steers the revolving family of stars."

Copernicus winds up this chapter, in which he clearly and simply has set forth the outlines of his new system, by briefly stating that this harmonious arrangement will explain to a careful observer why the retrograde arc of Jupiter is greater than that of Saturn and smaller than that of Mars, and why that of Venus is greater than that of Mercury; also why the outer planets are brightest in opposition, all these phenomena being caused by the motion of the earth. That nothing similar is seen among the fixed stars, proves their immense distance, in comparison to which even the annual orbit of the earth is a negligible quantity. That there is a very great space between Saturn and the fixed stars, is also, he thinks, proved by the twinkling of the latter, which marks the difference between immovable and moving bodies.

But in the eyes of Copernicus it was not sufficient to attribute to the earth a double motion, the rotation in 24 hours and the annual motion round the sun; he had still to account for the fact, that the axis of the earth, notwithstanding the annual motion, always points to the same spot on the celestial sphere. To modern minds, this is simply explained by saying that the axis remains parallel to its original position and therefore is not endowed with any separate motion[1]. But that would not have recommended itself as a proper explanation to the ancients. We have seen that they maintained that the moon does not rotate on its axis, since it always turns the same face to the earth (whereas we say that this proves that the period of rotation is equal to the period of revolution round the earth), and similarly they reckoned the anomaly in the epicycle from a

[1] Apart from the precessional motion, of which presently.

point in the latter which remained the most distant one (apo-centre) from the centre of the deferent, just as if the epicycle was a hoop with a long rod passing diametrically through it by which it was swung round the centre of the deferent. Coperni-cus therefore would have expected the axis of the earth to have continued during the year to be directed to a point a long way above the sun, as if the earth were the bob of a gigantic conical pendulum. This would have made the celestial pole in the course of a year describe a circle parallel to the ecliptic, and as it does nothing of the kind but remains fixed, Copernicus had to postulate a third motion of the earth, a "motion in decli-nation," as he calls it[1], whereby the axis of the earth describes the surface of a cone in a year, moving in the opposite direction to that of the earth's centre, i.e. from east to west. Hereby the axis continues to point in the same direction in space. But the period is not exactly a year, it is slightly less, and this slight difference produces a slow backward motion of the points of intersection of the ecliptic and the equator—the precession of the equinoxes. This was now at last correctly explained as a slow motion of the earth's axis and not as hitherto as a motion of the whole celestial sphere, and this almost reconciles us to the needless third motion of the earth, which certainly had its share in the unpopularity against which the Copernican system had to do battle for a long time, as it seemed bad enough to give the earth one motion—but three !

The slight difference between the periods of the orbital motion and of the axial motion of the earth would explain a steady retrocession of the equinoxes. But unluckily Copernicus shared the old error, the belief in the irregular motion of the equinoxes, as it did not occur to him that errors of observation were quite sufficient to account for the differences between the various values of the constant of precession resulting from observations made in antiquity and the middle ages. He made out that precession in the 400 years before Ptolemy had been slower than during the time between Ptolemy and Al Battani, and during the latter period quicker than since Al Battani. He also thought that the obliquity of the ecliptic

---

[1] Lib. i. cap. 11, p. 31.

showed signs of irregular ~~change~~, and he therefore designed an
hypothesis to account ~~for~~ both these (imaginary) phenomena.
He supposes two mo~~tio~~ns of the earth's axis at right angles to
each other, which he calls librations, because they, like the
motion of a pendulum, are quickest in the middle; one moving
the pole in a line through the pole of the ecliptic, whereby
the obliquity varies between the limits of 23° 52′ and 23° 28′

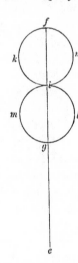

in a period of 3434 years, and another at
right angles to this, making the amount of
precession to vary in a period of 1717 years.
The combined effect is to make the pole
of the earth move in the circumference of
two small circles with radii of 6′ through
fkilgminf; *e* is the pole of the ecliptic, *ei*
the mean obliquity[1]. This geometrical representation is, however, only intended to give
an approximate idea of the phenomena, as it
cannot at the same time give values of the
variations of obliquity and of precession which
satisfy the observational data on which he
builds. The equinox in fact fluctuates as
much as 70′ to either side of the mean
position, while the annual precession varies
between the limits 50″·2015 ± 15″·3695. To avoid the inequality
Copernicus always counts his longitudes from the star γ Arietis,
and not from the equinox.

Being obliged on this occasion to give up the usual principle
of circular motion, Copernicus deems it necessary to prove
that a rectilinear motion may be produced by a combination
of two circular ones, as when a circle rolls on the interior
of another with a radius twice as great, in which case a point
on the circumference of the smaller circle will describe a
diameter of the greater one. We have seen that Nasir ed-din
Al Tûsi knew this theorem and made use of it in his
planetary theory. In his manuscript Copernicus had added,
but again struck out, the following sentence: "It is to be
noticed that if the two circles were unequal, other conditions

[1] Lib. III. cap. 2 et seq.

remaining unchanged, then they will not describe a right line
but a conic or cylindrical section which mathematicians call
an ellipse; but about this elsewhere." He probably noticed
that in general it would only be an hypocycloid looking like
an ellipse, and he therefore struck out this sentence[1]. All
the same it is interesting to see that he was aware that other
curves than circles and right lines could be produced by combinations of circular motions.

By giving an annual orbit to the sun and making it account
for the "second inequalities," Copernicus had laid the foundation of a system very much simpler than the Ptolemaic system.
But unfortunately he was compelled to mar the simplicity of
his work, because the heliocentric system was not sufficient to
explain the varying velocities of the planets in their orbits,
the "first inequalities." There was no help for it, he had to
make use of the excentrics and epicycles. As in the case of
the work of Ptolemy, we shall briefly describe the geometrical
constructions he employed.

As regards the motion of the earth round the sun Copernicus had of course nothing essential to add to the excentric
circle (or concentric circle with an epicycle) which Ptolemy had
used for the motion of the sun. He made the excentricity of
the orbit equal to 0·0323 and the longitude of the apogee
$= 96° 40'$[2]. Here again he did not make sufficient allowance
for the inaccuracy of Greek and Arabian observations. He
found that the excentricity had decreased and the longitude
of the apogee increased, but he imagined that these changes
had not progressed regularly. Though he acknowledged that
Al Zarkali's determination of the apogee must be erroneous[3],
it did not occur to him to doubt that there had been no change
between the times of Hipparchus and Ptolemy, nor that the
advance of the apogee had gone on increasing since that time.
He therefore thought it necessary to assume the following
motion of the line of apsides. The sun, the centre of the
world, is in S, round it the point A moves in a circle from west

[1] Lib. III. cap. 4, p. 166.
[2] Lib. III. cap. 16, p. 211.
[3] He knew of it from the *Epitome* of Regiomontanus, lib. III. prop. 13.

to east in about 53,000 years[1], while B, the centre of the earth's orbit, moves round A ˙ ⌐ a small circle in the opposite direction in 3434 years, th⌐ ⌐ame period as that of the variation

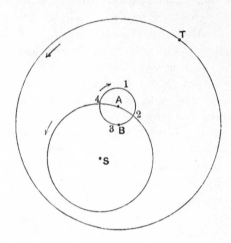

of the obliquity. If the radius of the earth's orbit BT be = 1, SA is = 0·0368 and $AB = 0·0047$. When B was at the point *1* on the small circle, the excentricity was a maximum, this took place about the year B.C. 64; at the point *3* (which B was to reach about a hundred years after the time of Copernicus), the excentricity would reach its minimum and the apogee would move fastest.

Rheticus, in his preliminary account (*Narratio prima*) adds a curious astrological commentary to his description of this motion of the apogee[2]. When the centre of the earth's orbit was at *1* and the excentricity was greatest, the Roman republic was inclining towards monarchy, and, as the excentricity declined, the Roman empire declined and vanished. When the excentricity reached its mean value, at *2*, Muhammedanism arose and another great empire took its beginning and was still increasing; but when the excentricity would reach its minimum, in the seventeenth century, this empire would, please God, rapidly

---

[1] This follows from the mean annual motion being 24″ 20‴ 14⁗ (lib. III.
p.⌐ap. 22, p. 222).
[2] *Narratio prima*, ed. of 1873, p. 453.
but ⌐⌐

noticed ⌐

‚se. When the excentricity again reaches its mean value
„ the second coming of Christ may be expected, for the
centre of the earth's orbit was at the same place at the
creation of the world; and this computation would not differ
much from the saying of Elias, that the world would last six
thousand years, in which time about two revolutions of this
rota fortunæ would take place. We should have liked to
hear what important events took place when the critical
points were passed during the first revolution. Nothing of
this theory of monarchies is mentioned by Copernicus himself,
but we cannot doubt that Rheticus would not have inserted
it in his account if he had not had it from his " D. Doctor
Praeceptor," as he always calls him[1].

The motion of the moon was by Copernicus represented by
constructions much simpler than those of Ptolemy. The equation of the centre he accounts for by an epicycle, but for the
second inequality he rejects the excentric deferent and uses
instead a second epicycle. The centre of the deferent is
therefore at d in the centre of the earth, and on its circumference the centre of the first epicycle moves from west to
east with the mean sidereal motion of the moon. The centre
of the second epicycle moves on the circumference of the first
one in the opposite direction with the mean anomalistic motion ($13°\ 3'\ 53''\ 56''''\cdot5$ per day, reckoned in the antique fashion
from the momentary apogee a), while the moon moves on the
second epicycle from west to east, twice round in every
lunation, being in e at every mean syzygy, and in f at every
mean quadrature[2]. By this arrangement the enormous changes
of parallax resulting from the constructions of Ptolemy were
avoided. Copernicus retained the ancient value of the sum
of the two inequalities $= 7°\ 40'$, and therefore put the radius
of the first epicycle $cb = 0\cdot1097$, and that of the second
$ae = 0\cdot0237$[3]. The greatest distance of the moon he found to
be $68\frac{1}{3}$, the smallest $52\frac{17}{60}$ semidiameters of the earth, both

---

[1] Rothmann, a firm adherent of Copernicus, in a letter to Tycho Brahe
written in 1587, attributes the idea to Rheticus, and asks how the excentricity
of the sun can have anything to do with the changes of empires. *Tychonis
Epist. astr.* p. 131.

[2] Lib. IV. cap. 3, p. 235.   [3] Ibid. cap. 8, p. 257.

occurring at quadrature[1]. The apparent diameter of the moon
therefore varies between 28′ 45″ and 37′ 34″, a great improvement indeed (as Copernicus remarks) on the theory of Ptolemy,
according to which the apparent diameter ought to be nearly
a degree at perigee[2].

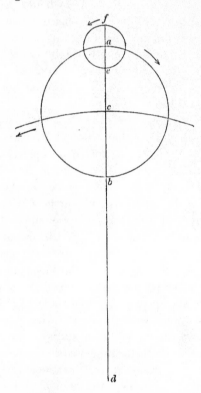

In the planetary theories Copernicus had the great advantage over Ptolemy, that he had (as regards the motion in
longitude) only the first inequality to deal with, the period of
which is the sidereal period of revolution. This Ptolemy had
accounted for by the excentric circle and the equant or circle
of uniform angular motion, the centre of the deferent or circle
of equal distances being half-way between the earth and the
centre of the equant. Copernicus might have adopted this

[1] Lib. iv. cap. 17, p. 278.   [2] Ibid. cap. 22, p. 285.

angement, but he considered that the principle of uniform circular motion had been violated by the introduction of the equant, and he had therefore to find some other explanation. For the outer planets this was comparatively easy. In the figure, *d* is the centre of the earth's orbit, to which point, as representing the mean motion of the sun (i.e. of the earth) Copernicus always referred the planetary motions. The centre of the excentric orbit of the planet is at *c*, while the planet moves on an epicycle in the same direction and with the same angular velocity with which its centre moves round the excentric. The radius *ae* of the epicycle is one-third of the excentricity *cd* of the deferent, in fact *cd* + *ae* is equal to Ptolemy's excentricity of the equant, so that instead of bisecting

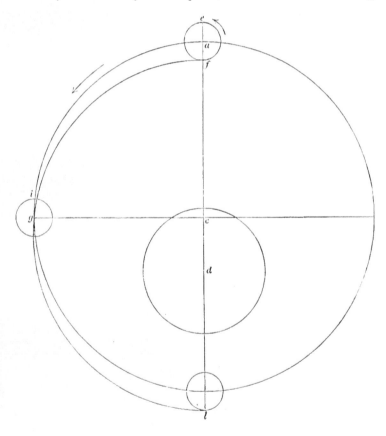

the excentricity as Ptolemy had done, Copernicus gave $\frac{3}{4}$ of it
to the deferent and let the epicycle account for the rest, but
the result is the same. The planet is at f when the centre of
the epicycle is at a, at i when the centre is at g, and so on;
and, as Copernicus points out, the planet will not describe a
circle, as i is outside a circle passing through f and l^1.

Instead of the excentric circle one of equal radius, but
with d as centre, might have been adopted². On this should
then have been moving an epicycle with radius cd and direct
motion, on this another epicycle with radius $\frac{1}{3}cd$, same period
but retrograde motion, on which finally the planet would move
directly with twice the velocity, so that whenever the centre of
the smaller epicycle was in the apsides of the greater, the
planet was in the perihelion of the smaller. This *epicycli
epicyclium* would have the same effect as the *eccentrepicyclum*
figured above, but Copernicus prefers the latter arrangement
as the simpler.

In the Copernican system the Ptolemaic epicycles of Venus
and Mercury became the orbits of the two planets round the
sun. But the greatest elongations of these planets are not
always equally great, a fact which is partly caused by the
excentricity of their orbits, partly by that of the earth's orbit.
In the case of Venus the phenomenon is simple enough, since
her own orbit has a very small excentricity; and Copernicus
therefore adopted a movable excentric after the manner of
Apollonius, i.e. he let the centre of the orbit of Venus move
round the mean centre of the planet's orbit in a small circle
with twice the angular velocity of the earth and in the same
direction. Whenever the earth passes the produced line of
apsides of Venus at a and b, the centre of the excentric is at
the point m of the small circle nearest to the mean sun, and
the radius of the small circle is one-third of the average excentricity, $dn = \frac{1}{3} cd^3$. But owing to the very great excentricity of
the orbit of Mercury ($\frac{1}{5}$, or more than twice that of Mars) this

---

[1] "Non describit circulum perfectum, sed quasi." Lib. v. cap. 4, p. 326.
To produce an excentric circle the motion on the epicycle should have been
retrograde; see above, Chapter VII. p. 154.

[2] This alone is given in the *Commentariolus*.

[3] Lib. v. cap. 22, p. 368.

.heory was not sufficient for that planet. The centre of the small circle is now at *n*, whenever the earth is at *a* or *b*, and the planet does not move on the excentric, but backwards and forwards on the line *k l* (the diameter of a small epicycle) which is always directed to the centre of the excentric, so that Mercury is at *k* every six months, when the earth is at *a* or *b* and the centre of the excentric at *n*, and at *l* whenever the mean heliocentric longitude of the earth differs 90° from the longitude of the apsides of Mercury, while *i* goes round the excentric in 88 days. "Therefore Mercury by its proper motion does not always describe the same circle, but very different ones according to the distance from the centre, the smallest when at *k*, the greatest when at *l*, the average one

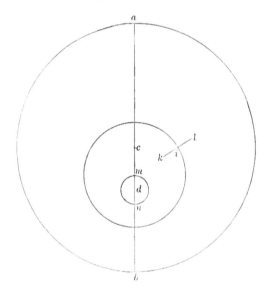

when at *i*¹, almost in the same manner as may be seen in the epicycle of the moon; but what in the case of the moon is done in the circumference, that occurs at Mercury in the diameter by reciprocal motions, composed of equal ones; and how this is done we have seen above when dealing with the

¹ Greatest 0·3953, smallest 0·3573, mean 0·3763 (*ca* = 1), *dn* = 0·0212, *cd* = 0·0736 (cap. 27, p. 382).

D. 22

precession of the equinoxes[1]." Copernicus gives no explanation as to why he deviates from the circular motion in this
particular case.

We have already mentioned that Copernicus was not a
persevering observer, but that he thought a few observed
places of each planet sufficient for determining the elements
of its orbit. He determined the excentricity and longitude
of apogee of the three outer planets from three oppositions
observed by Ptolemy as well as by three others observed by
himself, and he made the interesting discovery that the longitudes of the apogees had all increased much more than could
be accounted for by precession; and although he much
exaggerated the actual amount of the motions of the lines of
apsides, still the credit of this discovery cannot be denied
him[2]. Having found these two elements of the orbit it was
now a simple matter to determine the ratio of the semidiameter of the deferent of each planet to the semidiameter
of the earth's orbit, by a single observed place of the planet
outside an opposition. In thus giving the relative dimensions
of the whole system Copernicus scored heavily over Ptolemy,
as no geocentric system can give the smallest clue to the
distances of the planets, although, as we have seen, the actual
distances (in terms of the sun's distance) had in reality all
along lain hidden in the ratio of deferent-radius to epicycle-
radius found by Ptolemy. The distances of the inner planets
from the sun, the radii of their epicycles according to Ptolemy,
are readily found from observations of greatest elongation, and
here Copernicus had to rely solely on observations recorded by
Ptolemy. Indeed he quotes only one observation of Venus
made by himself (an occultation by the moon in 1529) and
none at all of Mercury, which planet he says has given him a
great deal of trouble, since one can rarely see it on account of

---

[1] Lib. v. cap. 25, p. 377.

[2] Copernicus determined the apogees, not the aphelia, of the planets, since
he let the lines of apsides pass through the centre of the earth's orbit and not
through the sun. The variations he found were for Saturn 1° in 100 years, for
Jupiter 1° in 300 years, for Mars 1° in 130 years (lib. v. caps. 7, 12, 16, pp.
339, 351, 360).

the vapours of the Vistula[1]. He had, however, obtained three observations of Mercury, one by Berhard Walther and two by Johannes Schoner, and he found also in the case of this planet a direct motion of the line of apsides[2].

The following are the mean distances of the planets from the sun found by Copernicus: they are almost identical with those resulting from the determinations of Ptolemy[3].

	Copernicus	True Value
Mercury	0·3763	0·3871
Venus	0·7193	0·7233
Earth	1·0000	1·0000
Mars	1·5198	1·5237
Jupiter	5·2192	5·2028
Saturn	9·1743	9·5389

As regards the earth's distance from the sun, Copernicus had to adopt the value of the solar parallax given by Hipparchus, only making a trifling correction to it resulting from the values of the apparent diameters of the sun and moon adopted by him. He makes the mean parallax $= 3' 1''$, and the mean distance 1142 semidiameters of the earth[4].

In his sixth book, the shortest of all, Copernicus deals with the latitudes of the planets, and this is the part of his work in which he keeps closest to Ptolemy, and in which the want of accurate observations is most strongly felt and prevents him from dispensing with needless complications. The orbits of the three outer planets are inclined to the ecliptic, but the angle of inclination is not constant but varies slightly in the synodic period, being greatest when the planet is in opposition, and smallest when it is in conjunction. The mean inclinations and their limits of variation are[5]:

$$\text{Saturn} \quad 2° \, 30' \pm 14'$$
$$\text{Jupiter} \quad 1° \, 30' \pm 12'$$
$$\text{Mars} \quad 1° \, 0' \pm 51'.$$

[1] He ought to have said " of the Frische Haff," but he probably thought nobody outside Prussia knew the name of this lagoon. He does *not* say that he had never *seen* Mercury. Lib. v. cap. 30, p. 387.

[2] He makes it out to be 1° in 63 years (cap. 30, p. 393), about ten times the actual amount.

[3] Lib. v. caps. 9, 14, 19, 21, 27, pp. 341, 353, 364, 367, 383.

[4] Lib. IV. cap. 21, p. 283.   [5] Lib. VI. cap. 3, pp. 421–422.

For Mercury and Venus the theory is fully as complicated as that of Ptolemy. For each of these planets the line of nodes falls in the line of apsides, and the greatest latitudes ought therefore to occur when the planet is 90° from the apogee; but they are subject to two kinds of fluctuations or "librations." The first has a period of half a year, so that whenever the mean place of the sun passes through the perigee or apogee of the planet the inclination is greatest. The second libration differs from the first by taking place round a moving axis, the planet always passing through it whenever the earth is 90° from the apsides; but when the apogee or perigee of the planet is turned towards the earth, Venus always deviates most to the north and Mercury to the south. Suppose, for instance, that the mean place of the sun falls in the apogee of Venus, and the planet happens to be in it, then the simple inclination and the first libration would produce no latitude, but the second one, which takes place round an axis at right angles to the line of apsides, produces the greatest deviation. But if Venus at that moment had been 90° from the apsides, then the axis of this libration would pass through the mean sun, and Venus would add to the northern " reflexion " the greatest " deviation," or decrease the southern by the same amount. Let $abcd$ be the orbit of the earth, and $flgk$ the excentric orbit of Venus or Mercury at its mean inclination to

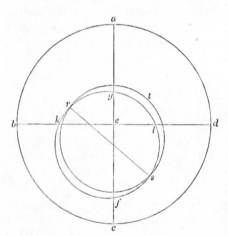

the former, fg being the line of nodes. When both the earth
and the planet are in the line ac the planet has no latitude,
but it will have some if in either of the two semicircles gkf
or flg, and this latitude is called obliquation or reflexion.
But when the earth is at b or d these latitudes in gkf or flg
are called declinations, "and they differ from the former in
name rather than in reality." But as the inclination was
found greater in the obliquation than in the declination, this
is supposed to be caused by another libration around fg as an
axis. Another "circle of deviation" is assumed, which is
inclined to $gkfl$, concentric with it in the case of Venus, but
excentric to it in the case of Mercury. Their line of intersection, rs, is a moving axis of this libration. When the
earth is at a or b the planet will reach its limit of deviation at
t, and while the earth moves away from a, the planet moves
away from t at the same rate, while the obliquity of the circle
of deviation decreases, and when the earth has reached b the
planet has reached the node r of this latitude. But at that
moment the two planes coincide and change positions, so that
the other semicircle of deviation, which hitherto lay to the
south, now will lie to the north, and Venus, which before was
in north latitude, will continue there, and will never by deviation be turned to the south. Similarly Mercury will only
deviate to the south. For both planets the period is a year[1].

The following are the numerical values adopted by Copernicus:

	Declination[2]	Maximum Obliquation[3]	Maximum Deviation[4]
Venus	2° 30′	Decl. + 59′	+ 10′
Mercury	6° 15′	Decl. + 45′	− 45′.

As if he thought he had perhaps adhered too closely to
Ptolemy, Copernicus also gives another explanation of the
deviation, assuming that the centre of the planet's orbit lies
outside the plane of the earth's orbit, always on the same side
of it, but at a distance varying in the period of a year.

On the whole this latitude theory was only a slight modifi-

---

[1] Lib. VI. cap. 2, pp. 417–419.   [2] Cap. 5, p. 424.
[3] Cap. 7, p. 429 sq.   [4] Cap. 8, p. 434.

cation of the theory of Ptolemy. It was natural that the
latter should find great difficulty in representing the latitudes,
since he had to let the line of nodes pass through the earth
instead of through the sun. But Copernicus had also erred,
though to a smaller extent, by letting it pass through the
centre of the earth's orbit. This displaced the nodes, so that
a planet was found to have some latitude when it ought to
have had none (or been in the ecliptic), and this amount of
latitude varied with the place of the earth in its orbit. For
the same reason the greatest north latitude of a planet would
turn out different from its greatest south latitude, and the
amount of the difference would also seem to vary with the
position of the earth. No wonder that it was necessary to
assume oscillations of the orbits.

With the latitude theory ends the immortal work *De
Revolutionibus*. Quite apart from the daring theory of the
earth's motion, which even its adversaries (at least those of them
whose opinion was worth anything) acknowledged to be worthy
of a great mind, the book at once placed its author on a level
with Hipparchus and Ptolemy. For the first time, since the
Syntaxis had been published, an astronomer had produced a
work fit to take the place of that time-honoured monument
of Greek science. This was not a mere commentary to the
Almagest, nor a mere sketch of a rival planetary theory such
as Al Betrugi and Fracastoro had given; this new book represented a complete overhauling of the contents of Ptolemy's
work, and supplied new theories and new tables of the
planetary motions, which were practically independent of the
doctrine of the earth's motion and might be used even by
those who were strongly opposed to that doctrine. The
principal elements of the planetary orbits had been determined anew, and though this was done on the basis of an
utterly insufficient number of new observations, this was a
defect which nobody seems to have remarked at the time.
Another and a more serious defect, partly caused by the want
of new observations, partly by an excessive confidence in the
accuracy of Ptolemy's observations, was that Copernicus in
many cases had kept too close to his great predecessor. The

man who had deposed the earth from its proud position as
the centre of the universe and had recognized it to be merely
one of the planets, had yet felt compelled to give it quite an
exceptional position in his new system. Though he had said
"in the midst of all stands the sun," he had in his planetary
theories assumed the centre of all movements to be the centre
of the earth's orbit, where the sun was not. And the year,
that is to say the period of revolution of the earth, was
intimately connected with the motion of the two inner planets,
both in longitude and latitude, and the same was the case
with the motion of the outer planets in latitude, so that the
earth was nearly as important a body in the new system as
in the old. Nor had the motion of the earth done much to
simplify the old theories, for though the objectionable equants
had disappeared, the system was still bristling with auxiliary
circles. This, however, does not seem to have been felt as a
fault by Copernicus, for he winds up his *Commentariolus* with
the following words: "Thus Mercury runs in all on seven
circles, Venus on five, the earth on three, and round it the
moon on four, lastly Mars, Jupiter and Saturn on five each.
Thus altogether 34 circles suffice to explain the whole construction of the world and the whole dance of the planets."

Kepler was not far wrong when he said that Copernicus
did not know how rich he was, and tried more to interpret
Ptolemy than nature (though he came nearer doing this than
anyone), since he failed to see how needless the variations of
latitude were, and attributed librations to the planes of the
excentrics, not following the motions in these excentrics, but
(*quod monstri simile sit*) the motion of a body that had nothing
to do with them, the earth[1].

Copernicus was, however, well aware that he had only
initiated the reform of astronomy, and that a great deal of
work would have to be done before it could be completed.
He said to Rheticus that he should be as delighted as
Pythagoras was when he had discovered his theorem, if he
could make his planetary theory agree with the observed

---

[1] *De motibus stellæ Martis*, cap. xiv. (Opera, iii. 234).

positions of the planets within 10′[1]. But the accuracy attained was very far indeed from reaching even that modest limit. Certainly it would have been possible, as observations multiplied and revealed errors in the theory, to have piled epicycle on epicycle to wipe out these errors, since this is just the same as expressing a function as a series of terms involving sines and cosines of angles proportional to the time elapsed since a certain epoch, which is what astronomers still do. But a fundamental error was the taking the centre of the earth's orbit as the centre of all motion, since this in the theory of Mars may cause very considerable errors in the geocentric longitude of the planet[2].

But if Copernicus did not produce what is now-a-days meant by "the Copernican system," let us not forget what he did do. He not only showed that the assumption of the annual motion of the earth round the sun would explain in a very simple manner the most glaring irregularities in the motions of the planets, but he built up a complete system of astronomy thereon, a system capable of being further developed as soon as an indefatigable observer had perceived the necessity of cross-examining the heavens in a persevering manner.

---

[1] *Ephemerides novæ a G. J. Rhetico* (Lipsiæ, 1550), p. 6.
[2] Kepler showed that though this would at most produce errors of 5′ in the heliocentric longitudes of Mars, the error might rise to 1° 3′ in the geocentric longitudes (*De mot. stellæ Martis*, cap. VI. Op. III. p. 199).

CHAPTER XIV.

TYCHO BRAHE AND HIS CONTEMPORARIES.

THE book of Copernicus was published in 1543, and Kepler's book on Mars, showing that the planetary orbits are ellipses, appeared in 1609. In the same year the telescope was first directed to the heavenly bodies and completely changed many of the prevailing notions as to their constitution. The period from 1543 to 1609 was a transition period, as the system of Copernicus had not yet been purified and strengthened by Kepler; but during that time the work was done which led to his great discoveries.

The book *De Revolutionibus* at once took its place as a worthy successor to the *Almagest* of Ptolemy, which had hitherto been the Alpha and Omega of astronomers. Erasmus Reinhold[1], who had already in 1542 hailed the forthcoming work as opening a new era, soon set about to prepare new tables of the celestial motions to take the place of the obsolete Alfonsine Tables, and they were published in 1551 under the title of *Tabulæ Prutenicæ*, so called in honour of the author's patron, Duke Albrecht of Prussia. The general arrangement is the same as that of the tables in the book of Copernicus, except that the intervals are smaller and the attempted accuracy greater (seconds being given where Copernicus had only given minutes), but the constants are determined anew from the data given by Ptolemy and Copernicus. Owing to the extreme scantiness of recent observations the tables were not very much better than those

---

[1] Born at Saalfeld in Thuringia in 1511, Professor in Wittenberg, died 1553. He is also the author of a valuable *Gründlicher und warer Bericht vom Feldmessen*, Saalfeld, 1574, published by his son; see *Nature*, LXVII. p. 42.

they superseded; still they represented a step forward, and
nothing better could be done until the work of Tycho and
Kepler had borne fruit. There was no occasion for Reinhold
to make a confession of scientific faith, and he gave no hint as
to whether the system of Copernicus was the physically true
one or not. All the same, the practical demonstration which he
supplied of the excellence of the mathematical part of the book
De Revolutionibus doubtless helped greatly to spread the fame
of the latter. In England the tables were already in 1556
utilized by a certain John Field for the preparation of an
Ephemeris for 1557, "juxta Copernici et Reinholdi canones,"
in the preface to which the author says that their writings are
established on true, certain and authentic demonstrations. In
an epistle prefixed to the same *Ephemeris*, John Dee, the well-
known English mathematician and astrologer, proclaims his
adherence to the system of Copernicus[1].

Probably the earliest pronouncement in England in favour
of the new system was made in 1551 by Robert Recorde, the
author of several books on arithmetic, in his *Pathway to Knowledge*, though in a guarded manner, as if he hardly thought the
world ripe for any such doctrine. In a dialogue between Master
and Scholar the former gives Ptolemy's reasons for asserting
that the earth "standeth in the myddle of the worlde," but adds
that "Eraclides Ponticus, a great philosopher, and two great
clerkes of Pythagoras schole, Philolaus and Ecphantus, were of
the contrary opinion, but also Nicias Syracusius and Aristarchus
Samius seem with strong arguments to approve it." After
saying that the matter is too difficult and must be deferred till
another time, the Master states that "Copernicus, a man of
greate learning, of muche experience and of wondrefull diligence
in obseruation, hathe renewed the opinion of Aristarchus Samius,
and affirmeth that the earthe not only moueth circularlye about
his own centre, but also may be, yea and is continually out of
the precise cètre 38 hundredth thousand miles; but bicause the

---

[1] Note by J. Hunter in *Month. Not. R. Astr. Soc.* III. p. 3; De Morgan,
"Notices of English Mathematical Writers between the Norman Conquest and
the year 1600," in *Companion to the Almanac*, 1837, p. 35. Field published
similar ephemerides for the years 1558, 1559, and 1560.

vnderstanding of that controuersy dependeth of profounder knowledge than in this introduction may be vttered conueniently,
I will let it passe tyll some other time[1]." While Field probably
only adopted the planetary theories of Copernicus, Recorde
seems to have been persuaded of the physical truth of the
earth's motion, or at least to have thought it very probable.
Another English mathematician who was of the same opinion
was Thomas Digges, author of an interesting book, *Alæ seu
scalæ mathematicæ,* which deals chiefly with the new star which
appeared in 1572[2]. In the preface he says that the Ptolemaic
system is like a set of head and limbs taken off different people,
which shows that the hypothesis is not the true one[3], and caused
Copernicus to use another one. He adds that the phenomena
are the same whether we assume the rotation of the starry
sphere or of the earth. In the concluding paragraph he expresses his regret that Copernicus is not alive now, as we might
in that case have attained to a perfect knowledge of the celestial
system. In 1592 Digges published a new edition of the *Pro-
gnostication everlasting,* a meteorological work by his father,
Leonard Digges. In an appendix added by himself he remarks,
after referring to the Ptolemaic system, that " In this our age,
one rare witte...hath by long study, paynfull practise, and rare
invention, delivered a new Theorick or Model of the world,
shewing that the Earth resteth not in the center of the
world...[4]." As this addition is furthermore headed " A Perfit
Description of the Cœlestiall Orbes, according to the most
ancie ; doctrine of the Pythagoreans : lately revived by Coper-
nicu   and by Geome^rical Demonstrations approved," there can
be i   doubt that Th mas Digges was a believer in the motion
of t : earth and not merely adopted the new system as a
wo   ng hypothesis.

      t the end of the sixteenth century we find another

[1] De Morgan, l. c.

[2] London, 1573, 4⁰. Digges had hoped to test the Copernican system by
trying whether the new star had an annual parallax, but he could not find any.
See my *Tycho Brahe,* p. 59.

[3] Copernicus had used this expression in his dedication to the Pope, p. 5.

[4] The whole passage is given by Whewell, *Hist. of the Induct. Sc.* i. p. 386
(3rd ed.).

Englishman of great renown accepting the doctrine of the earth's daily rotation as undoubtedly true. This is William Gilbert, the author of the epoch-making work *De Magnete*, published in London in the year 1600. He discusses very fully the "daily magnetic revolution" of the globe, which he considers not merely probable but certain, since nature ever acts with fewer rather than with many means, and because it is more according to reason that the one small body, the earth, should make a daily revolution than that the whole universe should be whirled around it. There is no reason whatever to assume the existence of spheres, without which it is absurd to imagine the stars to rush round the earth in twenty-four hours with an enormous velocity; in fact there can be no doubt that the stars, like the planets, are situated at various distances from us, and that many are so far off that the eye cannot perceive them. He attributes the cause of the diurnal motion to the magnetic energy of the earth, without, however, going into particulars; while he points out that the moon's orbit is a little more than twice $29\frac{1}{2}$ times the length of great circles on the earth, and as the moon's period is a little over $29\frac{1}{2}$ days, the moon and the earth agree in a twofold ratio of motion[1]. Though Gilbert repeatedly refers to the doctrine of Copernicus, he does not wish to enter into the question of the orbital motion in this book[2], but in a posthumous work, not published till 1651, he appears to hesitate between the systems of Tycho and Copernicus[3].

Contrary to what might have been expected, the Copernican doctrine does not appear to have had many followers in Germany in those early days. Among those who took it up was Christian Wursteisen, or Urstisius, of Basle (1544–1588), who is said by Galileo to have given lectures on the subject in Italy[4]. He

---

[1] Lib. VI. cap. 3.

[2] Edward Wright, a well-known English mathematician (d. 1615), wrote a preface to Gilbert's book, in which he says that there are many and great difficulties in the way of accepting the diurnal motion of all the spheres (if spheres there be), and considers it very probable that the earth rotates. He apparently does not believe in the annual motion.

[3] *De mundo nostro sublunari Philosophia nova.*

[4] Galileo lets one of the persons in his *Dialogue* (p. 143, Albéri's ed.) mention this, which has been misunderstood by some to mean that Galileo himself had attended this lecture.

wrote nothing about it, and in a lengthy commentary to Peurbach's *Theoricæ* he does not even allude to the new system, and
only mentions Copernicus a couple of times, though on one
occasion he calls him "a man of truly divine genius who in our
century has attempted the restoration of astronomy not without success[1]." But the book was probably written to order,
and the sale might have been injured by the introduction of
controversial matter. Michael Mästlin (1550–1631) was also
a Copernican, and as he was the teacher of Kepler he was
probably the first to instruct his great pupil in the details of
the new system. He saw Kepler's first work, *Mysterium Cosmographicum*, through the press, and added to it of his own accord
a new edition, the fourth, of the *Narratio Prima* of Rheticus[2].
In a preface added to the latter, Mästlin declares that the order
and magnitude of all the orbs are so disposed in the Copernican
hypothesis that nothing can be altered or transposed without
confusion to the whole Universe, "quin etiam omnis dubitatio
de situ et serie prout exclusa manet." He even thought of
publishing a new edition of the work of Copernicus and actually
wrote a preface, in which he strongly protested against the
condemnation of the Copernican system by the Congregation of
the Index; saying that nobody has refuted it by astronomical
or mathematical arguments, and that it is the old system of
Aristarchus, which Copernicus has solidly confirmed and proved
by unanswerable arguments and by the aid of geometry[3]. All
the same Mästlin wrote a text-book quite in the usual style
of the fifteenth and sixteenth centuries (*Epitome Astronomiæ*,
Tübingen, 1588), in which only the old theories are expounded.
But in his old age, when publishing a new edition of this work,
he added an appendix to the first book, at the end of which
(p. 95) he says of the rotation of the starry sphere that the
incomprehensible rapidity of this was doubtless not the last, if
indeed it was not the first, reason which gave Copernicus
occasion to think of other hypotheses and another arrangement

---

[1] *Quæstiones novæ in theoricas novas planetarum*, Basle, 1573, p. 46.

[2] The third edition (omitting the Encomium Borussiæ) was appended to the
second edition of the book, *De Revolutionibus*, Basle, 1566.

[3] Mästlin's plan was not carried out. The preface is printed by Frisch
from the original MS in *Kepleri Opera*, I. p. 56.

of the spheres corresponding more with reason, with nature, and with observations.

Another German Copernican was Christopher Rothmann, chief astronomer to Landgrave Wilhelm IV., of Hesse-Cassel. He was a constant correspondent of Tycho Brahe, and the two of them repeatedly discussed this subject in their letters, Rothmann defending himself with great ability against the arguments of Tycho, so that he must have had very strong convictions as to the truth of the Copernican doctrine[1]. He did not publish anything about it himself. A contemporary of his, Origanus or David Tost, a great astrologer and author of *Ephemerides*, accepted the daily rotation of the earth, which he supposed to be connected with its magnetic power, but otherwise he adopted the Tychonic system[2].

In Italy the spirit of humanism showed some signs of life still, and the time was fast approaching when experimental physics was to start as a revived science in that country. Giovanni Battista Benedetti (1530-90) was a precursor of Galileo in refuting Aristotelean errors as to motion, centrifugal tendency, &c.; he also preferred the "theory of Aristarchus, explained in a divine manner by Copernicus, against which the arguments of Aristotle are of no value," and he even went so far as to suggest that the planets were inhabited, since the centre of the lunar epicycle was not likely to be the only object of creation[3]. The rotation of the earth was also admitted by Francesco Patrizio (1530-97), a liberal-minded philosopher, on the ground that the stars, if really moving, would either have to be attached to a huge sphere, which he declares impossible on account of the enormous speed of its rotation, or they would move freely through space, which for the more distant ones would be equally impossible. The orbital motion he rejects, and even the Tychonic system does not find favour in his sight (he makes the absurd mistake of taking Tycho for a believer in solid celestial spheres), and he is rather behind his age in making

[1] *Tychonis Epist. astron.* pp. 188–192.
[2] *Origani Novæ Coel. Motuum Ephemerides* Frankfurt, 1609, T. I., dedication.
[3] *Diversarum speculationum math. et physicarum liber*, Turin, 1585, p. 195.

general remarks about the perfectly uniform motions of the planets, quite different from those we see[1]. Patrizio was an opponent of the Aristotelean physics solely because he was an admirer of Plato, and he can therefore hardly have contributed appreciably to pave the way for Galileo, as he altogether failed to see the value and necessity of observation and experiment.

While Patrizio as a Platonist could only accept as much of the new system as might be reconciled with the Platonic conception of the world, the revolutionary spirit of Giordano Bruno made him a red-hot Copernican. The infinity of the universe and the infinite multitude of worlds therein are specially insisted on by him, and in his book *De Immenso* he refutes at length the arguments of Aristotle against the infinity of the world, urging that it has no centre and that the rotation of the earth is the true cause of the apparent motion supposed to be produced by the *primum mobile*. The earth is a star like the moon and the planets. He praises the genius of Copernicus for its freedom from prejudice, though he regrets that he was more a student of mathematics than of nature and therefore unable to free himself from unsuitable principles. Evidently the planetary theories of Copernicus were not to the taste of Bruno, who did not in his speculations confine himself to what could be proved from observation and calculation but let his mind soar freely through space. Certainly some of his ideas have turned out to be true forecasts, e.g., that the earth is flattened at the poles, that the sun rotates, and that the fixed stars are suns[2]. He can, however, hardly be considered as a representative of his age either as regards his philosophical or his religious opinions. A hundred years earlier he would have been held in high respect in Rome, but in the year 1600 there was no room for him and he was burned as a heretic.

It is interesting, though useless, to speculate on what would have been the chances of immediate success of the work of Copernicus if it had appeared fifty years earlier. Among the

---

[1] *Nova de Universis Philosophia, libris L comprehensa*, Venice, 1593, *Kepleri Opera*, I. p. 225.

[2] For Giordano Bruno's opinions on astronomical matters see especially his book *De Monade*; the parts in which he eulogizes Copernicus are reprinted by Libri, *Hist. des sc. math. en Italie*, IV. pp. 416–435.

humanists there certainly was considerable freedom of thought, and they would not have been prejudiced against the new conception of the world because it upset the medieval notion of a set of planetary spheres inside the empyrean sphere, with places allotted for the hierarchy of angels. If one of the leaders of the Church (at least in Italy) at the beginning of the sixteenth century had been asked whether the idea of the earth moving through space was not clearly heretical, he would probably merely have smiled at the innocence of the enquirer and have answered in the words of Pomponazzi that a thing might be true in philosophy and yet false in theology. But the times had changed. The sun of the renaissance had set when, in 1527, the hordes of the Connetable of Bourbon sacked and desecrated Rome; the reformation had put an end to the religious and intellectual solidarity of the nations, and the contest between Rome and the Protestants absorbed the mental energy of Europe. During the second half of the sixteenth century science was therefore very little cultivated, and though astronomy and astrology attracted a fair number of students (among whom was one of the first rank), still theology was thought of first and last. And theology had come to mean the most literal acceptance of every word of Scripture; to the Protestants of necessity, since they denied the authority of Popes and Councils, to the Roman Catholics from a desire to define their doctrines more narrowly and to prove how unjustified had been the revolt against the Church of Rome. There was an end of all talk of Christian Renaissance and of all hope of reconciling faith and reason; a new spirit had arisen which claimed absolute control for Church authority. Neither side could therefore be expected to be very cordial to the new doctrine. Luther, in one of his Table Talks, had in his usual blunt way given his opinion of the " new astrologus " who would prove that the earth moves. " The fool will upset the whole science of astronomy, but as the Holy Scripture shows, it was the sun and not the earth which Joshua ordered to stand still." This is not very surprising, since Luther had always been a stranger to humanism, but it is more remarkable that the highly-cultured Melanchthon should give vent to more than one

sweeping condemnation of Copernicus. Already two years before
the publication of the book of Copernicus, Melanchthon wrote to
a correspondent that wise rulers ought to coerce such unbridled
licence of mind[1]. And in his *Initia doctrinæ physicæ*, published
in 1549, he goes fully into the matter in a chapter headed:
" Quis est motus mundi ? " First he appeals to the testimony
of our senses. Then he serves up the passages of the Old
Testament in which the earth is spoken of as resting or the sun
as moving. Finally he tries his hand at " physical arguments,"
of which the following is a specimen: " When a circle revolves
the centre remains unmoved ; but the earth is the centre of the
world, therefore it is unmoved[2]." A beautiful proof. It would
have been wiser if he had stuck to his Scriptural arguments or
to the *argumenta ad hominem* which he advocated in 1541.

While this was the attitude of the German reformers, it is
very curious to find a clerical voice crying in the wilderness,
trying to prove from Scripture that the earth does move. And
this was actually done in Spain, of all countries ! Didacus
a Stunica (de Stúñiga) published in 1584 at Salamanca a commentary on the book of Job, in the course of which he discussed
the passage, " Who shaketh the earth out of her place and the
pillars thereof tremble[3]." He maintained that it was much
easier to understand this passage in connection with the opinion
of the Pythagoreans which " in this our age Copernicus doth
demonstrate." The only argument he quotes in favour of
Copernicus is that his doctrine better explains the phenomena
of precession and why the sun is now forty thousand stadia
nearer to us tha it was held to be in times past. The passages
of Scripture ascribing motion to the sun really refer to the
motion of the earth which " by way of speech is assigned to
the sun, even by Copernicus himself and those who are his
followers."

---

[1] The wise rulers of Rome did that in 1633, so Protestants have no right to
blame them.

[2] Prowe, *Nic. Coppernicus*, I. 2, p. 232.

[3] This part of the commentary is translated by Th. Salusbury, *Math. Collections and Translations*, T. I. pp. 468–470 (London, 1661, fol.). I have not
seen the original. Diego de Stúñiga was an Augustinian monk and a Doctor of
Divinity of the University of Toledo ; he must not be mixed up with his name-
sake who wrote against Erasmus and who died in 1530.

The explanation of Stúñiga was not accepted by the defenders of literal interpretation; in fact the very passage he commented on was always quoted to show that the earth *has* a particular place of its own[1], and all the well-meaning writer got for his trouble was that his book was placed on the *Index* in the following century. It would be useless to set forth at length the arguments against the motion of the earth, borrowed from the Scriptures and made use of in the sixteenth and seventeenth centuries; they are much the same as those which the Fathers of the Church had used a thousand years earlier in defence of the Babylonian system of the world[2]. As yet the Church abstained, however, from taking any action to put down the new doctrine, probably because the latter was thought to be a mere academic subject for idle disputation, and not a matter which was likely to be taken up seriously by any sane person, so that there was no fear of this pernicious idea spreading any further on account of the supposed insuperable physical objections against the assumption of any motion of the earth. For some of these objections were awkward enough to answer in those days, while others rested on perfectly unproved assumptions[3]. The frequently repeated argument, that if the diurnal motion of the earth carried the air along with the earth, it would cause a terribly high wind, might seem futile to a man like Tycho Brahe[4]; still it was currently believed, and was even neatly expressed by George Buchanan, the

---

[1] Riccioli, *Alm. nov.* II. p. 480.

[2] They are given by Riccioli, l. c. pp. 479–495.

[3] We must here pass over the objections raised by people utterly ignorant of the rudiments of astronomy. A very glaring example of this kind of writing is F. Ingoli Ravennatis, *De situ et quiete terræ contra Copernici Systema Disputatio*, 1616, printed in Favaro's *Nuovi studi Galileani*, pp. 165–172, and in the national edition of Galileo's works, T. v. pp. 403–412. One of his arguments is that if the sun were in the centre it ought to have a greater parallax than the moon, because the farther bodies are from the Primum Mobile, in which their places are marked and "ubi notantur parallaxes," the greater is their parallax. The heaviest must be in the centre, for when wheat is sifted lumps of earth which are in the wheat are by the circular motion of the sieve brought into the middle, &c., &c. Kepler wrote him an answer, printed by Favaro, l. c. pp. 173–184, and Galileo another in 1624, Ed. naz. T. VI. pp. 509–561.

[4] *Epist. astr.* p. 74. Kepler in a letter to Fabricius says that this objection is like the wind, "nihil efficit nisi strepitum" (*Op.* III. p. 462).

well-known Scottish scholar and statesman, in his Latin poem
on the sphere[1]:

> Terra igitur nec sponte sua secedere mundi
> E media regione potest, nec viribus ullis
> In latus impelli potis est, tollive, premive:
> Cum sit nulla usquam tantæ violentia molis,
> Moliri quæ sede sua per vimque movere
> Congeriem terra possit. Nec rursus in orbem
> Se rotat, ut veterum falso pars magna Sophorum
> Crediderat, Samii jurata in verba magistri.
>
> .   .   .   .
>
> Ergo tam celeri motu si concita tellus
> Iret in occasum, rursusque rediret in orbem,
> Cuncta simul quateret secum, vastoque fragore
> Templa, ædes, miseris etiam cum civibus urbes
> Opprimeret subitæ strages inopina ruinæ.
> Ipsæ etiam volucres tranantes aera leni
> Remigio alarum, celeri vertigine terræ
> Abreptas gemerent sylvas, nidosque tenella
> Cum sobole et chara forsan cum conjuge, nec se
> Auderet zephiro solus committere turtur,
> Ne procul ablatos terra fugiente Hymenæos
> Et viduum longo luctu defleret amorem.

But there was another difficulty to which the most determined adherents of Copernicus could not give a satisfactory
answer. If the earth moved, it was said, an arrow shot vertically
upwards could never fall straight down again, but ought to fall
on a spot at a distance of many miles. For when it was answered
that the air was also moving and carried the arrow along with it,
the objector would reply that, even supposing the air moved (and
what was there to move it?), it would move much more slowly
than the earth, being very different in substance and quality,
so that the arrow would still be left behind, and for this reason
a man in a very high tower would always feel a strong wind.
Again, if a man should drop a stone from the top of a tower, it
could never reach the ground at a spot perpendicularly under
the place from which it was dropped. For if the air was moving

---

[1] *De Sphera*, lib. I. v. 320 et seq. (*Opera Omnia*, Lugd. Bat. 1725, vol. II.).
The poem was not published till after Buchanan's death in 1582. He probably
met Tycho Brahe in 1571, when he was sent to Denmark to try to persuade the
Danish Government to surrender Bothwell. See my *Tycho Brahe*, p. 100.

with the earth, it could not carry a heavy stone along at the
same rate; and even if the stone should move in a circle by its
own nature, like the earth, it could not move as fast as the
earth, because the latter is in its natural place, while the stone
is trying to reach its natural place by falling, as Copernicus
himself had acknowledged[1].

To this last objection a Copernican before Galileo could
generally not give the proper answer, for both sides were equally
ignorant as regards the laws of motion, and both sides had
recourse to the same kind of foggy talk as to what was natural
or not natural for a body to do, which Aristotle had indulged in
nearly two thousand years earlier. Tycho Brahe maintained
that this argument about the falling stone was unanswerable[2],
and it is very curious that not even he, who taught astronomers
to seek for the laws of planetary motion through observations,
seems to have thought of making the simple experiment of
dropping a pebble from the top of the mast of a swiftly moving
vessel. He might have done it scores of times while passing
backwards and forwards between his island and the shores of
the Sound; yet he boldly asserts that a bullet fired vertically
upwards from the deck of a moving ship will not fall down
straight again, as some people believe, but the faster the ship
moves the greater the distance will be[3]. Though Tycho was
fond of talking about the connection between heavenly and
earthly (i.e. chemical) research, it did not occur to him to verify
the truth of his assertion by experiment.

It is therefore not to be wondered at, that with the exception of the few men we have mentioned, nobody accepted the
Copernican doctrine as true during the first half century after
1543, though everybody used the Prutenic Tables. Some writers
even made use of rather strong language when alluding to it.
Thus Maurolico in a little book, *De Sphaera*, says that Copernicus,
who made the earth twirl round, should be tolerated, and is
more worthy of a whip than of a refutation. But perhaps he

[1] Lib. I. cap. 8, p. 23; see above, p. 324.
[2] *Epist. astron.* pp. 167 and 188.
[3] Ibid. p. 190. Galileo seems to have been the first who actually made the
experiment which Tycho neglected; see his reply to Ingoli, Ed. naz. vol. VI.
p. 509.

only means that as Copernicus made the earth spin round like
a child's top, he ought to have a whip given him to keep his toy
going[1]. Caspar Peucer, Professor at Wittenberg and son-in-law
of Melanchthon (by whom he had perhaps been prejudiced), says
in a book on astronomical hypotheses[2] "accommodated to the
observations of N. Copernicus and the tables founded by him,"
that he passes over the hypothesis of Copernicus, lest beginners
should be offended and disturbed thereby; and, in another place,
that the absurdity of Copernicus, so far from the truth, is
offensive.   Even Nicolas Müller, Professor in Groningen, who
in 1617 edited the third edition of the work of Copernicus, had
in two previous publications declared that he had never yet met
with any valid reason for rejecting the old system, which was
also supported by Scripture, and that he might more willingly
have followed Copernicus if he had left the earth in the middle
of the world and only given it a diurnal motion[3].   It is strange
that Müller and not a few others should have been willing to
concede the rotation of the earth but felt bound to reject its
orbital motion; for the alleged physical objections applied
equally to each.   But the support supposed to be given by
Scripture to the old system had possibly a good deal to do with
this distinction being made.   The reason given by Copernicans
for rejecting the diurnal rotation of the stars, that they would
have to travel with an absolutely incredible velocity, was after
all not a very weighty one.   It must be remembered that the
universe in those days seemed to be of quite moderate dimensions.   The distance of the sun, and consequently also the
distances of all the planets, were supposed to be twenty times
smaller than they are in reality; and even the most accurate
observer of the age, Tycho Brahe, would hardly have detected
an annual parallax of one minute in a star, if it had existed, so
that even the speed of a fixed star would not be so very
enormous.   In a text-book by an Englishman, Thomas Lydiat,
entitled *Prælectio astronomica*, the very sensible remark is made,

---

[1] As suggested by De Morgan, *Budget of Paradoxes*, p. 72.   Otherwise there
seems to be no sense in " tolerating " him.

[2] *Hypotheses astronomicæ*, Wittenberg, 1571.

[3] Riccioli, *Alm. nov.* II. p. 489; *Kepleri Opera*, VIII. p. 566.

that if Copernicus and his partisans had never seen any faster
motion than the flight of a bird, the speed of an arrow or a
cannon-ball would have seemed to them equally incredible[1].
This remark is almost the only original one in the book, which
is perfectly medieval in every respect, even in believing in the
water above the firmament, which is said to move along with
the latter, because water is never at rest except in a concave
vessel[2].

For many years after the publication of the work of
Copernicus astronomy was making no progress; opinion stood
against opinion, while planetary theory had hardly made any
advance since Ptolemy. The first warning that astronomy
would have to be cultivated in a totally different manner came
from the French mathematician, Pierre de la Ramée, or Petrus
Ramus, Professor of Philosophy and Rhetoric at the Collège
Royal at Paris, who had from his youth been a determined
opponent of the Aristotelean natural philosophy. He published
at Basle, in 1569, *Scholarum mathematicarum libri xxxi*, the
three first books of which contain a history of mathematics.
Dealing with the application of mathematics to astronomy, he
says in the second book that astronomy is nothing but an
arithmetical counting up of the celestial motions and a geometrical measuring of the dimensions of the celestial spheres.
Astronomy is involved and impeded by the many hypotheses
from which it can be liberated by mathematics. The Chaldeans
and Egyptians had possessed an astronomy without hypotheses,
but founded on observations; then Eudoxus invented the hypotheses of revolving spheres, which Aristotle and Kalippus

---

[1] *Prælectio astronomica de natura coeli & conditionibus elementorum...Item
Disquisitio de origine fontium*, London, 1605, p. 59. Only 75 pp. (out of 200)
are about astronomy.

[2] p. 51 sq. Other points of interest in the book are : The star of 1572 and
the comet of 1577 prove that changes can occur in the ethereal world just as
in the sublunary world, therefore there is no essential difference between them
(pp. 23–28). The stars are not attached to solid orbs but hang in the liquid
ether, which is the strongest kind of fire (pp. 28–39). The motion of the
planets towards the east is only a lagging behind in the common motion
towards the west; the turning of the sun at the solstices is caused by the denser
air (p. 60 sq.). Venus digresses more from the sun than Mercury owing to its
larger body being more obnoxious to the spreading of the solar rays (p. 73).

improved, while the Pythagoreans, in opposition to them, introduced epicycles and excentrics. Lately Copernicus, an astronomer not only comparable to the ancients but much to be admired, rejected all the old hypotheses and revived those admirable ones which demonstrate astronomy by the motion not of the stars but of the earth. If only Copernicus had proceeded without hypotheses, for it would have been easier to work out an astronomy corresponding to the true state of the stars than, like a giant, to move the earth; and it was to be hoped that some distinguished German philosopher would arise and found a new astronomy on careful observations by means of logic and mathematics, discarding all the notions of the ancients.

Travelling in Germany in 1569 or the beginning of 1570, Ramus met at Augsburg a young Dane, Tycho Brahe (1546–1601), who had already attracted some attention in Germany as an assiduous observer of the stars. In the course of a lengthy conversation, Ramus explained his views to Tycho (who has left us an account of the interview[1]), but the young man answered that astronomy without an hypothesis was an impossibility, for though the science must depend on numerical data and measures, the motions of the stars could only be represented by circles and other figures. Tycho evidently never saw that Ramus objected to the fundamental assumption of all previous systems, that the planets could only move in circular orbits or in orbits resulting from combinations of circles, and to the perfectly arbitrary assumption of Ptolemy that the centre of the deferent was half-way between the earth and the centre of the equant. Ramus wanted a man to start absolutely *de novo* and to find what kind of orbit would best satisfy a large number of observed places of a planet; and this was actually done thirty years later by Kepler. But Tycho had already, long before this meeting, perceived that the first desideratum of astronomy at that time was a long-continued course of observations of the planets; and he doubtless expressed his concurrence in the views expressed by Ramus in his recently-published book as to the necessity of founding astronomy on observations. This principle he kept in view during his whole life, and he did not confine himself to

[1] In a letter to Rothmann, *Epist. astr.* p. 60.

observing, but continued until his death to deduce important
results from his observations, a work which was brilliantly
carried on afterwards by his great successor, Kepler.   In this
place we have, however, specially to consider his attitude to the
burning question of the day and state his reasons for placing
himself in opposition to Copernicus, a man whose scientific
worth he was more competent to judge than any of his contemporaries, and of whom he always spoke with the greatest
veneration.

The difficulty of reconciling the motion of the earth with
certain passages of Scripture was with Tycho a real objection to
the new system[1].   But there were plenty of other objections.
First there were the difficulties of conceiving "the heavy and
sluggish earth" moving through space[2], and the immensity of
the distance which had to be assumed between the orbit of
Saturn and the fixed stars, since Tycho had found no trace
of annual parallax in the latter[3].   Then he fully participated in
the current belief that a stone falling from a tower would fall
very far from the foot of the tower if the earth either rotated or
travelled round the sun[4].   The "triple motion" of the earth
assumed by Copernicus seemed also difficult to conceive.   But
Tycho's chief objection, which he appears to have been the first
to put forward, was the following.   Until the invention of the
telescope had revealed the fact that the fixed stars, unlike the
planets, appear as mere luminous points and not as discs, the
most exaggerated ideas were prevalent as to their apparent
diameters, as we have already mentioned when dealing with
Arabian astronomy.   Tycho assumes the diameters to be: first
magnitude 120″, second 90″, third 65″, fourth 45″, fifth 30″,

---

[1] *Epist.* p. 148.   He says here that Moses must have known a good deal
about astronomy, since he calls the moon the lesser light, though the apparent
diameters of sun and moon are about equal.   The prophets must also be
assumed to have known more about astronomy than other people of their
time did.

[2] *De Mundi æth. rec. phæn.* p. 186.

[3] Distance from stars to Saturn 700 times the distance from the sun to
Saturn, *Epist.* p. 167.   Letter to Kepler, Dec. 1599, *Kepleri Opera*, VIII. p. 717.

[4] *Epist.* p. 167.   We have already mentioned his opinion about a shot fired
upwards from a ship in motion.

sixth 20″[1]. Now if the annual parallax of a star of the third magnitude was as great as one minute, the star would be as large as the annual orbit of the earth round the sun. And how big would the brightest stars have to be, and how enormously large would they be, if the annual parallax was still smaller?

All these objections Tycho set forth in various letters to Rothmann, but the latter did not consider them convincing. He replied with great common sense to the arguments based on the literal interpretation of Scripture, asking whether we ought perhaps to believe in the existence of the windows of heaven, mentioned in the account of the Deluge. As to the earth being too heavy to be in motion, he refers to the idea of Copernicus that gravity is nothing but a tendency implanted in all particles to group themselves into spherical bodies, and as the earth is in any case freely suspended in the ether like the planets, why should it not be in motion as the planets are[2]? Neither does the falling stone disturb him, and he replies that the stone as well as the tower participate in the motion of the earth both before and during the fall of the stone[3]. And why should it be absurd to assume an immense space to exist between the orbit of Saturn and the fixed stars, or to let a star of the third magnitude be as large as the earth's orbit? Are we to put limits to the Divine wisdom and power? As to the third motion, he points out that the earth is not attached to a solid orb which carries it round, but it is unsupported, and its axis simply keeps at the same angle with the axis of the zodiac. There is therefore no necessity for assuming a third motion; only the diurnal and the annual will suffice, and he acknowledges that Copernicus on this point has expressed himself rather obscurely[4].

It is very creditable to Tycho Brahe that he printed all

[1] *Astr. inst. Progymn.* pp. 481–482.

[2] *Epist. astr.* p. 129.

[3] Compare Kepler's arguments (*De stella Martis*, Introduction, and letter to David Fabricius, *Opera*, III. pp. 152 and 458. He says the case would be different if the stone were at a distance comparable to the earth's diameter. Gilbert (*De Magnete*, VI. 5) expresses himself less clearly: heavy bodies are united to the earth by their heaviness and advance with it in the general movement; the motion of a falling body is not a composite one, the resultant of a motion of coacervation and a circular motion, but is simple and direct.

[4] *Epist.* pp. 185–187.

these replies to his own attack on the Copernican system; but he takes care to weaken the effect by adding a note filling five closely printed pages[1], in which he states that he pleaded his cause so well with Rothmann during a visit of a month's duration which the latter paid him in 1590, that this generally very obstinate man began to waver, and finally declared himself defeated. Tycho repeats his own arguments at length in this note, but there is no need to repeat them here; with regard to the fall of a stone or the range of a cannon fired off in succession towards the east and towards the west, he refuses to believe that a body can be endowed with two motions at the same time (for one would disturb the other), and that the thin air should be able to carry a heavy stone along in its alleged circular motion—which Rothmann, by the way, had not said it did.

But though Tycho maintained that the earth was at rest, he did not accept the Ptolemaic system. In three letters written in the years 1587 to 1589 he states that he was induced to give it up when he found from morning and evening observations of Mars at opposition (between November 1582 and April 1583) that this planet was nearer to the earth than the sun was, while according to Ptolemy it ought to be more distant than the sun[2]. Now, Tycho did not determine the solar parallax anew (as he did every other astronomical constant) but accepted the ancient value of 3'; did he then find a parallax of Mars greater than 3'? He did not, for Kepler was unable to find any sensible parallax from Tycho's observations; but to his surprise he found from Tycho's manuscripts that some pupil or assistant (as he suggests, by a misunderstanding) had computed the parallax of Mars from the planetary elements of Copernicus and found it greater than that of the sun[3]. That Tycho should have fallen into the error of believing that his observations gave a larger parallax of Mars than of the sun becomes the more remarkable,

[1] *Epist. astron.* pp. 188–192.

[2] *Epist. astron.* p. 42 (to the Landgrave); Weistritz, *Leben des T. v. Brahe*, I. p. 243 (letter to Peucer, 1588), and *Epist. astr.* p. 149 (to Rothmann).

[3] Kepler, *De stella Martis*, cap. XI. *Opera*, III. p. 219 and p. 474. In his *Progymn.* I. p. 414, Tycho says that the outer planets have scarcely perceptible parallaxes, but that he had found by an exquisite instrument that Mars was at opposition nearer than the sun. Comp. ibid. p. 661.

when we find that in 1584 he had declared that these very same observations gave a parallax very much smaller than that of the sun, which showed that the Copernican system was wrong[1]! Anyhow, he afterwards believed the reverse, and therefore rejected the Ptolemaic system, and (he adds in his letter of 1589 to Rothmann) he remarked that comets when in opposition did not become retrograde like the planets, for which reason he thought he had to reject the Copernican system also, so that there was nothing to do but to design a new one.

In the eighth chapter of his book on the comet of 1577 (where the parallax of Mars is not mentioned) Tycho describes his own system, which he says he had found "as if by inspiration" four years before the book was written, that is, in 1583[2]. The earth is the centre of the universe and the centre of the orbits of the moon and the sun, as well as of the sphere of the fixed stars, which latter revolves round it in twenty-four hours, carrying all the planets with it. The sun is the centre of the orbits of the five planets, of which Mercury and Venus move in orbits whose radii are smaller than that of the solar orbit, while the orbits of Mars, Jupiter, and Saturn encircle the earth. In order that the distance of Mars at opposition may be smaller than that of the sun, the semidiameter of the orbit of Mars is a little smaller than the diameter of the solar orbit, so that the two orbits intersect each other, but as they are only imaginary lines and not impenetrable spheres, there is nothing absurd in this.

This system is in reality absolutely identical with the system of Copernicus, and all computations of the places of planets are

---

[1] "Cum tamen longe minores fuisse, creberrimis, exquisitissimisque et sibi invicem correspondentibus observationibus eas deprehendimus; ut ob id tota Martis sphaera ulterius removeatur a nobis quam ipse Sol." Letter to Brucæus of Rostock, *T. B. et ad eum doct. vir. epist.* p. 76.

[2] In a letter dated the 31st Jan. 1576, Tycho's friend Joh. Pratensis asks him to instruct the writer about the hypotheses of Ptolemy and Copernicus, whether one of them is to be accepted, "an vero potius Ptolemaica assumptio sit castiganda, prout a te factum est, et Copernicea ad stabilitatem Terrae convertenda, uti etiam insinuasti tuoque sic nobis praeluxisti ingenio." Ibid. p. 20. Query, had Tycho really thought out his system before 1576? If so, why did he not say so later on? This letter only exists in a copy made for the purpose of publication during the last years of Tycho's life. Is it perhaps fictitious?

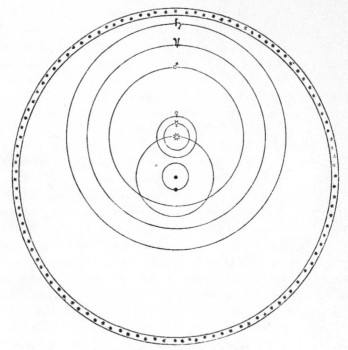

The Tychonic System of the World.

the same for the two systems[1].   As it leaves the earth at rest,
the Tychonic system might serve as a stepping-stone from the
Ptolemaic to the Copernican system, and one might have
expected it to have been proposed before the latter. It may
have occurred to Copernicus in his younger days, but if so, he
did not rest content with it, but proceeded at once to its logical
sequel, the heliocentric system.   The planetary theories of
Copernicus could, of course, be applied unaltered to the new
system, and it had been Tycho's intention, had he lived longer,
to have utilised his own observations to the preparation of new

---

[1] All the same, Tycho, who was very proud of his system, would not allow
it to be called a modification of that of Copernicus. He writes to Rothmann in
1589: " Occasionem vero has Hypotheses construendi non desumsi ex inversis
Copernianis, etsi tu tale quidpiam unquam cogitasti, mihi id, ut satis nosti, non
innotuit, neque simile quid unquam ex Rhetico vel Reinholdo colligere licuit."
After which he describes his observations of Mars in 1582. *Epist.* p. 149.

elements of the orbits in a great work to be called *Theatrum astronomicum*. In his *Progymnasmata* he only gives a sketch of the theory of Saturn for the purpose of finding the greatest distance of Saturn from the earth, adopting the "epicycli epicyclium" of Copernicus[1]. He thus makes out that the greatest distance of Saturn from the earth is 12,300 semidiameters of the earth, and as he objects to a great void between the orbit of Saturn and the fixed stars, he places these at a distance of 14,000, and the new star of 1572 at least at 13,000 semidiameters. This new star, which in so many ways determined the direction of Tycho's studies, caused him to speculate on the nature of the celestial bodies. He believed the star to have been formed of "celestial matter," not differing from that of which the other stars are composed, except that it was not of such perfection or solid composition as that forming the stars of permanent duration, which was the cause of its rapid dissolution. The matter of which it was formed was taken from the Milky Way, close to the edge of which the star was situated[2]. Before the invention of the telescope it was indeed very natural to assume the Milky Way to be of a nebulous character, and Tycho's idea is therefore not discreditable to him. He did not think the substance of the stars to be the same as that of our earth, but rather to stand in the same relation to it as the soul to the body. In opposition to Rothmann, Tycho did not believe that the celestial space was filled with thin air[3].

The discussion of the motion of the comet of 1577 gave Tycho the opportunity of promulgating his system. During the who e of the Middle Ages the prevailing Aristotelean notion of the a mospheric origin and nature of comets had prevented proper attention being paid to these bodies, and Regiomontanus was the first to attempt to determine the distance of a comet. Owing to the want of good instruments he failed in his attempt, as he only made out that the comet of 1472 could not have a parallax greater than 6°. Though comets from that time began to be regularly observed, Tycho Brahe was the first to prove conclusively that comets have very small parallaxes and

---

[1] *Progymn.* I. p. 477. See above, Chapter XIII. p. 336.
[2] *Progymn.* I. p. 787 sq.   [3] *Epist.* p. 138.

are therefore much farther off than the moon, the orbit of which was still considered the limit of the elementary world[1]. The comet of 1577 was also the first of which an attempt was made to calculate the orbit; Tycho found as the result of his calculations that the comet moved round the sun in a circular orbit outside that of Venus, the greatest elongation from the sun being 60° and the motion retrograde. He was unable to represent the observed places by a uniform motion in this orbit and was obliged to assume an irregular motion, to account for which he remarks that an epicycle might be introduced, but as the inequality was only 5′, he did not deem it necessary to go so far in refining the theory of a transient body like a comet; besides, it is probable that comets, which only last a short time, do not move with the same regularity as the planets do. As an alternative he suggests that the figure of the orbit may not have been " exactly circular but somewhat oblong, like the figure commonly called oval[2]." This is certainly the first time that an astronomer suggested that a celestial body might move in an orbit differing from a circle, without distinctly saying that the curve was the resultant of several circular motions. Mästlin also worked out the orbit of this comet and, like Tycho, found a circle round the sun outside the orbit of Venus, but he accounted for the irregularity of the motion by introducing a small circle of libration, along the diameter of which the comet moved to and fro. Tycho did not approve of this idea because orbits were not really existing objects; but years afterwards, when he worked out his lunar theory, he found himself unable to do without this and similar expedients[3].

The book *De Mundi ætherei recentioribus phænomenis liber secundus* was ready from the press in 1588, and though not

[1] Cardan had already in 1550 concluded from the absence of parallax that comets could not be bodies in the atmosphere, but he does not appear to have stated how he had found this. Pingré, *Cométographie*, I. p. 70.

[2] " Sive igitur cometa hic noster non undequaque et exquisite rotundum ad solem circuitum sed aliquantulum oblongiorem, in modum figuræ quam Ovadam vulgo vocant, confecerit..." *De Mundi æth. rec. phæn.* p. 194.

[3] Ibid. p. 266. For further particulars about this comet, the tail of which Tycho believed to be turned away from Venus and not from the sun, see my *Tycho Brahe*, p. 158 sq.

regularly published till 1603, some copies were at once distributed to friends and correspondents. Not a few of these copies
are still in existence, having the original title-page and colophon
of 1588. In this way the Tychonic system of the world became
known at once, and a Scotchman, Duncan Liddel, lectured on it
at Rostock already in 1589 or 1590 and afterwards even claimed
to have found it independently himself[1]. But a worse competitor
appeared immediately in the person of Nicolai Reymers, called
Ursus, a native of Holstein. This person had been in the service
of a Danish nobleman, with whom he went to Hveen in 1584,
and less than two years afterwards he appeared at Cassel, where
he described as his own discovery a system exactly like Tycho's,
except that it admitted the rotation of the earth. The Landgrave of Hesse, an enthusiastic astronomer, was so pleased with
the idea, that he got his instrument-maker, the celebrated
mathematician Bürgi, to make a model of it. Tycho only heard
of this when his own book reached Cassel, but very soon afterwards a little book was published in which Reymers had set
forth the new system. The title is *Nicolai Raymari Vrsi Dith-
marsi Fundamentum Astronomicum*, Strassburg, 1588 (4 + 40 ff.,
4to, with two plates). Most of the book treats of trigonometry,
but the last chapter is "On observing the motions of the
planets, wherein about our new hypotheses," describing the
new system without once mentioning Tycho's name. The latter
concluded that Reymers had stolen the idea from him, but
when his accusation to this effect appeared in print in his
publish d correspondence with Rothmann, Reymers replied in
a very scurrilous book, *De astronomicis hypothesibus* (Prague,
1597), and the squabble went on till Reymers' death in 1600[2].

There is, however, not the least proof of the alleged plagiarism.
The idea of the Tychonic system was so obvious a corollary to
the Copernican system that it almost of necessity must have
occurred independently to several people ; and Reymers, who
was certainly an able mathematician, may very well have
thought of it himself. He cannot have been a believer in the
ordinary objections to the earth's motion, since he accepted the

---

[1] *Tycho Brahe*, pp. 137 and 181.
[2] For particulars see *Tycho Brahe*, pp. 183, 273, 288, 304.

earth's rotation; indeed, it is difficult to see why anyone who did this should object to the orbital motion, unless perhaps on theological grounds. From the diagram in Reymers' book we see that he did not perceive the necessity of letting (with Copernicus and Tycho) the distance of Mars at opposition be less than that of the sun, for he does not let the two orbits intersect each other.

Unlike Copernicus, Tycho had at his disposal a great mass of observations, made during many years according to a well-considered plan of following the sun, moon, and planets right round the heavens and not merely observing them occasionally at opposition or other interesting points of their orbits. Hereby he succeeded in making the first important step forward since Ptolemy as regards the motion of the moon, so that at his death all the great lunar perturbations were known with the one exception of the secular acceleration of the mean motion, which could only be discovered by the comparison of observations made in the course of centuries. The motion in longitude he represented in a manner different from that of Copernicus and agreeing better with the observed positions. He places the centre of the deferent (radius $= 1$) on a small circle with radius $0\cdot02174$, in the circumference of which the earth is placed, so that the centre of the deferent is in the earth in the syzygies and farthest from it in the quadratures. There are two epicycles with radii $0\cdot058$ and $0\cdot029$, the period in the former being the anomalistic month, and the moon moving in the latter twice as fast and in the opposite direction, so that at apogee the moon is $0\cdot029$ outside the deferent, at perigee $0\cdot087$ inside it. The effect of the two epicycles gives the maximum of the first inequality equal to $4° 59' 30''$, while the circle through the earth gives the second one $= 1° 14' 45''$, rather nearer the truth than Ptolemy's value. The third inequality or variation had been discovered by Tycho before he left Denmark and was announced by him in 1598, but he did not attempt to account for it by adding another epicycle. He merely let the centre of the first epicycle oscillate (librate) backwards and forwards on the deferent to the extent of $40'\cdot5$ on each side of its mean position, the latter moving along the deferent with the moon's

mean motion in anomaly, and the centre of the epicycle being
in its mean position at the syzygies and quadratures, and farthest
from it at the octants, the period of a complete libration being
half a synodical revolution[1]. At the same time Tycho's observations showed the existence of another inequality in longitude,
the fourth one, of which the solar year was the period, so that
the observed place was behind the computed one, while the sun
moved from perigee to apogee, and before it during the other
six months. This was remarked at the latest during Tycho's
stay at Wittenberg (between December, 1598, and the beginning
of May, 1599), but it was difficult to find a convenient way of
introducing this inequality in the already complicated theory.
As the period of the phenomenon was a year, Tycho (or rather
his disciple Longomontanus) ultimately allowed for it by correcting the equation of time, or rather using a value differing
from the ordinary one by $8^m 13^s$ multiplied by sine of the solar
anomaly, though this leaves $5'$ or $6'$ of the irregularity unaccounted for[2].

Tycho Brahe's discoveries as regards the lunar motion in
latitude were as important as those he made of inequalities
in longitude. Having first noticed when discussing his observations of the comet of 1577 that the value of the inclination
of the lunar orbit to the ecliptic adopted since the days of
Hipparchus ($5°$) was too small, the examination of all his
observations finally showed him that the inclination varied
between $4° 58' 30''$ and $5° 17' 30''$, while the retrograde motion
of the nodes was found not to be uniform, so that the true
places of the nodes were sometimes as much as $1° 46'$ before
or behind the mean ones. This inequality of the nodes had
not been detected by the ancients because it disappears at the
time of an eclipse, when the moon is both at the node and in
syzygy. Tycho explains this and the change of inclination by
assuming that the true pole of the lunar orbit describes a
circle with a radius of $9' 30''$ round the mean pole, so that the

---

[1] *Tycho Brahe*, p. 338. About Sédillot and Abu 'l Wefa see above,
Chapter XI. p. 252.

[2] Ibid. pp. 306 and 310.

inclination reaches its minimum at syzygy and its maximum at quadrature.

Tycho's numerous observations of the planets were destined in the hands of Kepler to put the finishing touch to the work of Copernicus by revealing the true nature of the planetary orbits. But he did not rest content with the mere accumulation of material, but had already in 1590 (or earlier) commenced to draw some conclusions from the comparison of his results with the tabular places of the planets. In that year Giovanni Antonio Magini of Bologna, a rather well-known astronomer in his day, wrote to Tycho that he suspected the excentricity of Mars to be periodically variable. In his reply Tycho stated that he had found this difficulty not only in the case of Mars but also to a less extent in the theories of the other planets, and that he wanted to observe oppositions of Mars all round the zodiac in order to investigate the phenomenon fully[1]. In a letter of 1591 to the Landgrave of Hesse, Tycho alludes to this again as "another inequality arising from the solar excentricity," and in a letter to Kepler of April 1, 1598, he goes further by saying that not only is the ratio of the semidiameters of planetary epicycles not as simple as imagined by Copernicus, but that the annual orbit of the earth (according to Copernicus) or the epicycle of Mars (according to Ptolemy) seemed to vary in size[2]. This was the first step towards the discovery of the elliptic orbit, and it was correctly interpreted by Kepler as proving that the excentricity of the solar orbit (which Tycho had found equal to 0·03584) was only half as great as hitherto supposed, so that the motion was not simply uniform with regard to the centre of the orbit, but with regard to a *punctum æquans* as in the Ptolemaic theory of the other planets[3]. Observations of the sun alone could never have revealed the insufficiency of the simple excentric circle. During the last year of his life, on the completion of his lunar theory, Tycho had commenced to investigate

---

[1] *Carteggio inedito di Ticone Brahe...con G. A. Magini*, Bologna, 1886, pp. 393 and 397.

[2] *Epist. astron.* p. 206; *Astr. inst. Mechanica*, fol. G 3 verso; *Kepleri Opera*, I. p. 44, III. p. 267.

[3] As regards the motion of the sun we may mention that Tycho found the longitude of the apogee = 95° 30′ with an annual motion of 45″.

the motions of the planets, in which work Kepler became associated with him, but in October, 1601, Tycho's death set Kepler free to prosecute the work in his own way.

Though Tycho had rejected the motion of the earth, he had, mathematically speaking, adopted the system of Copernicus, and by proving comets to be celestial bodies he had finally put an end to the idea of solid spheres, whereby he greatly increased the chance of success of the new system. In his writings Kepler repeatedly claims for Tycho the merit of having " destroyed the reality of the orbs[1]." Another ancient error which Tycho practically abolished was the belief in the irregular motion of the equinoxes, which he showed to have been caused solely by errors of observation[2]. Though Kepler was rather inclined to admit some slight irregularity in the amount of annual precession, trepidation with its cumbersome machinery may now be said to have disappeared from the history of astronomy.

[1] For instance, *De stella Martis*, cap. XXXIII. *Opera*, III. p. 301.

[2] When the first chapter of the *Progymnasmata* was written (in 1588) Tycho must have believed in the irregularity, since he attributed to it the different values of the length of the year found from observations made at different epochs (p. 38). Later on he saw more clearly how errors of observation would produce such discrepancies (p. 253 sq.).

CHAPTER XV.

KEPLER.

In January, 1599, Mästlin, having heard from his former pupil, Johann Kepler, of the difficulties which Tycho Brahe had encountered in determining the excentricities of the planets, wrote in reply that Tycho had hardly left a shadow of what had hitherto been taken for astronomical science, and that only one thing was certain, which was that mankind knew nothing of astronomical matters[1].

The great practical astronomer had indeed thoroughly shown the insufficiency of all previous theories, but he had at the same time increased the accuracy of observed positions so vastly, that it would now be possible to produce a satisfactory theory and, better still, to determine the actual orbit in space in which each planet was travelling, a feat never yet attained. The material for the investigation was ready, thanks to Tycho, and the mathematician to make use of it was also ready; it was the very man to whom Mästlin had addressed those despairing words and who had already made a very promising début in the scientific world.

Kepler was born on December 27, 1571, in Würtemberg and studied from 1589 at the University of Tübingen, where he through Mästlin became acquainted with the doctrine of Copernicus and convinced himself that it represented the true system of the world. He had originally intended to enter the Church, but as Lutheran divinity or rather the very narrow-minded spirit then prevailing among its ministers was not to his taste, he accepted in 1594 the post of "provincial

---

[1] *Kepleri Opera*, I. p. 48.

mathematician" of Styria, and from thenceforth devoted his life to science. Already in 1596 appeared his first great work, which he, feeling that it was only a forerunner of still greater works, entitled *Prodromus Dissertationum Cosmographicarum continens Mysterium Cosmographicum*. Though it does not reveal the secret of the arrangement of the planetary orbits, as its author fondly hoped it did, the book contains Kepler's first great discovery. The reasons for abandoning the Ptolemaic in favour of the Copernican system are set forth in the first chapter with remarkable lucidity. By two very instructive diagrams he shows that the Ptolemaic epicycles of the outer planets are seen exactly under the same angle from the earth as the orbit of the earth is from a point in each of the outer planetary orbits, and he shows how this explains why Mars has an epicycle of such enormous size, while that of Jupiter is much smaller and that of Saturn smaller still, though their excentrics are much larger than that of Mars. The Ptolemaic system could assign no cause for this curious arrangement, nor for the strange fact that the three planets when in opposition to the sun should be in the perigees of their epicycles. Neither could it explain why the periods of the inner planets in their excentrics should be equal to that of the sun, nor give any reason why the sun and moon never became retrograde. All these facts are so simply explained by the doctrine of the earth's annual motion, while Copernicus is also able to account for p cession without requiring " that monstrous, huge and starless ninth sphere of the Alphonsines." Certainly, it is difficult to see how anyone could read this chapter and still remain an adherent of the Ptolemaic system.

What Kepler aimed at throughout his whole life was to find a law binding the members of the solar system together, as regards the distribution of their orbits through space and their motions, knowing which law he expected it would be possible to compute all the particulars about any planet if the elements of one orbit were known. The first instalment is given in the *Mysterium Cosmographicum*, which deals with the problem of finding the law connecting the relative distances of the planets. In the preface he tells the reader how he was led to what he supposes to be a great discovery. He thought there must be a reason why the number, distances and velocities of the moving bodies have the values which had been found by observation; and the hope of finding it was strengthened by the manner in which the resting things, the sun, the fixed stars and the intermediate space corresponded to God the Father, the Son and the Holy Ghost. He tried whether one sphere might be twice, three times, four times as great as another; he tried to insert a planet between Mars and Jupiter, and another between Mercury and Venus (supposed to be too small to be perceived by us), and when he even then failed to find some simple ratio between the distances from the sun, he tried whether these were proportional to some trigonometrical function. Chance led him at last to seek the law of the distances by geometry. While describing in the course of a lecture (on the 9/19 July, 1595) the cycles of the great conjunctions of planets and how the conjunctions pass from one "trigon" of the zodiac to another, the diagram he had drawn to illustrate this[1] brought the five regular solids to his mind, and it struck him that these and not plane figures would be properly belonging to the orbs of space (*inter solidos orbes*). Between the six planetary spheres there are five intervals, and adopting for the semidiameters of the spheres the values given by Copernicus, Kepler found that the five solids fitted between the spheres in the following order :

Saturn,
Cube,
Jupiter,
Tetrahedron,
Mars,
Dodecahedron,
Earth,
Icosahedron,
Venus,
Octahedron,
Mercury.

[1] *Myst. Cosmogr.*, præfatio (*Opera*, I. p. 108) ; about trigons see *Tycho Brahe*, p. 49.

The sphere of Saturn is circumscribed to a cube in which the sphere of Jupiter is inscribed; the latter is circumscribed to the tetrahedron, and so on. But as the orbits of the planets are not concentric but excentric circles, it became necessary (with the Arabs and Peurbach) to give to each sphere a thickness sufficient to afford room for the excentric orbit between the inner and outer surface. In the Middle Ages, as we have seen, the outer surface of one sphere could be assumed to touch the inner surface of the sphere next outside it, because the Ptolemaic system afforded no clue to the relative distances of the planets. But in the Copernican system the relative dimensions of the spheres cannot be arbitrarily chosen, they are determinate quantities, leaving plenty of room between the spheres, and the question was now: How far did the dimensions of the spheres, resulting from the distances and excentricities according to Copernicus, fit the dimensions of the five regular solids computed from them, so that the inner surface of a sphere coincided with the sphere circumscribed to the solid next below, and the outer surface with the inscribed sphere of the solid next above? The following table shows the result of Kepler's computations[1]:—

Semidiameter of inner surface = 1000	Semidiameter of outer surface	
	Computed	Acc. to Copernicus
Saturn	Jupiter 577	635
Jupiter	Mars 333	333
Mars	Earth 795	757
Earth	Venus 795	794
Venus	Mercury 577 or 707	723

The second value for Mercury is the semidiameter of the circle inscribed in the square formed by the four middle sides of the octahedron. If the thickness of the earth-sphere be

[1] Ibid. cap. xiv. *Op.* i. p. 151. It must not be forgotten that to Kepler these spheres were only mathematical conceptions, not actually existing bodies.

increased by including the lunar orbit, the figures in the last
column become for Venus 847 and for the earth 801. The
agreement between the computed values and those of Copernicus is fairly satisfactory except in the case of Jupiter, " at
which nobody will wonder, considering the great distance."
Kepler adds that it is easy to see how great would have
been the difference if the arrangement had been contrary
to the nature of the heavens, that is if God had not at the
Creation had these proportions in view, for the agreement
cannot be accidental. There must be a reason for everything,
and Kepler is quite ready to explain why the five regular
solids have been arranged in this particular order. They are
of two kinds, primary (cube, tetrahedron, dodecahedron) and
secondary (icosahedron and octahedron), differing in various
ways. The earth as the dwelling-place of man created in
the image of God was worthy to be placed between the
two kinds of solids; the cube is the outermost because it is
the most important, being the only one generated by its base
and indicating at its angles the three dimensions of space. For
the order of the other bodies he gives a great many reasons,
one more fantastic than the other[1]. But we must pass over
all these curious details as well as over his ninth chapter,
in which the astrological qualities of the five planets are
derived from the nature of the five solids.

Though Kepler's solution of the "cosmographic mystery"
has turned out to be a failure, it was natural enough that
he should have commenced his work by looking for some
relation between the distances of the planets from the sun,
and it is rather curious that he did not stumble on the series
erroneously known as Bode's law. Perhaps he might have
found it if he had not so early become enamoured of the
five solids, and he remained true to his first celestial love
all his life[2].

The agreement between the theory and the numerical data
of Copernicus was not perfect, and the next question for

---

[1] Chapters III.–VIII. *Op.* I. p. 127 sq.

[2] He published a second edition of the book in 1621 unaltered, but with
notes added to each chapter.

Kepler to consider was now how it could be improved. He reminds the reader[1] that the work of Copernicus was not cosmographic but astronomical; that is, it was of little importance to him whether he erred a little as to the true proportion of the spheres, if he could only by the observations find figures suitable for computing the motion and places of the planets. There is therefore nothing to prevent anybody from correcting his figures, so long as the equations of time suffer little or no change. What chiefly interested Kepler in the investigation he had in hand were the excentricities, on which the thickness of the spheres depended. It struck him now that although Copernicus beyond a doubt had placed the sun in the centre of the universe, still "as an aid to calculation and in order not to confuse the reader by diverging too much from Ptolemy" he had referred everything not to the centre of the sun but to the centre of the earth's orbit. Through this point therefore not only the line of nodes of each planet, but also the line of apsides pass in the theory of Copernicus, so that the excentricities are reckoned from a point, the distance of which from the sun measures the earth's excentricity. To follow Copernicus in this matter therefore means to give the earth no excentricity and its sphere no thickness, so that the centres of the faces of the dodecahedron and the vertices of the icosahedron fall in the same spherical surface, reducing the dimensions of the system more than observations would allow. Kepler consulted Mästlin, who willingly undertook to calculate the changes which the adoption of the sun as the centre would entail in the data of Copernicus. Naturally the changes turned out to be very considerable; thus the longitude of the aphelion of Venus was found to differ by about three signs of the zodiac (90°) from the apogee, while the new distance of Saturn differed from the old one by the whole amount of the earth's excentricity.

Kepler next gives a table of the annual parallaxes of the planets in aphelion first (1) computed from his theory, excluding the lunar orbit from the earth-sphere, next (2) according to the distances from the sun (Copernicus), and thirdly (3) computed from his theory, increasing the earth-sphere by the

[1] Cap. xv. *Opera*, I. p. 153.

lunar orbit. The differences are very considerable[1], and the positions of planets calculated from the new theory would therefore differ materially from those calculated by the Prutenic tables. But this does not alarm Kepler as to the truth of his theory. In a masterly written chapter[2] he discusses the shortcomings of the Copernican theory and the Prutenic tables, which often differed several degrees from the observed pl᠎ es of the planets, and he shows in particular that the excentri᠎ ties given by Copernicus are of no value. Copernicus believed that the excentricities of Mars and Venus had changed, whereas it turned out on referring them to the sun that there was no change in them. Mästlin called Kepler's attention to the utterances of Copernicus reported by Rheticus, which showed how well aware the great master had been of the insufficiency of the data on which he had built, and which he attributed to three causes : first, that some of the observations of the ancients had not been honestly reported but had been modified to suit their theories; secondly, that the star-places of the ancients might be as much as 10′ in error; and thirdly, that there were no comparatively recent observations extant such as those Ptolemy had had at his disposal. Kepler therefore calmly awaited the judgment of astronomers.

Finally Kepler endeavours to find "the proportions of the motions to the orbits[3]." As the periods of revolution are not proportional to the distances from the sun we must either assume that the "animæ motrices" more distant from the sun are feebler, or that there is only one anima motrix in the centre of all the orbits, that is, in the sun, which acts more strongly on the nearer bodies than on the more distant ones. He decides in favour of the latter assumption. He considers it probable that this force is inversely proportional to the circle over which it has to be spread, so that it diminishes as the

---

[1] Especially for Mars 40° 9′, 37° 22′, 37° 52′, and Venus 49° 36′, 47° 51′, 45° 33′. End of cap. xv. *Op.* i. p. 157. In his book on Mars, cap. xxix. (*Op.* iii. p. 291), Kepler mentions that the figures under the third supposition when the earth's excentricity is halved come very close to the truth.

[2] Cap. xviii. : " De discordia προσθαφαιρεσεων ex corporibus a Copernicanis in genere et de astronomiæ subtilitate." *Opera*, i. pp. 164–168.

[3] Cap. xx. *Op.* i. p. 173.

distance increases. At the same time the period increases with
the length of the circumference, "therefore the greater distance
from the sun acts twice to increase the period, and conversely
half the increase of period is proportional to the increase of
distance." For instance, the period of Mercury is 88 days and
that of Venus is $224\frac{2}{3}$ days, so that half the increase of period
is $68\frac{1}{3}$; therefore $88 : 156\frac{1}{3} :: $ distance of Mercury : distance of
Venus. Starting from Saturn, Kepler finds the following ratios
of distances :

				Copernicus
Jupiter	:	Saturn	0·574	0·572
Mars	:	Jupiter	0·274	0·290
Earth	:	Mars	0·694	0·658
Venus	:	Earth	0·762	0·719
Mercury	:	Venus	0·563	0·500

"We have come nearer to the truth," says Kepler. But
twenty-two years were to pass before he found the true law.
It is interesting to notice that already in 1596 he had
recognized that the motion of the planets must be controlled
or caused by a force emanating from the sun, and that he
already then made the erroneous assumption from which he
never swerved, that the effect of this force is inversely proportional to the distance from the sun.

Although the principal idea of the *Mysterium Cosmographicum* was erroneous, we owe a great debt of gratitude to that
work, since it represents the first step in cleansing the Copernican system from the remnants of Ptolemaic theory which
still clung to it. The greatest desire of Kepler was now to
obtain more correct values of the mean distances and excentricities in order to prove his theory to be absolutely true, and
the only place in the world where this information could be got
was in the observatory of Tycho Brahe. The great distance of
Gratz from Denmark might have deterred Kepler from joining
Tycho on his island, but fortunately for the advance of science
Tycho, who had quarrelled with many influential people in
Denmark, and possibly feared that his great treasure of
observations might be taken from him on the plea that it
had been gathered at the public expense and therefore was

public property[1], left Denmark in 1597 and settled in Bohemia two years later. Driven from Styria by religious persecution, Kepler went to Prague in January, 1600, and in the following year was appointed collaborator to Tycho, whom he succeeded in October, 1601, as Imperial mathematician. Though he ॒ s up to August, 1601, frequently interrupted by journeys to Styria to settle his private affairs, and by illness, he soon began to make good progress in investigating the motion of the most troublesome planet.

When Kepler joined Tycho at the Castle of Benatky in February, 1600, Mars had just been in opposition to the sun, and a table of the oppositions observed since 1580 had been prepared and a theory had been worked out which represented the longitudes in opposition very well, the remaining errors being only $2'$[2]. But the latitudes and the annual parallaxes could not be represented by the theory, and Kepler began to consider whether the theory might not after all be wrong, though it represented the longitudes in opposition so well. There were several particulars in the theory, to which Kepler objected. In the first place Tycho had like Copernicus referred the motion of the planet to the mean place of the sun. This principle Kepler had rejected in his book, as it implied motion round a mathematical point instead of round the great body of the sun. But there was also a practical objection to the principle. From the observations at opposition the time had been deduced, when Mars differed $180°$ from the mean longitude of the sun, and the motion of the sun (or rather the earth) had therefore to be assumed as a known quantity. To a certain extent, therefore, the great advantage in using oppositions (that the observed longitudes were equal to the heliocentric longitudes) was lost, and the " first inequality " was not determined independently of the " second " one, caused by the motion of the earth or, in the Tychonic system, of the sun. In the case of Mars the longitude at " mean opposition " might differ more than $5°$ from that at true

[1] *Tycho Brahe*, p. 252.

[2] Kepler, *De stella Martis*, cap. VIII. *Opera*, III. 210. The semidiameter of the greater epicycle was 0·1638, that of the smaller 0·0378, or in Ptolemy's theory the excentricity of the equant = 0·2016.

opposition, a very serious difference. Eventually Kepler was
able to persuade Tycho and Longomontanus to adopt the
apparent place of the sun in the lunar theory, while his own
continued researches on Mars more and more showed him the
necessity of referring the planet's motion to the true place
of the sun. The other objection raised by Kepler to Tycho's
theory of Mars was that the annual orbit of the sun had
been assumed to be a simple excentric circle (as in the theories
of Ptolemy and Copernicus) with an excentricity $= 0.03584$.
In the *Mysterium Cosmographicum* Kepler had expressed the
opinion that all the planets, including the earth, would be
found to move exactly in the same manner. He now pointed
out to Tycho that the apparent alternate shrinkage and
expansion of the earth's (or the sun's) annual orbit, which
Tycho had found in 1591, was simply caused by the fact
that the motion in this orbit was not uniform with regard
to the centre, but with regard to a *punctum æquans* exactly as
in the planetary theories of Ptolemy. In this case it is easy to
see that the annual parallax or difference between the heliocentric and geocentric longitude of a planet will vary with
its position with regard to the earth's line of apsides[1]. If Mars
be in the prolongation of this line and be observed from two
points at equal distances on both sides of the line, then the
parallaxes will be equal, no matter where in that line the
punctum æquans be situated. But if Mars be about 90° from
the earth's apsides and be observed from the apsides or from
two points in mean anomalies α and $180° - \alpha$, the parallaxes will
not be equal unless the *punctum æquans* be in the centre of the
orbit, but they will differ more or less according as the earth is
nearer to or farther from its apsides. Tycho had apparently
suspected that this was the true explanation of the strange
phenomenon[2], but as he wished his book (*Progymnasmata*)
published without further delay, the bisection of the solar
excentricity was not introduced into it, but was merely alluded

[1] Kepler, *De stella Martis*, cap. xxii. *Op.* iii. p. 267.

[2] Kepler wrote in July, 1600, to Herwart von Hohenburg about the equant
of the sun : " Et hoc est quod Tycho quasi sub aenigmatis involucro (ut inter-
dum solet) ad me perscripserat de variabili quantitate orbis annui." *Opera*,
iii. p. 24.

to by Kepler in the appendix with which he wound up the book after Tycho's death[1].

Tycho Brahe died on October 24, 1601, and on his deathbed he begged Kepler to carry out the contemplated reform of theoretical astronomy on the basis of the Tychonic system instead of the Copernican. Though Kepler's work of reform eventually led to the firm establishment of the latter system, yet he conscientiously demonstrated the theory of Mars according to the three systems of Ptolemy, Tycho and Copernicus, remembering the last wish of the great practical astronomer[2], whose admirable foresight had provided an inexhaustible treasury of observations made under all conceivable conditions. Already before Tycho's death Kepler had made good progress with the work on Mars[3], and four years later it was finished. We shall now follow his investigations in the order in which he records them himself.

Having first shown that no certain conclusion could be drawn from Tycho's observations with regard to the horizontal parallax of Mars, except that it did not exceed 4′ and was probably very much smaller[4], Kepler proceeds to find those elements of the orbit which can be determined separately. The longitude of the ascending node he found by searching Tycho's ledgers for observations of the planet at times when it had no latitude and then calculating its heliocentric longitude by Tycho's theory. Six observations of this kind gave him the longitude of the ascending node $= 46\frac{1}{3}°$ [5]. The inclination of the orbit to the plane of the ecliptic he next determined by three different methods. First, by picking out observations of Mars at 90° from the nodes made at a time when the distance from the earth to Mars is equal to the distance of Mars from the sun, when the observed latitude is equal to the inclination. Secondly, by taking the planet at a time when it was in quadrature to the sun, while the earth and the sun were both in the line of nodes; again the observed latitude is equal to the

---

[1] *Progymn.* p. 821.　　　　[2] *De stella Martis*, cap. VI. *Opera*, III. p. 193.

[3] See his letter to Magini, dated June 1, 1601, describing the use of four oppositions for finding the apsides. *Opera*, III. pp. 5 and 40.

[4] Cap. XI. *Op.* III. p. 219. Compare above, p. 362.

[5] Cap. XII. *Op.* III. p. 225.

inclination.   Thirdly, by the method of Copernicus, using
latitudes observed at opposition.   The first and third method
assume the ratio of the dimensions of the orbits to be known,
while the second method is quite independent of any previous
theory, and Kepler succeeded in finding four observations
satisfying the conditions of the second method.   In these
various ways he found the inclination $= 1°\ 50'$ and proved
that the plane of the orbit passes through the sun and that
the inclination is constant, so that the oscillations of the orbit
hitherto deemed necessary had no real existence.   It is when
proclaiming this important discovery that he makes the remark
that Copernicus did not know of his own riches[1].

The next and most important step was to determine the
position of the line of apsides (longitude of the aphelion), the
excentricity and the mean anomaly at some date or other.   To
determine these three quantities Ptolemy required only three
oppositions, because he assumed the bisection of the excentricity (in the figure $CA = CS$), but as Kepler was determined
to follow Copernicus and Tycho in making no assumption
of that kind, he had to use four oppositions.   From the
ten oppositions observed by Tycho (to which he was able to
add two observed by himself in 1602 and 1604) he selected
those of 1587, 1591, 1593, and 1595 and deduced from them
the time of true opposition.   In the figure D, G, F, E are
the four observed places of Mars, S the sun, C the centre

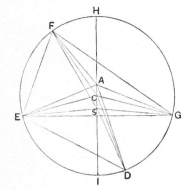

of the circular orbit, A the
punctum æquans, HI the line
of apsides.   The position of
this line and the mean anomaly of the first opposition, that
is, the angles HSF and HAF,
were in the first instance
borrowed from Tycho's theory.
The observations gave directly
the heliocentric longitudes,
that is, the angles at S between the lines SF, SE, SD

[1] Caps. xiii.–xiv. *Op.* iii. pp. 228–234.

and SG, while the angles at A, the differences of mean anomaly, were known, as the period of sidereal revolution gave the mean motion. From the triangles ASF, ASE, ASD and ASG, in which the angles at AS are known, the distances SF, SE, SD and SG are now computed, expressed in parts of $1S$. From the triangles SFE and SFG the angle F of the quadrilateral $FEDG$ is found, and similarly the three other angles E, D, G. If now the four points F, E, D, G lie on the circumference of a circle, we should have

$$F + D = G + E = 180°.$$

When this condition has been fulfilled, we have to find whether the centre of the circle lies on the line AS. In the triangle SFG we can compute the length of FG, since we know the angle at S and the two other sides ; in the isosceles triangle FCG we now know FG and the angle FCG, the latter being equal to twice FEG (or twice the sum of FES and SEG), we can therefore find the two radii in parts of AS, and the angle CFG. Furthermore the angle $SFC = SFG - CFG$, therefore we can in the triangle CSF find the side CS and the angle CSF, and we ought to have

$$CSF = HSF.$$

It is therefore necessary to alter the assumed direction of HI or the angles HSF and HAF (the true and the mean anomaly of the first opposition) until both conditions are fulfilled, that is, until the four points lie on a circle the centre of which is in the line joining S and A.

Kepler truly says that if the reader finds this account of the method tedious, he should pity the author, who made at least seventy trials of it, and he should not wonder that upwards of five years were devoted to Mars, although nearly the whole of the year 1603 was spent in optical research[1]. The result of the seventy trials was (radius of circle $= 1$)

Longitude of aphelion 28° 48′ 55″ of Leo[2] (1587)

$$AC = 0·07232 \qquad CS = 0·11332.$$

---

[1] Cap. xvi. *Op.* iii. p. 245.

[2] The motion of the nodes and that of the aphelion were determined by comparison with Ptolemy, through the medium of the star Regulus, in order to be independent of Ptolemy's precession. Cap. xvii. *Op.* iii. p. 250.

This theory Kepler afterwards called the "vicarious hypothesis." It represented the longitudes of the twelve oppositions
very well, the greatest residual error being 2′ 12″ which Kepler
held to be chiefly due to errors of observation, since the apparent
diameter of Mars when nearest the earth seemed very considerable[1]. And yet the theory turned out to be false, and Kepler
believed that when Ptolemy adopted the bisection of the
excentricity $(AC = CS)$ he must have met with a similar
experience, which probably also induced Tycho to lay the
theory of Mars aside and take up that of the moon instead.
Kepler tested his theory by the latitudes of the oppositions of
1585 and 1593, when Mars was near the limits of greatest
north and south latitude and at the same time near the
aphelion and the perihelion. Using Tycho's solar theory
unaltered he found that the excentricity came out $= 0·08000$ or
$= 0·09943$, according as true or mean oppositions were used,
very different from 0·11332, but not differing very much from
$\frac{1}{2}(AC + CS) = 0·09282$. He therefore tried the effect of putting
$AC = CS = 0·09282$, but this turned out to be a bad move,
for while places about 90° from the apsides were well represented, those in anomalies 45°, 135°, etc., differed about 8′.
We see now, says Kepler, why Ptolemy acquiesced in the
bisection of the excentricity, for 8′ was well within the limit of
accuracy to his observations (10′); but to us Divine goodness
has given a most diligent observer in Tycho Brahe, and it
is therefore right that we should with a grateful mind make
use of this gift to find the true celestial motions[2]. Another
proof that the vicarious hypothesis was wrong was supplied by
examining longitudes outside oppositions but near the apsides.
These also gave an excentricity of about 0·09. The vicarious
hypothesis, which had cost such an immense amount of labour,
was thus a perfect failure. And this showed that either the
orbit was not a circle, or, if it was, that there was not a
fixed point within it, seen from which the planet moved
uniformly, but that the *punctum æquans* would have to

---

[1] Cap. xviii. *Op.* iii. p. 254.
[2] Cap. xix. *Op.* iii. p. 258.

# #

oscillate backwards and forwards in the line of apsides, which could not be the effect of any natural cause[1].

Having thus proved the impossibility of forming a correct theory from oppositions alone, Kepler saw the necessity of attacking the problem in a more general manner instead of like his predecessors investigating the first and the second inequalities quite separately. He determined to tackle the second inequality first, by examining more rigorously the annual orbit of the earth. In the *Mysterium Cosmographicum*[2] he had tried to explain that a planet moves fastest at the perihelion and slowest at the aphelion because at these points it is nearest to and farthest from the sun and therefore respectively most and least under the influence of some power emanating from the sun. But he had acknowledged that if this explanation were correct, the earth ought to move exactly in the same manner as the planets; and yet nobody had attributed an equant to the annual orbit or made it anything but a simple excentric circle. It was therefore a great pleasure to him when it flashed on him (*dictabat mihi genius*, as he says) that the apparent change of the diameter of the annual orbit must be caused by the fact that the centre of equal distances and the centre of equal angular motion were not coincident in the case of the earth any more than in the orbits of the planets. But this had now to be rigorously proved.

Having first proved the reality of the alleged phenomenon by means of two observations of Mars in the same heliocentric longitude, made on two occasions when the differences of heliocentric longitudes of the planet and the earth were equal, by showing that the parallaxes instead of being equal differed by 1° 14'·5,[3] Kepler determined the excentricity of the earth's orbit by means of observations of Mars in one point of its orbit, taken from several points of the earth's orbit. In the triangle between the sun (S), the earth (E) and the projection of Mars on the plane of the ecliptic (M) the angles at S and E were known, the heliocentric longitude of Mars being taken either from Tycho's or from the vicarious theory; from them the ratio

[1] Cap. xx. *Op.* iii. pp. 259–262.   [2] Cap. xxii. *Op.* i. pp. 182–183.
[3] Cap. xxii. *Op.* iii. p. 271.

of the sides SE to SM was found. Similarly the ratio of other radii vectores to SM could be found by picking out other observations of Mars taken after the lapse of exactly one or more periods of sidereal revolution, and it was then a simple geometrical problem to find the radius of the circle, the distance of S from the centre and the direction of the diameter through S, i.e. the line of apsides. From the same observations and in the same manner the distance of the punctum æquans from the centre of the circle was determined, and both this and the distance of the sun from the centre were found to be equal to about 0·01800 (the radius being $= 1$), or practically half of Tycho's excentricity, so that Kepler's strong suspicion, that the latter ought to be bisected and that the earth moved exactly according to the same principles as the planets, had been completely verified[1]. The smaller value of the excentricity agreed perfectly with the very small variation in the sun's apparent diameter in the course of a year, while the difference between the equation of the centre computed by the old and by the new theory was found to be insensible, amounting at most to a few seconds[2].

This confirmation of Kepler's idea of the similarity of the motion of the earth and of the planets naturally led him to resume the suggestion made in the *Mysterium Cosmographicum*, that this motion is caused by a force emanating from the sun; and as the effect of any such force must necessarily vary in some way or other with the distance from the sun, he was led to speculate on the variation of a planet's velocity throughout its orbit. Hereby he eventually succeeded in getting rid of the Ptolemaic equant and substituted for it the law which subsequently became known as the second law of Kepler, though it really was discovered before the first one. As the orbits of the planets are situated nearly in one plane, that of the ecliptic, Kepler supposed the force (*virtus*) only to act in the planes of

---

[1] Caps. XXIII.–XXVIII. pp. 272–290. In chapter XXVIII. Kepler tests the result by finding the heliocentric longitude and distance from the sun of Mars from various combinations, on the assumption of $e = 0·018$ for the earth.

[2] Caps. XXIX. and XXXI. pp. 291 and 296. About the sun's diameter compare *Astronomiæ Pars Optica*, cap. XI. *Op.* II. p. 343.

the orbits and consequently to be simply inversely proportional
to the distance. The same will be the case with the velocity in
the orbit, and consequently the short time which the pla et takes
to pass over a very small arc of the orbit is proportiona₁ to the
radius vector. Kepler proves this for the neighbourhood of the
apsides in the Ptolemaic excentric circle[1] and assumes without
further examination that it holds good for any point of the
orbit; and even later on, when he recognized the orbits to be
elliptic, he took it for granted that the proof still held. We
know now that he was wrong in this, since the velocity at any
point is proportional to the perpendicular from the focus to the
tangent at the point in question, so that Kepler's theorem is
only true at the apsides where the radius vector is perpendicular
to the tangent. But the flaw in Kepler's reasoning is curiously
counteracted by another one in deducing his law. As the time
spent in passing over a very small arc is proportional to the
radius vector, the sum of the times spent in passing over the
sum of minute arcs making up a finite arc of the orbit will be
proportional to the sum of all the radii vectores, that is (he
thinks) to the area of the sector described by the radius vector.
Here is the second flaw, since a sum of an infinite number of
lines side by side does not make an area, a fact of which Kepler
was quite aware. Still, the theory of gravitation has proved
the truth of the celebrated second law of Kepler, that the time
of describing an arc of the orbit is proportional to the area of
the sector swept over by the radius vector. But the way in
which Kepler deduced the law was anything but unobjection-
able. He never found out the error of his law of distances, but
he knew that the sum of a number of radii vectores did not
correctly measure the area of a sector[2]; yet when he found that
the mean anomalies could be computed accurately by his second
law, so as to agree with the observations, not only for the earth's
orbit, to which he had in the first instance only applied it, but
also for the elliptical orbit of Mars, he justly considered it to be
fully established[3]. At first Mars, however, continued as troublesome as ever, for when Kepler from observations taken near

[1] Cap. XXXII. *Op.* III. p. 297.   [2] Cap. XL. *Op.* III. p. 324.
[3] Cap. LIX. p. 401.

the perihelion and aphelion deduced new values of the elements,
the comparison with observed places in other parts of the orbit
again showed outstanding errors which in the octants rose
to 8'[1].

The last result led Kepler to suspect that the form of the
orbit was not circular and showed him the necessity of proceeding without any preconceived notion as to its figure. But
the figure could be determined if the distances of Mars from
the sun in various parts of the orbit could be found. He therefore computed three distances both from the circular hypothesis
and from the observations, with the following result:

Date	Distance from aphelion	From circular hypothesis	From observations	Circle minus observations
1590 Oct. 31	9° 37′	1·66605	1·66255	+0·00350
1590 Dec. 31	36° 43′	1·63883	1·63100	+0·00783
1595 Oct. 25	104° 25′	1·48539	1·47750	+0·00789

As the observed distances were all smaller than those
resulting from the excentric circle, the natural conclusion was,
that the orbit is not a circle but a curve which, except at the
apsides, lies wholly within the circle. This would also explain
why the application of the law of areas gave the result that the
planet seemed to move too fast near the apsides and too slowly
at mean distance, since the sector-areas of the circle would
everywhere, except close to the apsides, be greater than those
of a curve lying inside the circle. Kepler therefore concluded
"that the orbit of the planet is not a circle but of an oval
figure[2]." At the apsides this oval coincides with the circle and
in anomalies 90° and 270° it deviates most from the circle, the
oval being egg-shaped, broader at the aphelion and more pointed
at the perihelion. In order to explain this remarkable form of
orbit Kepler resolved the motion of the planet into one on an
excentric circle and one on an epicycle. He supposed the
planet to possess some power of resisting the force emanating

[1] He found the mean distance$=1\cdot52640$, excentricity (bisected)$=0\cdot09264$,
one $=4^\circ 28° 39' 46''$, cap. XLII. p. 333. The difference between ellipse
$e^2 \sin 2a$, or for $a=45°$ and $e=0\cdot09264$, $7'\cdot4$.

[2] *Op.* III. pp. 335–337.

from the sun which drives it along, so that it describes an epicycle by a retrograde motion, and he assumed that the planet moves unequally (according to the second law) on the excentric, but uniformly on the epicycle[1].

Kepler had thus finally broken with the assumption of the circular orbit, but he now encountered great difficulties in dealing with the ovoid orbit and its quadrature, so that he was obliged to have recourse to approximate methods. The vicarious theory would give with sufficient accuracy for this purpose the heliocentric longitude, that is, the direction of the radius vector; it remained to determine its length. The line from perihelion to aphelion (IH) was therefore first divided unequally, so that $AC = 0\cdot07232$ and $SC = 0\cdot11332$, S being the sun. Then the angle HAM is made equal to the mean anomaly, and from C the line CM' is drawn, equal in length to the mean distance of Mars. SM' will then be the true heliocentric direction of Mars.

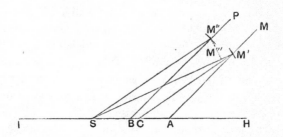

Next, AS is bisected in B and BP is drawn parallel to AM, so that HBP is the mean anomaly, then a circle round B will give the distance. On BP BM'' is marked off equal to the mean distance, then SM'' is the true length of the radius vector, and if we make $SM''' = SM''$, then M''' (situated on the line SM') will be the true place of the planet[2].

For finding the areas of the oval-sectors Kepler substituted for the oval an ellipse, the greatest breadth of the lunula between it and the excentric circle being $0\cdot00858$[3]. This also gave errors of about 7' in the octants, but with signs opposite

---

[1] Cap. xlv. *Op.* iii. p. 337.   [2] Cap. xlvi. *Op.* iii. p. 341.
[3] This is the square of $0\cdot09264$. Cap. xlvii. p. 347. The sun is not in one of the foci of this auxiliary ellipse.

to those of the excentric circle, showing that the true orbit was somewhere between the circle and the auxiliary ellipse[1]. After a number of other fruitless experiments he computed twenty-two distances of Mars from the sun by means of the new hypothesis. This computation showed that he had determined the line of apsides correctly, and proved in the most conclusive manner that it really passed through the body of the sun, as he had always maintained, and not through the mean sun. But the distances turned out to be too small, the difference being 0·00660 about the place of mean distance[2]. The true orbit was therefore clearly proved to be situated between the circle and the oval.

The breadth of the lunula between the orbit of Mars and the excentric circle at last supplied the long-sought clue to the mystery of the planet's motion. It amounted to 0·00660, the semidiameter of the circle being 1·52350, or 0·00432 if the semidiameter is = 1. This is very nearly equal to 0·00429 or half the breadth of the lunula of the oval theory. By mere chance, as he acknowledges, Kepler noticed that 1·00429 is equal to the secant of the greatest optical equation of Mars, that is, the secant of the angle (5° 18′) of which the tangent is equal to the excentricity. "I awoke as if from sleep, a new light broke on me." At the mean distances the optical equation is a maximum, and there the shortening of the distances was found to be greatest, being the excess of 1·00429 over unity; this result Kepler extended to all points of the orbit, substituting everywhere for the radius vector of the excentric circle the same quantity multiplied by cosine of the optical equation, or the distantia diametralis, as he calls it. The comparison of a number of distances computed by this rule with those resulting from Tycho's observations showed that this assumption was perfectly justified[3]. Thus the great discovery was made

---

[1] Cap. xlvii. p. 350, where a table is given, containing for the excentric anomalies 45°, 90°, 135°, the true anomalies by single and bisected *e*, by the vicarious hypothesis and the "physical" theory, circular and elliptic. The auxiliary ellipse agrees pretty closely with Ptolemy's theory.

[2] Cap. lv. p. 384.

[3] Cap. lvi. *Op.* iii. p. 384.

that the radius vector of Mars is always represented by the
equation

$$r = a + ae \cos E,$$

where a is the mean distance and E is the excentric anᵣ naly,
counted according to ancient custom from the aphelion, while
ae is the distance between the sun and the centre of the orbit.
Though the goal was really reached, Kepler at the last moment
created fresh trouble for himself. The diminution of the radius
vector as the planet moves away from the aphelion suggested a
libration of the planet along the diameter of an epicycle moving
on a circle concentric with the sun. But while this would represent the above equation, that is, the length of the radius
vector, an attempt to compute the corresponding true anomaly
in this way left errors of $4'$ or $5'$[1]. This compelled Kepler to
return to the ellipse, which he had already employed as a substitute for the oval, and he finally proves[2], that an ellipse with
the sun in one of the foci gives the length of the radius vector
in accordance with the above equation, while its direction is
given by

$$r \cos v = ae + a \cos E.$$

The great problem was solved at last, the problem which
had baffled the genius of Eudoxus and had been a stumblingblock to the Alexandrian astronomers, to such an extent that
Pliny had called Mars the "inobservabile sidus." The numerous
observations made by Tycho Brahe, with a degree of accuracy
never before attained, had in the skilful hand of Kepler revealed
the unexpected fact that Mars describes an ellipse, in one of the
foci of which the sun is situated, and that the radius vector of
the planet sweeps over equal areas in equal times. And the
genius and astounding patience of Kepler had proved that not
only did this new theory satisfy the observations, but that no
other hypothesis could be made to agree with the observations,
as every proposed alternative left outstanding errors, such as it
was impossible to ascribe to errors of observation. Kepler had
therefore, unlike all his predecessors, not merely put forward a
new hypothesis which might do as well as another to enable a

[1] Cap. LVIII. p. 399.
[2] Caps. LIX.-LX. pp. 401–411, being the end of Part IV. of the book.

computer to construct tables of the planet's motion; he had found the actual orbit in which the planet travels through space. In the fifth and last part of his book on Mars he finally shows how perfectly the new theory represents the observed latitudes[1]. The longitudes had been bad enough to previous theorists, but the latitudes had been simply hopeless, driving astronomers to the most unreasonable assumptions such as oscillations of the orbit. Now that the true nature of the orbit had been found and it had been proved that its plane intersected that of the earth's orbit in a line passing through the sun, everything became clear and the many hitherto inexplicable phenomena were at once accounted for. Among these was the fact that the latitude is not always a maximum exactly at the time of opposition, and Kepler quotes both Tycho's manuscripts and conversations held with him to show the anxiety this had caused the great observer. Now it became simply a question of finding whether the sine of the heliocentric latitude or the distance between Mars and the earth varied most rapidly, and another source of annoyance to theoretical astronomers had thus been removed[2].

The discovery of the elliptic orbit of Mars was an absolutely new departure, as the principle of uniform circular motion had been abandoned; a principle which from the earliest times had been considered self-evident and inviolable, though Ptolemy had tacitly dropped it when introducing the equant. To the enquiring mind of Kepler it became therefore a necessity to endeavour to explain, why the planet describes an elliptic and not a circular orbit.

In the *Mysterium Cosmographicum* Kepler had supposed the existence of an *anima motrix* in the sun, and this idea he now develops further. This force emanates from the sun, but unlike light it does not spread in all directions but only in the plane close to which the planes of all the planetary orbits are situated, so that it simply diminishes as the distance increases. The velocity of a planet in its orbit therefore varies inversely as the distance, which idea, as we have seen, led him to the discovery of his second law. But the rule could not be extended

[1] Cap. LXII. *Op.* III. p. 415.   [2] Cap. LXVI. p. 422.

from one orbit to another, as the periods of revolution would then be proportional to the squares of the distances. The solar force produces revolution because the sun rotates on its axis and thereby swings the straight lines, along which the force proceeds, around with itself from west to east. The result is, that a circular stream or vortex is produced which carries the planets along with it[1], though in different periods owing to the different amount of resistance made by each planet, which depends on its mass. The solar equator is naturally assumed to coincide with the ecliptic, and the period of rotation of the sun is estimated in a very curious manner. The periods of the planets nearer to the sun are shorter than those of the planets farther away from it, so that the rotation-period of the sun must be less than 88 days, the time of revolution of Mercury. The semidiameters of the sun and of the orbit of Mercury he supposes to be in the same ratio to each other as those of the earth and the lunar orbit, and therefore the periods ought to be in the same ratio, which gives the period of rotation of the sun equal to about three days[2]. Kepler had to acknowledge a few years later[3], when the discovery of sunspots had followed soon after the invention of the telescope, that this determination as well as the assumption about the position of the solar equator were equally erroneous.

The vortices caused by the sun would carry the planets round in circular orbits concentric with the sun, and it was therefore necessary to look for some force capable of changing this circular motion into an elliptic one. Already before the publication of Gilbert's book on the magnet Kepler had become greatly interested in magnetism, as we learn from letters written in 1599 to the Bavarian Chancellor Herwart von Hohenburg; and he made repeated attempts to find the position of the earth's magnetic poles by means of the few determinations of magnetic declination then available. First

---

[1] "Flumen est species immateriata virtutis in sole magneticæ." *De stella Martis*, cap. LVII. *Op.* III. p. 387.

[2] Ibid. cap. XXXIV. p. 306.

[3] In a letter to Wackher, written in 1612 after reading Scheiner's announcement of his determination of the period of rotation of the sun and the position of the sun's equator. *Opera*, II. p. 780.

he thought that the north pole was 23° 28′ distant from the
magnetic pole, and that the latter indicated the place where
the rotation-pole had been at the time of the Creation, since
when the two poles had gradually drifted asunder, the earth's
equator becoming inclined to the ecliptic. Afterwards he
concluded from the observations made by the Dutch expedition
to Novaja Zemlja that the two poles were only $6\frac{1}{2}$° apart,
which seemed to him to agree well with the theory of Domenico
Maria da Novara that the position of the earth's axis had
changed 1° 10′ since the days of Ptolemy, which in the 5600
years elapsed since the Creation would amount to more than
5°[1]. Though Kepler, after the publication of Gilbert's book,
and as he obtained access to more measures of declination,
perceived the impossibility of fixing the position of the magnetic
pole by such measures, he continued to be deeply interested in
magnetism, the phenomena of which he believed to furnish
the explanation of the elliptic motion of the planets[2].

Every planet, according to Kepler, has a magnetic axis
which always points in the same direction and remains parallel
to itself, just as the rotation
axis of the earth does without
requiring the "third motion"
postulated by Copernicus[3].
He justifies the assumption
that the planets are "huge
round magnets" by referring
to the fact that Gilbert had
proved this to be the case
with the earth, which, according to Copernicus, is one
of the planets[4]. One of the

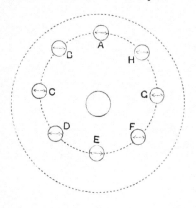

---

[1] *Ungedruckte wissenschaftliche Correspondenz zwischen Johann Kepler und
Herwart von Hohenburg.* Edirt von P. Anschütz, Prag, 1886, pp. 28 and 59.

[2] The following theory is partly given in the *Commentary on Mars*, partly in
Kepler's later text-book, *Epitome Astronomiæ Copernicanæ*, published in three
parts in 1618, 1620, 1621.

[3] See *Opera*, I. p. 121, "motus iste revera motus non est, quies potius
:enda"; compare III. p. 388 and p. 447.

[4] *De stella Martis*, cap. LVII. *Op.* III. p. 387.

magnetic poles of a planet seeks the sun, while the other is repelled by it. Let us begin by considering the osition of the planet at *A*, where the magnetic poles are equidistant from the sun; the sun will neither attract nor repel the planet, but will simply move the planet along. But by this motion the planet is successively brought to the positions *B, C, D, E*, and the pole which is "friendly to the sun" (*soli amica*) is turned towards the sun, while the hostile one (*discors*) is turned away from it. In this part of the orbit the planet is therefore attracted by the sun, and continues to approach it until the position *E*, where the attraction and repulsion again balance each other. When the planet has passed *E* the hostile pole is turned towards the sun, and the planet will therefore in the second half of the orbit be repelled from the sun and the distance will increase until the aphelion is reached at *A*. The axis does, however, not remain

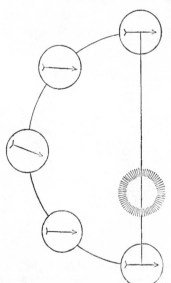

rigorously parallel to the original direction, but owing to the action of the sun it suffers a slight deflection, so that it is pointed exactly to the sun when the planet is at its mean distance. In the upper quadrant, from the aphelion to the mean distance, the sun produces this "inclination" during a longer interval but with a weaker force, in the lower quadrant during a shorter period but with a stronger force, *fitque compensatio perfecta*. If the magnetic axis after a whole revolution has not quite come back to its original position a slow motion of the line of apsides is the result, such as is indeed observed to exist[1]. The amount of excentricity is

[1] *De stella Martis*, cap. LVII. (*Op.* III. p. 389), *Epitome*, lib. IV. pars III. (*Op.* VI. pp. 377–380). The two figures are taken from the latter account.

different in the various planetary orbits owing to the different
intensity of the magnetism in each planet, but the final cause
is that the ratio of the greatest to the smallest velocity has to
be an harmonious one[1].

The planets would move in the plane of the solar equator,
that is, the ecliptic (and have no latitudes), except for the
inclination of the axis of each planet to the ethereal current
produced by the sun, whereby the planet is deflected just as a
ship is by its rudder.

Kepler's explanation of the planetary motions is the first
serious attempt to interpret the mechanism of the solar system.
He did not advance beyond the notions of mechanics current
in the sixteenth century, since he supposes that a constantly
acting force is necessary to keep up the motion of a planet,
and that the planet would stop where it was if the force
were to cease acting. The active force is magnetism pure
and simple, and Kepler never tired of emphasizing this whenever an opportunity offered. Thus the thesis No. 51 of his
little polemical book *Tertius Interveniens* (1610) states that
" the planets are magnets and are driven round by the sun by
magnetic force, but the sun alone is alive," and in the commentary to this thesis occurs the following passage, which we
shall quote in the original, quaint language : " Für mein Person,
sage ich, dass die Sternkugeln diese Art haben, dass sie an
einem jeden Ort dess Himmels, da sie jedesmals angetroffen
werden, stillstehen würden, wann sie nicht getrieben werden
solten. Sie werden aber getrieben per speciem immateriatam
Solis, in gyrum rapidissime circumactam. Item werden sie
getrieben von jhrer selbst eygnen Magnetischen Krafft, durch
welche sie einhalb der Sonnen zuschiffen, andertheils von der
Sonnen hinweg ziehlen. Die Sonn aber allein hat in jhr selbst
ein virtutem animalem, durch welche sie informiert, liecht
gemacht, vnd wie ein Kugel am Drähstock beständiglich
vmbgetrieben wirdt, durch welchen Trieb sie auch jhre speciem
immateriatam ad extremitates usque mundi diffusam in gleicher
Zeit hervmb gehen macht, vnd also successive alle Planeten

---

[1] *Epitome, Op.* vi. p. 379.

mit hervmb zeucht. Mehrere scientia animalis wirdt zu den himmlischen bewegungen nicht erfordert[1]."

Though Kepler deserves credit for having attempted to find the cause of orbital motion, he cannot be called a fore-runner of Newton. His force is not directed to the sun but is tangential, *non est attractoria sed promotoria*, as he said in a letter to Mästlin in March, 1605[2]. If the sun did not rotate the planets would not revolve, and similarly if the earth did not rotate the moon would not revolve round it; but the moon has no rotation because there is no satellite moving round it, "rotation was therefore omitted in the case of the moon, being unnecessary[3]." But Jupiter and Saturn must rotate on their axes, as they have satellites[4]. The rotation of a planet is partly caused by the solar force, but chiefly by a force inherent in the planet; thus the earth rotates 365 times during one revolution, of which the sun is responsible for five, as the earth but for its influence would only rotate 360 times[5].

Thus we see that gravity had no place in Kepler's theory of celestial mechanics. Yet his ideas about gravity are very superior to those prevailing since the days of Aristotle. As usual we find the germ of his ideas in the *Mysterium Cosmographicum*: "No point, no centre is heavy; but everything of the same nature as a body will tend to it; nor does the centre acquire weight by drawing other things to itself or by being sought by them, any more than a magnet gets heavier while drawing iron[6]." In the introduction to the book on Mars "the true doctrine about gravity" is summed up in the following

---

[1] *Opera*, I. pp. 554 and 590. A few years previously he had hesitated to identify the motive force altogether with magnetism, as the earth is also affected by the planetary aspects, "percipit igitur Terra aliquid, quod sola ratio percipit. Magnetes vero a nullo rationis subjecto seu objecto moventur." Letter to Brengger of Nov. 1607, *Op.* II. p. 589. But this is really nothing but quibbling, for in other letters he plainly calls the solar force "effluvium magneticum" (letter to the English astrological writer Sir Christopher Heydon in May, 1605, *Op.* III. p. 37), or uses similar expressions.

[2] *Op.* III. p. 57.

[3] *Epitome*, lib. IV. part 2, *Op.* VI. p. 362.

[4] No satellites of Saturn had yet been discovered, but the mysterious "appendages" (the ring imperfectly seen) were often spoken of as satellites.

[5] Ibid. p. 359.　　　　　　　　[6] Cap. XVI. *Op.* I. p. 159.

axioms[1]. Every bodily substance will rest in any place in which it is placed isolated, outside the reach of the power of a body of the same kind (*extra orbem virtutis cognati corporis*). Gravity is the mutual tendency of cognate bodies to join each other (of which kind the magnetic force is), so that the earth draws a stone much more than the stone draws the earth. Supposing that the earth were in the centre of the world, heavy bodies would not seek the centre of the world as such, but the centre of a round, cognate body, the earth; and wherever the earth is transported heavy bodies will always seek it; but if the earth were not round they would not from all sides seek the middle of it, but would from different sides be carried to different points. If two stones were situated anywhere in space near each other, but outside the reach of a third cognate body, they would after the manner of two magnetic bodies come together at an intermediate point, each approaching the other in proportion to the attracting mass. And if the earth and the moon were not kept in their orbits by their animal force (*vi animali*), the earth would ascend towards the moon one fifty-fourth part of the distance, while the moon would descend the rest of the way and join the earth, provided that the two bodies are of the same density. If the earth ceased to attract the water all the seas would rise and flow over the moon. On the other hand the *virtus tractoria* of the moon reaches as far as the earth and produces the tides, as to which and their influence in forming bays and islands Kepler has a good deal to say, which we must pass over here.

It is remarkable how firmly Kepler clung to the close analogy between gravity and magnetism on the one hand, and between the motive force of the sun and magnetism on the other hand. And yet he failed to see the identity of gravity and the force which keeps the planets in their orbits. This is the more remarkable when we notice that he, in the notes to his *Somnium*, written between 1620 and 1630, expressly attributes the tides to "the bodies of the sun and moon attracting the waters of the sea with a certain force similar to

---

[1] *Op.* III. p. 151.

the magnetic[1]." He acknowledged then that the a raction of the sun (and not merely the tangential force emai..ting from it) could reach as far as the earth. But there he stopped, and he could not have got any further without recasting his general conception as to the cause of motion of a body.

Kepler was a very prolific writer, not only of books but of letters, and as he wrote very openly about his work and his correspondence has fortunately been preserved, we are able to trace the progress of his researches from year to year. Thus he wrote to Herwart in July, 1600, that he had already found two of the suggestions in the *Mysterium Cosmographicum* confirmed, the use of the true place of the sun instead of the mean one and the existence of an equant in the solar orbit[2]. On the 1st of June, 1601, he wrote to Magini, describing the use of four oppositions and explaining that the irregularities of the inferior planets arise from the earth's motion, that the inclinations are constant, and that the theories of all the seven planets must be similar[3]. During the first half of 1602 he found that the orbit of Mars is oval; but the year 1603 was almost altogether devoted to his book on Optics, and the work on Mars was not taken up again till the beginning of 1604, as we learn from a letter he wrote to Longomontanus in the following year, adding that he did not yet clearly see the cause of the oval orbit except that it was due to a force propagated from the sun; he had then written fifty-one chapters[4]. Chapters LIII.–LVIII. were written about May, 1605, and before the end of the year chapter LX. had been written, completing the theory of the motion in longitude, but the remaining ten chapters, on the latitudes, do not seem to have been written till 1606. In December, 1606, the Emperor granted 400 florins towards the printing, and these were actually paid, which was

---

[1] *Op.* VIII. p. 61, note 202. In note 66 (p. 47) he gives the following definition : "Gravitatem ego definio virtute magneticae simili, attractionis mutuae. Hujus vero attractionis major vis est in corporibus inter se vicinis, quam in remotis. Fortius igitur resistunt divulsioni unius ab altero, cum adhuc sunt vicina invicem."

[2] *Op.* III. p. 24. In Oct. 1602 Kepler wrote to Herwart that he saw the sun in the theory of Mars "tanquam in specula" (p. 29).

[3] *Op.* III. pp. 5 and 37.   [4] *Op.* III. p. 34 sq.

frequently not the case with Imperial grants at that time. The manuscript was sent to the printer in September, 1607, and the book was ready from the press in July or August, 1609[1], under the title, *Astronomia nova αἰτιολογητος seu Physica Coelestis, tradita commentariis de motibus stellae Martis. Ex observationibus G.V. Tychonis Brahe.* In the history of astronomy there are only two other works of equal importance, the book *De Revolutionibus* of Copernicus and the *Principia* of Newton. The "astronomy without hypotheses" demanded by Ramus had at last been produced, and well might Kepler (on the back of the title-page) proclaim that if Ramus had still been alive he should himself have claimed the reward offered for the achievement, the surrender of the professorship held by Ramus.

Among the correspondents of Kepler the one most worthy to exchange ideas with him was David Fabricius, Protestant clergyman at Resterhave, from 1603 at Osteel in East Friesland, a very able observer who had spent some time with Tycho at Wandsbeck in Holstein in 1598, and had visited him again at Prague in June, 1601, for a couple of weeks during the absence of Kepler[2]. Their correspondence extended over the years 1602 to 1609, and Kepler kept Fabricius fully informed of the progress of his work on Mars[3]. Fabricius was an adherent of the Tychonic system, and never perceived that this is practically the same as the Copernican system, so that he was always trying to make out the absolute motion of Mars with reference to the earth, instead of resting content with investigating its heliocentric motion. In 1602 Kepler gave him an account of the vicarious hypothesis, and in July, 1603,

---

[1] Ibid. pp. 9–11. There is no publisher's name on the title-page, and the book was only sold privately.

[2] David Fabricius was born at Esens in East Friesland in 1564, and was murdered by one of his parishioners at Osteel on May 7, 1617. He discovered the variable star Mira Ceti in Aug. 1596, and at once wrote to Tycho to announce it (letter printed in *Vierteljahrsschrift d. a. G.* iv. p. 290). Kepler considered him an observer second to Tycho only. His son Johann was the first observer of sun-spots.

[3] The letters are among the Kepler MSS at Pulkova. Most of them were first printed in Apelt's delightful book, *Die Reformation der Sternkunde* (Jena, 1852) ; more completely by Frisch, *Kepleri Opera*, iii. pp. 61–133.

he transcribed for the benefit of Fabricius h. method of
determining distances into the Tychonic system, and communicated the important discovery that the orbit of Mars is an
oval. He even told him of the approximate method employed
for finding the direction and length of the radius vector and
the failure to represent the true anomalies (*desperata res
erat*)[1]. Several letters followed, in which Fabricius raised
various objections to the results obtained by Kepler both as
regards the sun and Mars, to which Kepler replied at length in
February, 1604. A pause followed then, until Fabricius, in a
letter dated October 27, 1604, announced that he had found,
from a comparison with his own observations, that the oval
hypothesis gave the radius vector of Mars too small at mean
distance[2]. Kepler mentions this in chapter LV. of his book,
and generously adds that Fabricius thus very nearly anticipated him in finding the true theory[3]. But this is really going
too far, for even if Kepler had abandoned the investigation,
Fabricius would never have discovered the elliptic form of the
orbit, which indeed he called absurd and never would accept,
when Kepler had found it. And creditable as it is to Fabricius
to have perceived the insufficiency of the oval hypothesis,
Kepler had found the same before receiving his letter, and was
able on December 18, 1604, to tell Fabricius that the path of
Mars is a perfect ellipse[4].

To the conservative mind of Fabricius it appeared impossible to give up the ancient principle of combinations of circular
motion, and he therefore designed a theory of his own to avoid
recognizing the elliptic motion[5]. If on the circumference of a
circle an epicycle moves along, while the planet moves on the

---

[1] *Op.* III. p. 82.

[2] Ibid. p. 95.

[3] Ibid. p. 384. This is the only reference to him in the book on Mars.

[4] Ibid. p. 96. In a long letter finished on Oct. 11, 1605, Kepler describes
at length how he found the ellipse and explains his magnetic theory. Ibid.
pp. 99–105.

[5] "Per ovalitatem vel ellipsin tuam tollis circularitatem et aequalitatem
motuum, quod mihi inprimis penitius consideranti absurdum videtur. Coelum
ut rotundum est ita circulares et maxime circa suum centrum regulares et
aequales motus habet." Letter of Jan. 20, 1607; *Op.* III. p. 108.

epicycle in the opposite direction with twice the velocity, the planet will describe an ellipse. Fabricius preferred transforming this construction by letting the centre of an excentric circle make oscillations (librations) in its own plane in a straight line perpendicular to the line of apsides. In this way he represented elliptic motion, but not motion in accordance with Kepler's second law, the true anomaly not corresponding to the correct mean anomaly. Fabricius set forth his theory in a letter of February, 1608, and a note dated October 2, 1608, but he never published any account of it, and Kepler only refers to it in a few lines in his *Epitome Astronomiae Copernicanae*[1]. But among the students of theoretical astronomy in the seventeenth century David Fabricius deserves an honoured place, although he is one of the last representatives of a principle then about to be finally abandoned.

Kepler had intended to write a systematic treatise on astronomy, like the Syntaxis of Ptolemy, doing for the other planets what he had already done for Mars. The book was to be called *Hipparchus* in honour of the great astronomer. But although he had made considerable progress with the part dealing with the moon's motion, various circumstances made him alter his plan, and he wrote instead a more elementary text-book, *Epitome Astronomiae Copernicanae*, in three parts, of which the first was published in 1618 at Linz, to which town Kepler had moved in 1612 as " Provincial mathematician "; the two remaining parts followed in 1620 and 1621. In this work the two first laws of Kepler, which in the first instance had only been proved for Mars, were assumed to extend to the other planets. With regard to the moon he found the introduction of the elliptic motion in its theory very troublesome owing to the variability of the excentricity, and he made in the course of years many changes in the way in which he represented the observed longitudes[2]. He must indeed often

---

[1] Lib. v. pars I. *Op.* vI. p. 414. Kepler says of the theory : " Nec enim mera aequabilitas motuum, nec praecisio omnimoda obtinetur, nec operae compendium fit, et causae motuum occultantur abneganturque."

[2] Frisch has collected the existing fragments of *Hipparchus* and all the various MSS on the moon in *Opera*, Vol. III. pp. 511–717.

have envied his predecessors, who could introduce an epicycle
to account for every new inequality. Independently of Tycho
Kepler had discovered the annual equation of the moon. The
solar eclipse of March 7 (N.S.), 1598, as well as the lunar
eclipse in February and the Paschal full moon, occurred more
than an hour later than the calendar computed by him had
announced, while the lunar eclipse in August of the same year
occurred earlier than expected. In the calendar for 1599 he
therefore suggested that the moon's period with regard to the
sun is in winter a little longer than in summer. Having in
January, 1599, been invited by Herwart von Hohenburg to
explain the matter more fully, Kepler in his reply suggested
that the moon might be retarded in its motion by a force
emanating from the sun, which would be greatest in winter,
when the earth and the moon are nearer to the sun than they
are in summer. The cause of the phenomenon might also be
that the speed of the earth's rotation depended on the distance
of the earth from the sun, and is a little quicker in winter, so
that the moon in winter seems to take longer time than in
summer to pass through equal arcs[1]. This idea is further
worked out in the *Epitome*[2]. The correction he applied in the
same manner as Tycho had done by using a different equation
of time for the moon, but the amount of the annual equation,
which Tycho had given as 4'·5, was by Kepler estimated at its
correct value, 11'[3].

The new planetary tables, the *Tabulæ Rudolphinæ*, on
which Kepler had worked for many years, were published in
1627 at Ulm, having been printed under the personal supervision of Kepler who for that purpose had left Linz for Ulm at
the end of the previous year. It is characteristic of the noble
mind of the author, that he states on the title-page that the
tables contain the restoration of astronomy, conceived and
carried out "a Phoenice illo astronomorum TYCHONE."

---

[1] *Op.* i. p. 409. For another letter of April, 1599, see *Ungedruckte wissenschaftliche Correspondenz*, p. 11.

[2] *Op.* vi. p. 359 sq.

[3] He knew this already in 1606, see *Op.* viii. p. 627. Compare *Tab.
Rudolph.* cap. xv. (*Op.* vi. p. 571), where the maximum is given as 21ᵐ 40ˢ.

But long before the completion of this work Kepler's genius had scored another triumph by the discovery of the third law of planetary motion. This is contained in his work *Harmonices Mundi libri v*, published at Linz in 1619, a continuation of the *Mysterium Cosmographicum*, completing to the author's satisfaction the chain of ideas about the harmony of the world which had occupied his mind since his youth.

It will be remembered that Kepler's chief desire, when he joined Tycho Brahe in Bohemia, was to obtain the means of computing more accurate values of the mean distances and excentricities of the planets in order to test his theory of the five regular polyhedra. When after many years' patient labour he had computed the distances from Tycho's observations, it turned out that the theory was only approximately correct, as the neighbouring planetary spheres did not accurately coincide with the spheres inscribed in and circumscribed to the corresponding polyhedra. Kepler concluded from this that the distances of the planets from the sun were not taken from the regular solids only, which idea seemed to be confirmed by the circumstance that the maximum and minimum distances of two planets give four ratios, so that there are in the whole planetary system twenty ratios of distances of successive planets, while the solids only supply five. The deviation of the construction of the world from the five solids, and the change of distance during a revolution of a planet, are consequences of the "harmony of the world," and this harmony must be sought for in the greatest and smallest distances of the planets from the sun, since it is the form of the orbit, i.e. the excentricity, of which it is required to find the law. The distances themselves did not on examination show any sign of producing harmony, which showed that it had to be sought in the motions (*in ipsis motibus, non in intervallis*), that is, in the angular velocities seen from the common fountain of motion, the sun. The following table gives for each planet the heliocentric angular velocity (daily motion) at aphelion and perihelion[1].

---

[1] *Harm. Mundi*, lib. v. cap. 4, *Op.* v. p. 287.

Compass

Divergent Interval	Convergent Interval
$\dfrac{a}{d}=\dfrac{1}{3}$,	$\dfrac{b}{c}=\dfrac{1}{2}$
$\dfrac{c}{f}=\dfrac{1}{8}$,	$\dfrac{d}{e}=\dfrac{5}{24}$
$\dfrac{e}{h}=\dfrac{5}{12}$,	$\dfrac{f}{g}=\dfrac{2}{3}$
$\dfrac{g}{k}=\dfrac{3}{5}$,	$\dfrac{h}{i}=\dfrac{5}{8}$
$\dfrac{i}{m}=\dfrac{1}{4}$,	$\dfrac{k}{l}=\dfrac{3}{5}$

♄ $\left\{\begin{array}{l} 1'\,46''\ a \\ 2'\,15''\ b \end{array}\right\}$ $\dfrac{1'\,48''}{2'\,15''}=\dfrac{4}{5}$, major third

♃ $\left\{\begin{array}{l} 4'\,30''\ c \\ 5'\,30''\ d \end{array}\right\}$ $\dfrac{4'\,35''}{5'\,30''}=\dfrac{5}{6}$, minor third

♂ $\left\{\begin{array}{l} 26'\,14''\ e \\ 38'\,\ 1''\ f \end{array}\right\}$ $\dfrac{25'\,21''}{38'\,\ 1''}=\dfrac{2}{3}$, fifth

♁ $\left\{\begin{array}{l} 57'\,\ 3''\ g \\ 61'\,18''\ h \end{array}\right\}$ $\dfrac{57'\,28''}{61'\,18''}=\dfrac{15}{16}$, semitone

♀ $\left\{\begin{array}{l} 94'\,50''\ i \\ 97'\,37''\ k \end{array}\right\}$ $\dfrac{94'\,50''}{98'\,47''}=\dfrac{24}{25}$, diesis

☿ $\left\{\begin{array}{l} 164'\,\ 0''\ l \\ 384'\,\ 0''\ m \end{array}\right\}$ $\dfrac{164'\,\ 0''}{394'\,\ 0''}=\dfrac{5}{12}$, octave with minor third

We may note here at the outset that the harmony is to Kepler only a mathematical conception; he does not imagine that there really is any "music of the spheres": "Iam soni in coelo nulli existunt, nec tam turbulentus est motus, ut ex attritu aurae coelestis eliciatur stridor[1]." The daily heliocentric angular velocity expressed in seconds is considered to represent the vibration number of a certain tone, but as the velocity changes in the course of a revolution the tone will not remain the same but will run through a musical interval, the length of which depends on the excentricity and can easily be determined, if the smallest velocity is considered to be the number of vibrations which the key-note makes in the unit of time. But the position of the interval on the key-board must depend in some way on the absolute length of the radius vector, and it was therefore necessary to find whether there was any law connecting the mean motion (or the period of revolution) with the

[1] Ibid. p. 286.

mean distance, as, if there was, it would be possible to compute the mean distance from the harmony of the heavens. This computed distance ought then to agree with the observed distance. After many trials Kepler found on May 15, 1618, his celebrated third law, that the squares of the periods of revolution of any two planets are proportional to the cubes of their mean distances from the sun[1]. This law he soon found to apply not only to the planets but also to the four recently discovered satellites of Jupiter[2].

There are now three ways in which consonance may appear in the planetary motions. First: the ratio of the slowest motion at aphelion to the quickest motion at perihelion is the interval due to the excentricity of the planet's orbit. The above table shows that the intervals are nearly perfectly consonant, as the dissonance is less than a semitone, except in the cases of the earth and Venus, owing to their small excentricities. Secondly, the extremes of motion of two successive planets may be compared to each other in a twofold manner, as the interval may either be taken from the lowest tone (motion at aphelion) of the outer planet to the highest tone (perihelion) of the one next below it, or from the highest tone of the outer to the lowest of the inner planet. The former Kepler calls the divergent, the latter the convergent interval, and the above table shows almost perfect consonance in both, except in the case of the interval between Mars and Jupiter, which agrees with the tetrahedron and not with the musical theory. Thirdly, there may be a consonance of all the six planets.

In order to find to which octaves the lowest and highest tone of each planet belong, the figures expressing the greatest and smallest angular velocity of each planet must be divided by some power of 2 in order to produce ratios smaller than 1 : 2, that is, inside an octave. The exponent of 2 employed will then indicate to what octave the tone belongs[3].

---

[1] " Sed res est certissima exactissimaque, quod proportio, quae est inter binorum quorumcunque planetarum tempora periodica, sit praecise sesqui-altera proportionis mediarum distantiarum, id est orbium ipsorum." *Harm. Mundi*, v. 3, *Op.* v. p. 279.

[2] *Epitome*, IV. 2, *Op.* VI. p. 361.

[3] *Harm. Mundi*, v. cap. 5, p. 291.

Velocity of Saturn	in aph.	divided by	$2^0 = 1'\,46''$
,, ,,	per.	,,	$2^0 = 2'\,15''$
,, Jupiter	aph.	,,	$2^1 = 2'\,15''$
,, ,,	per.	,,	$2^1 = 2'\,45''$
,, Mars	aph.	,,	$2^3 = 3'\,17''$
,, ,,	per.	,,	$2^4 = 2'\,23''$
,, Earth	aph.	,,	$2^5 = 1'\,47''$
,, ,,	per.	,,	$2^5 = 1'\,55''$
,, Venus	aph.	,,	$2^5 = 2'\,58''$
,, ,,	per.	,,	$2^5 = 3'\,\ 3''$
,, Mercury	aph.	,,	$2^6 = 2'\,34''$
,, ,,	per.	,,	$2^7 = 3'\,\ 0''$

Putting the aphelion velocity of Saturn equal to G, the lowest tone of the earth will also be G, because the two tones are represented by the figures $1'\,46''$ and $1'\,47''$, practically identical, but it will be G treble, five octaves higher. The figure for the highest tone of Mercury is $3'\,0''$, very nearly $\frac{5}{3}$ of $1'\,47''$, the tone is E^v, seven octaves and a major sixth above the lowest tone of Saturn. The tunes played by the planets are therefore[1]:

Saturn   Jupiter   Mars

Earth   Venus   Mercury

This gives an interesting view of the excentricities of the planets, as one sees at a glance the great difference between the

[1] Lib. v. cap. 6, *Op.* v. p. 294; Apelt, *Johann Keppler's astronomische Weltansicht*, p. 93.

almost circular orbit of Venus and the very considerable excentricity of Mercury. The well-known gap between the orbits
of Mars and Jupiter is also striking.

It would lead us too far if we were to show how Kepler
tempered the intervals of the six planets in order to produce in
the concert made by them all together the most perfect consonance[1]. He finally obtains the following ratios of the smallest
and greatest velocities[2]:

Saturn	64 : 81
Jupiter	6561 : 8000
Mars	25 : 36
Earth	2916 : 3125
Venus	243 : 250
Mercury	5 : 12

With these new values he computes new excentricities, mean
motions, and by his third law mean distances. The agreement
of these values of the distances with the observed ones is as
follows[3], the earth's mean distance from the sun being assumed
= 1000.

	From Harmony		From Tycho's Observations		Semidiam. of	
	Aphelion	Perihelion	Aphelion	Perihelion	sphere inscr. in	
Saturn	10118	8994	10052	8968	Cube	5194
Jupiter	5464	4948	5451	4949	Tetrah.	1649
Mars	1661	1384	1665	1382	Dodecah.	1100
Earth	1017	983	1018	982	Icosah.	781
Venus	726	716	729	719	Octah.	413
M...	476	308	470	307	Square in do.	336

The last column of this table shows how far the theory of
the five regular solids agrees with the harmony. The faces of
the cube come down a little lower than the mean distance
of Jupiter, while the faces of the octahedron do not quite reach
the mean distance of Mercury; those of the tetrahedron intersect the outer sphere of Mars, but the faces of the dodecahedron

---

[1] For full particulars see Apelt, p. 95 sq.
[2] Lib. v. cap. 9, *Op.* v. p. 318.   [3] Ibid. pp. 319 and 320.

and the icosahedron do not reach the outer spl res of the earth
and Venus. This is supposed to show that the ratios of the
orbits concluded from the regular solids are not directly, but
only indirectly through the harmony, represented in the actual
planetary orbits.

Many writers have expressed their deep regret that Kepler
should have spent so much time on wild speculations and filled
his books with all sorts of mystic fancies. But this is founded
on a misconception of Kepler's object in making his investigations of the cosmographic mystery and the harmony of the
world, for even in his wildest speculations he took as his base
carefully observed facts and he aimed at and obtained results of
great practical value. To the attempt at solving the " mystery "
of the solar system we owe the brilliant discovery, that the
planes of all the planetary orbits pass through the centre of the
sun, a law which ought to be called his first law, the failure of
detecting which had greatly contributed to render the work of
Copernicus incomplete. To his determination to build up his
system of polyhedra on the solid rock of thoroughly reliable and
systematically-made observations was due the perseverance with
which he clung to his post under Tycho, of whom he said (probably with some justice) that he was a man one could not live
with without exposing oneself to the greatest insults[1]. To his
continued work in the same direction we owe the first and second
law, and to the work on the harmony we owe the third. There
is thus the most intimate connection between his speculations
and his great achievements; without the former we should
never have had the latter.

We have yet to say a few words about Kepler's notions concerning the other celestial bodies. Though he emancipated
himself in so many ways from the opinions of the ancients, he
shared their opinion that the fixed stars form part of a solid
sphere, in the centre of which the sun is situated. The idea,
held by Giordano Bruno, that the stars are suns, surrounded by
planets, he regards as improbable, as our sun if removed to the
same distance would be much brighter than the fixed stars,

---

[1] And yet he is never tired of singing Tycho's praises in his books, and
speaks of him with the greatest respect even when differing from him.

... latter are moons or earths[3]." D.
there is no mention of suns. The starry sphere is there described as being only two German miles in thickness, so that
the stars are very nearly at the same distance from the sun[4].
This distance he makes out to be sixty million semidiameters of
the earth, assuming the distance of Saturn to be the geometrical
mean between the distance of the stars and the semidiameter
of the sun, and assuming the latter equal to 15 semidiameters
of the earth, its parallax being at most 1', a great step in
advance[5]. The Milky Way is concentric to the sun, as it
divides the heavens into two hemispheres and appears of nearly
the same breadth everywhere, so that the earth must be nearly
at its centre. The Milky Way is therefore on the inner surface
of the starry sphere.

The interior of the sphere is filled with ethereal air (*aura
ætherea*), through which the planets move. Occasionally this
air or ether becomes condensed so as to be opaque to the light
of the sun and stars, and this ether-cloud, which we call a
comet, receives an impulse from the rays of the sun and is
driven to move through space in a rectilinear path, swimming
in the ether as a whale or monster does in the sea. But the
matter forming the comet is gradually destroyed by the sunlight and is pushed away in the direction of the solar rays,
forming a tail, and in this way the comet is soon dissolved.
Though the comet while it lasts moves in a straight line (with
gradually increasing velocity), the motion appears to us to be

[1] *Epitome*, iv. 1, *Op.* vi. p. 335.
[2] *Ad Vitellionem Paralip.* vi. 12, *Op.* ii. p. 293, also i. p. 424. Among his
arguments is that Venus does not show phases.
[3] "Dissertatio cum Nuncio Sidereo," *Op.* ii. p. 500. Already in 1607 he had
in a letter agreed with Bruno and Tycho in assuming the planets to be like the
earth and inhabited, *Op.* ii. p. 591.
[4] *Epitome*, iv. 1, *Op.* vi. p. 334.   [5] *Ibid.* p. 332.

and the icosahedron do not reach the outer spheres of the earth
and Venus. This is supposed to show that the ratios of the
orbits concluded from the regular solids are not directly, but
only indirectly through the harmony, represented in the actual
planetary orbits.

..,Many writers.¹ celestial phenomena, as well as in Kepler
the Aristotelean doctrine of the immutability of everything in
the ethereal region. The absence of parallax in comets, and
the appearance of the new star of 1572 had furnished Tycho
with plenty of weapons against this doctrine, and Kepler called
attention to other phenomena which also indicated changes of
celestial matter, such as the unusual haze or fog of 1547 and
the light seen round the sun (the corona) during the total
eclipse of October 12, 1605. The new star in Ophiuchus
in the year 1604 gave him another proof of celestial change; he
suggested that it was composed of matter run' together from
the starry sphere, which, when the star vanished, flowed back
into the sphere again².

The publication of the Rudolphine Tables in 1627 is the
closing act of Kepler's fruitful life. He died on November 15, 1630, having completely succeeded in purifying the
system of Copernicus from the remnants of Alexandrian notions
with which its author had been unable to dispense. The solar
system was now fully revealed in all its simplicity, and the
single members thereof had for the first time been linked
together by the law connecting the distances with the periods
of revolution.

---

[1] *Ad Vitell. Par.* x. *Op.* ii. p. 339; *De Cometis libelli tres, Op.* vii. p. 53 sq.;
letter to Herwart in 1602, *Op.* iii. p. 28.

[2] *De stella nova in pede Serpentarii*, cap. xxiii. *Op.* ii. p. 693. The Aristotelean doctrine of the "sublunary" nature of comets was taken up by an
Italian writer, Scipione Chiaramonti, in his *Antitycho*, but Kepler exposed his
ignorance very unmercifully in his *Tychonis Brahei Dani Hyperaspistes* (1625),
Opera, Vol. vii.

CHAPTER XVI.

CONCLUSION.

THE system of Copernicus had been perfected by Kepler, and all that remained to be done was to persuade astronomers and physicists that the motion of the earth was physically possible, and to explain the reason why the earth and planets moved in accordance with Kepler's laws. To give a detailed account of how the earth's motion was gradually accepted, and how Newton's great discovery of the law of universal gravitation accounted for Kepler's laws, would be to write the history of the whole science of astronomy during the seventeenth century and does not come within the plan of this book. We shall only in a few words sketch the progress of the belief in the earth's motion and the feeble attempts at proposing modifications of existing theories up to the time of Newton.

A few months before Kepler's book on Mars came out the newly-invented telescope had been directed to the stars, and in the spring of the following year (1610) Galileo published his *Sidereus Nuncius*, giving the first account of the wonderful discoveries made with the new instrument, especially of the mountains in the moon and the four satellites of Jupiter. At the end of the little book Galileo, who had already for many years been an adherent of the Copernican system[1], publicly

---

[1] When he wrote his *Sermones de motu gravium* (in Pisa, before 1592) he seems to have been an adherent of the Ptolemaic system, as he says that rest is more agreeable to the earth than motion. But on Aug. 4, 1597, he wrote to Kepler that he had "for many years" been a follower of Copernicus, though he had not hitherto dared to defend the new system in public. *Kepleri Opera*, I. p. 40.

declared in its favour, pointing out the an ogy between the earth and the celestial bodies, and remarking that the discovery of four moons attending Jupiter during its motion round the sun put an end to the difficulty of the moon alone forming an exception to the general rule by moving round a planet instead of round the sun. Before the end of the year 1610 the discovery of sun-spots had supplied a new and very striking proof of the fallacy of the Aristotelean doctrine of the immutability of all things celestial, while the discovery of the phases of Venus deprived the opponents of Copernicus of a favourite weapon. But above all it was of the greatest importance that the fixed stars in the telescope appeared as mere luminous points, so that the apparent diameters of several minutes attributed to them by all previous observers were proved to have no existence. This swept away the very serious objection raised by Tycho that a star having no annual parallax and yet showing a considerable apparent diameter must be incredibly large.

No wonder that an old supporter of the Ptolemaic system, and hitherto a most determined opponent of Copernicus, Christopher Clavius, in the last edition of his commentary to Sacrobosco (1611) remarked that astronomers would have to look out for a system which would agree with the new discoveries, as the old one would not serve them any longer[1]. But the common objections referring to a stone dropped from a tower or a cannon ball fired in the direction north and south were still brought forward with confidence to disprove the rotation of the earth, and in refuting them Galileo rendered important service. The three laws of motion were not enunciated by him, as often assumed by popular writers, for he never fully grasped the principle of inertia and failed to realise the continued motion in a straight line of a body left to itself, while he supposed a body describing a circle to continue to do so for ever if not acted on by any force. Although he therefore never shook himself quite free from Aristotelean ideas, maintaining the perfection of circular motion and even letting a

---

[1] Kepler drew attention to this utterance in the preface to his *Epitome Op.* VI. p. 117. Clavius died in February, 1612.

falling body describe an arc of a circle through the earth's centre, there can be no doubt that his popular explanations must have strongly impressed many a wavering reader. But his infatuation for circular motion went so far that he quite ignored the fact that the planets do not move round the sun in concentric orbits. In the whole of his celebrated *Dialogo sopra i due massimi sistemi del Mondo, Tolemaico e Copernicano*, there is no allusion to the elliptic orbits; he even says (near the end of the " fourth day ") that we are not yet able to decide how the orbits of the single planets are constituted, " as a proof of which Mars may be mentioned which now-a-days gives astronomers so much trouble; even the theory of the moon has been set forth in very different ways after that Copernicus had considerably altered that of Ptolemy[1]."

Planetary theory was thus left altogether untouched by Galileo, nor was his opinion as regards the nature of comets what might have been expected from so determined an opponent of Aristotelean physics. Tycho had conclusively proved that they are celestial bodies, but Galileo does not seem altogether satisfied that they have no parallax, and believes them to be vapours originally risen from the earth and refracting the light in a peculiar manner. His opinion on this matter was therefore not very different from that of Scipione Chiaramonti, who in his book *Antitycho* (1621) had upheld the Aristotelean doctrine of the sublunary nature of comets; but on the other hand Galileo fully agreed with Tycho in pronouncing new stars to be celestial bodies. It was not until Hevelius had again shown from accurate observations that comets are much farther off than the moon that the opponents to their character of heavenly bodies were finally silenced some sixty years after

[1] Galileo (just before making this statement) remarks that the angular velocity of the moon must be greater at new moon than at full moon, since the moon when nearer to the sun describes a smaller orbit with reference to the sun. He compares the sun to the point of suspension of a pendulum, and earth and moon to two weights attached to the rod of the pendulum, the one representing the moon being placed at different distances from the point of suspension. Just as this alters the period of vibration of the pendulum, so (he concludes) does the earth move slower at the time of full moon than at that of new moon. This is a curious anticipation of the idea of planetary perturbation.

Tycho's death, and not till 1681 that the paral〉lic form of their orbits with the sun at the focus was discovered by Dörfel.

The persecution of Galileo by the Pope and the Inquisition for having (notwithstanding a previous warning) discussed the motion of the earth in a manner favourable to Copernicus, has been described so often that there is no need to give an account of it in this place. It looks like an act of Nemesis, that the Roman authorities should have taken special umbrage at the curious and altogether erroneous theory of the tides which Galileo propounded in the fourth day of his *Dialogo*, rejecting the ancient idea that they are caused by the moon and maintaining that they are quite incompatible with the Ptolemaic system. Perhaps his adversaries feared that " there might be something in it " and became exasperated in consequence. On the other hand, Galileo had hardly dealt quite fairly with his opponents in pretending that the Ptolemaic system was the only alternative to that of Copernicus. In the whole book there is no allusion whatever to the Tychonic system, although it is scarcely too much to say that about the year 1630 nobody, whose opinion was worth caring about, preferred the Ptolemaic to the Tychonic system.

The Copernican system had from the beginning been viewed with extreme dislike by theologians, both Roman Catholic and Protestant. We have seen in a previous chapter how strongly Luther and Melanchthon expressed themselves about it, and from a letter written to Kepler in 1598 by Hafenreffer, Professor of Divinity at Tübingen, it appears that the theory of the earth's motion was not in good odour among the theologians there[1]. But as yet the theory had not been directly forbidden anywhere, probably because Osiander's preface to the book of Copernicus (supposed to have been written by the author himself) disarmed the opponents by representing the theory as a mere means of computation. The fate of Giordano Bruno can hardly have been influenced by his advocacy of the earth's motion, for he had set forth a sufficient number of startling ideas to provide stakes for many scores of heretics. But the invention of the telescope and the analogy it

[1] *Kepleri Opera*, I. p. 37.

revealed between the earth and the ;
assume an entirely different aspect.^s to have adopted the
hypothesis which did not concern mank^or at Oxford (1617–
a question of the actual position in th^d the geometry of
dwelling-place of man, whether it was ^ard) merely adopted
the most important part of Creation ^ond focus[1]. Until
significant part thereof. In the course ^a necessary conse-
theologians had retreated step by step ^heory found some
of the Fathers of the Church; the Babyl^on account of its
world favoured by them had given way ^that the orbit of
antipodes and other abominations had been t^it, in which the
when impious hands tried to push the earth o^fixed points or
place in the centre of the world and make it ^was as bad as
stars, though they were moved by angels and th^became only
devil as central occupant, the theologians turned^ber that the
Action was taken within a few years of the in^dly accepted
telescope; on February 24, 1616, the consultors of the Inquisition at Rome declared the doctrine of the earth's motion to
be heretical, and on the 5th March following the "Sacred
Congregation" solemnly suspended the book of Copernicus and
the commentary to Job by Didacus a Stunica[1] "until they are
corrected" (*donec corrigantur*), and altogether damned and
forbade the recently published book by a Carmelite Father,
Foscarini, in which an attempt was made to show that the
motion of the earth is in accordance with Scripture[2]. This was
followed in 1620 by the issue of a *Monitum Sacræ Congregationis
ad Nicolai Copernici lectorem*, in which instructions are given
about the alterations to be made in the book *De Revolutionibus*
before it may be reprinted. These are not very numerous, and

---

[1] See above, p. 353.

[2] The text of the *Decretum* is given by von Gebler, *Die Acten des Galilei'schen Processes*, Stuttgart, 1877, p. 50. The title of Foscarini's book is *Lettera del R. Padre Maestro Paolo Antonio Foscarini, Carmelitano, sopra l' opinione de' Pittagorici, e del Copernico, della mobilità della Terra, e stabilità del Sole e il nuovo Pittagorico Sistema del Mondo. In Napoli per Lazzaro Scorrigio*, 1615. An English translation is given in Salusbury's *Mathematical Collections and Translations*, Vol. I. (1661), pp. 471–503. This book also contains translations of Galileo's *Dialogue* and of his letter to the Grand Duchess Christina of Tuscany, an attempt to reconcile the Bible and the Fathers with the Copernican system.

416　　　　　　　　　*Cc* which the motio of the earth is
Tycho's death, and not till ᴸe whole of chapteʳ VIII. of the first
orbits with the sun at the ᵈ likewise the disrespectful allusion to
　The persecution of G ᵉditor was ever found to publish a
for having (notwithstanc ᵉ book of Copernicus. As the decree of
motion of the earth in ᵉʳ books teaching the same thing," the
been described so often ᵃlso put on the *Index* in 1633. In the
of it in this place. ᵉ *Index* the clause forbidding "all other
the Roman authorit, omitted, but the book of Copernicus,
at the curious and Galileo's *Dialogue* and some other books
which Galileo proⱼⁱll 1822, so that the edition of 1835 is the
rejecting the anciⱼ y are not mentioned. By that time it had
and maintaining f geology to suffer from the *odium theologicum,*
Ptolemaic systeⁿ ᵃe theory of organic evolution came to occupy
might be some; theological mind once held by the Copernican

On

In Prote.. nt countries no serious attempt was made to put
down the doctrine of the earth's motion, perhaps because it
would not have looked well to imitate the action of the hated
Inquisition; but wherever the power of the Roman Curia
could reach, philosophers had to submit, though some of them
did it very unwillingly. Among these was Pierre Gassendi
(1592–1655), who in his numerous writings often praises the
Copernican system, and says that he would have preferred it if
it had not been pronounced contrary to Scripture, for which
reason he was obliged to adopt the Tychonic system. He
made the experiment of dropping a stone from the top of
the mast of a ship in motion and concluded, justly enough, that
the result neither proved nor disproved the earth's motion[2].
All the same he fell foul of his countryman, Morin, a very
violent anti-Copernican, who devoted one of his polemical

---

[1] This document is printed by Riccioli, *Almag. Nov.* II. pp. 496–97.

[2] *De motu impresso a motore translato epistolæ duæ*, Paris, 1642. He says (p. 156) that it is only "some cardinals" who have declared the earth to be at rest, and this is not an *articulum fidei*, but it must all the same have the greatest weight with believers. In his book, *De proportione qua gravia decidentia accelerantur* (1646), he says: "Videlicet ego Ecclesiæ alumnus, ita me totum ipsi devoveo, ut quicquid illa improbat, ipse anathema conclamem" (p. 286).

writings (*Alae telluris fractae*, Paris... to have adopted the Gassendi. Of another renowned astr...ssor at Oxford (1617– the Jesuit Giovanni Battista Riccioli...ised the geometry of difficult to say what his private opin...board) merely adopted great treatise on astronomy, *Almagestu*...second focus[1]. Until folio volumes (Bologna, 1651), an in... is a necessary conse- historian of astronomy, he gives twenty... theory found some refutes) in favour of the earth's moti...tly on account of its against it, many of the objections being v...ed that the orbit of to facts which have no bearing on the qu...e it, in which the speaks very highly of Copernicus and th...wo fixed points or system, and the arguments from Scripture ...ry was as bad as well as the action of the Curia, are evident...soon became only weight with him. Yet he produces an argum...ember that the which he attributes great weight[1]. If a body f...rvedly accepted of a tower under the equator of the earth (stan... Horrox (1619– in four seconds pass through spaces proportiona...

1, 3, 5, 7 ; but if the earth were rotating he mak...nade to modify four spaces would be about equal and that the b...matician and strike the ground with more force than it wou...tution of the after one second; therefore the earth does not ...ld its ground argument was shown to be fallacious by a well-kn...much longer matician, Stefano degli Angeli, and a lively controv...Descartes to between him, Borelli and Riccioli and the latter's ...Copernican Manfredi and Zerilli[2]. Riccioli adopted the Tychon...abjuration with a slight modification; while he accepts the ...to conflict Mercury, Venus and Mars round the sun, he lets Jup... what he Saturn move round the earth, because they have sat... earth in their own, those of Saturn being his "laterones" or appe...t was to i.e. the ring imperfectly seen and not yet recognized as...without The other three planets are satellites of the sun. He di...so that consider the first law of Kepler to have been proved, ... agreement between theory and observation was no proof...

---

[1] *Alm. Nov.* II. p. 409.

[2] For titles see the Catalogue of the Crawford Library of the Ro... Observatory, Edinburgh.

[3] *Alm. Nov.* I. p. 529. Riccioli's own planetary theory is wonderfully com... plicated, the excentricity being variable in the direction of the line of apsid... and the semidiameter of the epicycle varying in the same period.

There were of course till some opponents o. the Copernican system who did not reject it from fear of the Church. The only one of any distinction was Longomontanus (1562–1647), the principal disciple of Tycho Brahe, who wrote a treatise which he appropriately called *Astronomia Danica*, since it was mainly founded on the work of Tycho, whose system he adopted, though he admitted the rotation of the earth. He rejected the elliptic orbits of Kepler, and his standpoint was altogether that of the sixteenth century.

But the opposition of the Church did not retard the progress of astronomy, though no doubt it made it difficult for the Copernican system to become recognized outside the sphere of professional astronomers. Slowly but surely the idea of the earth's motion gained ground. There were, however, still not a few astronomers who, though followers of Copernicus, did not accept the planetary theories of Kepler altogether. Among these was Philip Lansberg (1561–1632), who published planetary tables, founded on an epicyclic theory, which were much used among astronomers, though they were very inferior to the Rudolphine Tables[1]. The second law of Kepler was objected to by Ismael Boulliaud (1605–1694), who, in his *Astronomia Philolaica* (Paris, 1645), substituted for it an extraordinary theory. He supposed the ellipse to be a section of an oblique cone, on the axis of which the focus not occupied by the sun is situated, while the angular velocity is uniform with regard to the axis of the cone, being measured in circular sections parallel to the base of the cone[2]. In addition to being utterly unreasonable (for why should the planets climb round imaginary cones?) the theory is a very poor substitute for Kepler's, as the true anomaly is very badly represented except when the excentricity is very small. Neither would it

---

[1] *Philippi Lansbergii Tabulæ cœlestium motuum perpetuæ*, Middelburg, 1632. (P. probably owed a great deal of the good repute they enjoyed for some time at e circumstance that they by a fluke represented the transit of Venus in 1639 fairly well, while the Rudolphine Tables threw Venus quite off the sun's disc.

[2] "Ut omnibus æqualis motus partibus respondeant singulæ partes appa-[les, ita tamen ut Aphelio minores circuli æqualis motus conveniant, Perihelio maiores." *Astron. Phil.* p. 26.

have been any advantage to astronomers to have adopted the theory of Seth Ward, Savilian Professor at Oxford (1617–1689), who in two little books criticised the geometry of Boulliaud, and (throwing his cones overboard) merely adopted uniform motion with regard to the second focus[1]. Until Newton proved that Kepler's second law is a necessary consequence of the law of gravitation, Ward's theory found some admirers in England, and it was apparently on account of its obvious defects that J. D. Cassini suggested that the orbit of a planet is not an ellipse but a curve like it, in which the rectangle of the distances of a point from two fixed points or foci is a constant quantity[2]. But this theory was as bad as the one it was to displace, and both of them soon became only historical curiosities. How sad it is to remember that the first astronomer of note who fully and unreservedly accepted the great results of Kepler's work, Jeremiah Horrox (1619–1641), only reached the age of twenty-two years.

While these various fruitless attempts were made to modify the planetary theory of Kepler, a great mathematician and philosopher set up a general theory of the constitution of the universe, which, owing to its author's celebrity, held its ground in his native land for upwards of a hundred years, much longer than it deserved. It had been the intention of Descartes to prepare a work "On the world," founded on the Copernican system; but when he heard of Galileo's trial and abjuration he gave up the idea, as he had no desire to get into conflict with the Church. But some years later he found what he considered to be a way out of the difficulty, since the earth in his system was not to move freely through space, but was to be carried round the sun in a vortex of matter without changing its place relatively to neighbouring particles, so that

---

[1] *In Ismaelis Bullialdi Astronomiæ Philolaicæ Fundamenta Inquisitio brevis*, Oxford, 1653, and *Astronomia geometrica*, London, 1656. Boulliaud replied to the former in his *Astronomiæ Philolaicæ Fundamenta clarius explicata*, Paris, 1657. Ward became Bishop of Exeter in 1662, of Salisbury 1667.

[2] *De l'origine et du progrès de l'astronomie* (1693), *Mém. de l'Acad. R. des Sciences*, 1666–1699, T. VIII. p. 43. As a native of Italy Cassini was afraid to pronounce publicly in favour of the earth's motion, even after his removal to Paris. Pingré, *Cométographie*, I. p. 116.

it might (by a stretch of imagination) be said to be at rest.
His account of the origin and present state of the solar system
is contained in his *Principia Philosophiæ*, which appeared at
Amsterdam in 1644. He assumes space to be full of matter
which in the beginning was set in motion by God, the result
being an immense number of vortices of particles of various
size and shape, which by friction have their corners rubbed off.
Hereby two kinds of matter are produced in each vortex, small
spheres which continue to move round the centre of motion
with a tendency to recede from it, and fine dust which gradually
settles at the centre and forms a star or sun, while some of it,
the particles which have become channelled and twisted when
making their way through the vortex, form sun-spots. These
may either dissolve after a while, or they may gradually form
a crust all over the surface of the star, which then may wander
from one vortex to another as a comet, or may settle perma-
nently in some part of the vortex which has a velocity equal
to its own and form a planet. Sometimes feebler vortices are
gathered in by neighbouring stronger ones, and in this way the
origin of the moon and satellites is explained.

The vortex theory of Descartes does not account for any of
the peculiarities of the planetary orbits, and is indeed pure
speculation unsupported by any facts. It was an outcome of
the natural desire to explain the motion of the planets round
the sun, why they neither wander off altogether nor fall into
the sun, but it could only with difficulty account for the non-
circular form of the orbits, in which respect it was inferior to
Kepler's magnetic vortex theory.   Another attempt at a general
theory of the solar system was made by Giovanni Alfonso
Borelli (1608–1679) in a book which professes to deal with
the satellites of Jupiter only, perhaps to avoid saying anything
about the motion of the earth[1]. He assumes that the planets
have a natural tendency to approach to the sun (and the
satellites to their central body), while the circular motion gives
them a tendency to fly away from it, and these opposing forces
must to a certain extent counterbalance each other. The

---

[1] *Theoricae mediceorum planetarum*, Florentiæ, 1666, reviewed by E. Gold-
beck in an essay, *Die Gravitationshypothese bei Galilei und Borelli*, Berlin, 1897.

former is a constant force, the latter is inversely proportional
to the distance. As to the motion in the orbit, Borelli, like
Kepler, connects it with the rotation of the sun, as the rays of
the sun's light catch the planet and drive it along; but in the
case of Jupiter, which is not self-luminous, he merely calls
them "moving rays." To explain the oval form of the orbit
he can only say that at the aphelion the tendency to approach
the sun gets the upper hand, so that the planet gradually
comes nearer to the sun. Hereby its velocity is increased and
also the centrifugal force, which is inversely proportional to the
radius vector, until the two forces become equal, after which
the centrifugal force makes itself most felt and again increases
the distance from the sun until the aphelion is reached.

Both in the way of mathematical theory and in speculation
men of science thus vainly tried during the fifty years following
that of Kepler's death to improve or modify his results. When
they tried to substitute other rules for his two first laws they
failed utterly, and when they speculated on the origin and
cause of planetary motion they only produced theories just as
vague as his were. All the same they did not labour in vain,
as they accustomed themselves and others to recognize the
Copernican system as a physical fact, and helped it to become
more and more accepted by educated people at large. An
interesting proof of the gradual change of feeling as regards
the earth's motion is afforded by the utterances of some
prominent men in England in the beginning and middle of
the century. In several places in his writings Francis Bacon
speaks of the Copernican system without in any way doing
justice to it, and as if it were of no greater authority than the
notion of the Ionians, that the planets describe spirals from
east to west[1]. On the other hand, we find that John Wilkins,
afterwards a brother-in-law of Cromwell, and still later Bishop
of Chester, in 1640 published "A Discourse concerning a New
Planet, tending to prove that (it is probable) our earth is one

---

[1] *Novum Organum*, II. 36, also other passages quoted by Whewell, *Hist.
Induct. Sc.* 3rd ed. Vol. I. pp. 296 and 388. Gilbert and Harvey were two other
investigators of whom Bacon did not think much. He was ready to teach
scientific men how to work, but he was singularly unlucky when laying down
the law about the work done by those inferior creatures.

of the planets"; while Milton in the ⅃ radise Lost speaks sympathetically about the new system. In England, where no astronomer of any importance except the short-lived Horrox had yet arisen, the ground was thus by degrees being prepared for the man whose work was to confirm the truth of Kepler's laws, and show them not to be arbitrary freaks of Nature but necessary consequences of a great law binding the whole universe together. From Thales to Kepler philosophers had searched for the true planetary system; Kepler had completed the search; Isaac Newton was to prove that the system found by him not only agreed with observation, but that no other system was possible.

INDEX.

Daït, in Egyptian cosmology 4
Dante, cosmology of 235
Declination of planets 341, magnetic 395
Deferent 153
Demokritus, cosmology 26, 115, 189
Derkyllides 11, 70, 99, 151
Descartes, vortex theory 421
Dicuil 225
Diels 10
Digges, Leonard and Thomas 347
Dikæarchus 173
Diodorus of Tarsus 212
Diogenes of Apollonia 33
    ,,   ,, Babylon 168
Diogenes Laertius 9, 49, 50, 82, 88, 124
Diophantus 206
Diverse, circle of the 64
Divine nature of stars 13
Doshiri, Egyptian name of Mars 5
Doxographic Writers 10
Dyugatih, Indian name of Jupiter 241

Earth, spherical form, Parmenides 20, Pythagoras 38, Plato 55, Aristotle 117
Earth, rotation of 49, 72, 125, 139, 140, 272, 285, 292, 323, 318 sq., 368
Earth, dimensions 118, 171 sq., 243, 249, 257
Earth, motion of, Philolaus 41, Aristarchus 136, arguments against 192, 355 sq., Copernicus 312, 324
Earth, third motion of 328, 361, 395
Earth, orbit, motions referred to centre of 343, 344, 377, excentricity of 381, 386
Eclipses, cause of 5, 13, 15, 17, 19, 23, 25, 28, 47
Ecliptic 93
Egyptian cosmology 3
"Egyptian" system 129
Ekphantus, rotation of earth 50, 124
Eleatic school 18
Elements 23, 85, 257
Elliptic orbits 392
Elysium 7
Empedokles, cosmology 23
"Enoptron" of Eudoxus 94

Epicycles, theory of 66, 148, 197, 201
Epikurus 23, 171
"Epinomis" 84
"Epitome Astronomiæ Copernicanæ" 403, 417
Equant 197
Eratosthenes 161, size of earth 175
Erebus 7
Eridu 1
Erus, myth of 56
Ether 120, 158, 411
Euclid 88
Eudemus, history of astronomy 12, 149
Eudoxus 87 sq., 115, 168, 173, 205
Euktemon 93, 106
Eusebius 10
Evection 195, 368
Excentricity of planets 332, 370, 378, of earth 377
Excentrics, movable 144, 154

Fabricius, David and Johann 401
Fates, in Plato's Republic 60
Fathers of the Church 207 sq.
Fergil, *see* Virgil
Field, John 346
Fire, central 41 sq., 147
  ,, above the air 120
Fracastoro 296 sq.

Gabir ben Aflah 247, 254, parallax of planets 261, inferior planets 267
Galenus (Pseudo-) 11, 139
Galileo, mentions Wursteisen 348, Sidereus Nuncius 411, 413, follower of Copernicus 413 sq., develops mechanics 414, dialogue 415, persecution 416
Gallus, C. Sulpicius 181
Gassendi 418
Gauzahar-sphere 260
Geminus 106, 130 sq., 143, 150, 169, 177, 203
Genesis, cosmology of 3, 209 sq., 229
Georgios of Trebizond 296
Gerbert 226
Gherardo of Cremona 296
Giese, Tiedemann 318
Gilbert, on rotation of earth 348

CAMBRIDGE: PRINTED BY JOHN CLAY, M.A. AT THE UNIVERSITY PRESS.